COON MOUNTAIN CONTROVERSIES

COON MOUNTAIN CONTROVERSIES

METEOR CRATER AND THE
DEVELOPMENT OF IMPACT THEORY

William Graves Hoyt

The University of Arizona Press / Tucson

THE UNIVERSITY OF ARIZONA PRESS

Copyright © 1987
The Arizona Board of Regents
All Rights Reserved

This book was set in 10½ x 13 Linotron 202 Berkeley Old Style
Manufactured in the U.S.A.

Library of Congress Cataloging-in-Publication Data

Hoyt, William Graves.
　Coon Mountain controversies.
　　Bibliography: p.
　　Includes index.
　　1. Meteor Crater (Ariz.)　2. Meteor craters.
3. Lunar craters.　I. Title.　II. Title: Impact
theory.
QB756.M4H69　1987　　551.3　　87-5009
ISBN 0-8165-0968-9 (alk. paper)

British Library Cataloguing in Publication data are available.

Contents

List of Illustrations	vii
Foreword by Eugene M. Shoemaker	xi
Introduction	1
1. The Craters of the Moon	7
2. Iron and Diamonds	31
3. Moonlets and Other Matters	54
4. Staking a Claim	73
5. Proving the Claim	100
6. Under the Southern Cliffs	127
7. Coon Mountain and the Moon	154

CONTENTS

8. The New Meteoric Hypothesis	184
9. Blue Sky Laws and Bureaucrats	214
10. The Final Shaft	238
11. The Great Debate	264
12. The Aftermath	297
13. Meteor Crater—The Later Years	318
14. Impact: A Cosmic Process	344
Notes to the Chapters	367
Photo Credits	424
Index	427

Illustrations

	following page
The moon	68
Franz von Paula Gruithuisen	
Nasmyth and Carpenter's "Fountain" and resulting crater	
R. A. Proctor	
A. E. Foote	
Willard D. Johnson	
Coon Mountain from the south	
Wide-angle view of Meteor Crater	
Grove Karl Gilbert	
Gilbert's lunar crater forms	
Gilbert's sketch of random and "moonlet" impact	
Gilbert's sketch of Mare Imbrium "sculpture"	
W. H. Pickering	
Holsinger at Volz's Trading Post	180

viii ILLUSTRATIONS

following page

Crater wall showing tilted strata — 180

Shaft house

Silica flour, south rim of crater

Map of distribution of meteoric material around crater

J. C. Branner

G. P. Merrill

Map of distribution of ejecta from Meteor Crater

W. F. Magie

N. H. Darton

Stone museum on crater rim

Aerial view of Meteor Crater

W. W. Campbell

T. J. J. See — 292

See's sketch of impact process

A. W. Bickerton

A. C. Gifford

Gifford's crater profiles

D. M. Barringer

G. R. Agassiz

D. M. Barringer, Jr.

Meteor Crater cross-section

H. L. Fairchild

F. R. Moulton

Elihu Thomson

E. P. Adams

Henry Norris Russell

P. L. Webster — 356

Percy W. Bridgman

following page
356

H. H. Nininger

E. M. Shoemaker

Boon and Albritton's sketch of crater structure

Robert S. Dietz

Ralph B. Baldwin

Baldwin's chart of diameter/depth

Shoemaker's concept of sequence of events in formation of Meteor Crater

FOREWORD

I first met Bill Hoyt when the headquarters of the U.S. Geological Survey's Astrogeology Branch was transferred to Flagstaff, Arizona, in 1963. Bill was then managing editor of the *Arizona Daily Sun,* and he frequently called us in connection with stories on the exploration of the Moon. When Bill interviewed me over the phone, I could hear him typing the story at the other end of the line. I knew full well that there would be no time for him to rewrite that story before it hit the press. To my amazement, every story was invariably lucid and scientifically accurate. Bill had an uncanny ability to grasp both the principles and the details of a scientific problem and to translate that understanding into lively, readable prose without ever compromising the scientific rigor. Moreover, he accomplished this feat in "real time" during the interview. It very quickly became evident that we were dealing with one of the outstanding science writers among practicing journalists. And he was editor of our hometown newspaper!

It was a matter of great satisfaction to many of us that Bill was later able to devote himself full-time to research on the history of science, as a resident scholar at Lowell Observatory. When he told me that he was planning to work on the history of Meteor Crater, I promptly placed at his disposal a small collection of papers I had gathered in anticipation of writing a brief history myself. There was no doubt that Bill would do a far superior job.

What he produced, however, totally surprised me with its depth. Hoyt developed the history in a richness and detail that far transcended my own knowledge and gave me a new appreciation for my scientific predecessors at Meteor Crater.

Coon Mountain Controversies represents a landmark account of the early history of the scientific study of impact craters. A major new paradigm has arisen in the last few decades concerning the impact origin of craters found on the Earth, the Moon, and the other solid bodies in the solar system. This paradigm is the primary empirical support for the prevailing cosmogonic view that the Earth and the other terrestrial planets, as well as Uranus, Neptune, and the cores of Jupiter and Saturn, were formed by accretion of smaller solid bodies. Hardly any scientist working on these problems today, however, has much knowledge of the history of the ideas and the research that preceded the acceptance of the paradigm.

Meteor Crater, Arizona, is the centerpiece in the evolution of the new paradigm, as it is the best exposed and the best preserved known impact crater on Earth in the diameter range greater than one kilometer. It remains the largest known crater with associated meteorites, and it drew the attention of some remarkable individuals. Hoyt has chronicled their involvement with the crater perceptively. The key role and contributions and also the errors of Daniel Moreau Barringer are described with meticulous scholarship.

What makes this book a lasting contribution is that Hoyt understood the subject very thoroughly and was able to make an accurate judgment of the important contributions and intellectual influence of the various actors in the drama. Hoyt has placed the story of Meteor Crater in a perspective that few others would have achieved. He made thorough use of the relatively abundant primary sources of historical data. At the same time, his grasp of the subject enabled him to winnow the significant facts and events from the trivial detail. As an original piece of historical research, this is his opus magnum.

Certainly this book will be of general interest to geologists and planetary scientists. As an account of the growth and evolution of a scientific idea, it is also of great interest as a case history of how science actually works, as opposed to idealized and misleading concepts of how science "ought to work." I think that will be one of its chief values to lay readers and specialists in other disciplines. The interplay of individuals, each with their own approaches and prejudices, and the dogged persistence of Barringer in his attempt to bring about a successful mining venture not only makes good reading but illuminates the human aspects of the scientific enterprise.

Eugene M. Shoemaker

INTRODUCTION

> The goddess of learning is fabled to have sprung full blown from the brain of Zeus, but it is seldom that a scientific conception is born in its final form, or owns a single parent. More often, it is the product of a series of minds, each in turn modifying the ideas of those that came before, and providing material for those that come after.
>
> G. P. THOMPSON
> 1937 Nobel Prize Lecture

COON MOUNTAIN

Coon Mountain, or Coon Butte as it was sometimes called, appears from a distance and from any direction as a long, low ridge rising from a broad, arid, canyon-cut plain that is a part of the Little Colorado River drainage system in north-central Arizona. How and when the name originated is not known, but for many years it was in general use in the area itself as well as in the scientific papers that described this unusual, if undistinguished-looking feature.[1]

The name is a misnomer. Aside from the particularizing adjective, Coon Mountain is not properly a mountain or a butte at all. It is a depression some 570 feet deep and 4,000 feet in diameter, collared by a rim that rises from 120 to 160 feet in height and provides the low topographic relief that deceives the casual observer on the surrounding plain. It is not surprising that George Russell Agassiz, as a young Harvard graduate in the 1880s, herded cattle around this "queer-looking butte" for a year without recognizing its true nature.[2] And while it "presents a very peculiar appearance," as mining engineer and entrepreneur Daniel Moreau Barringer later pointed out, "it does not suggest . . . the existence within itself of a large crater."[3]

For Coon Mountain is a large crater, a fact that becomes evident only to those who climb to the top of its boulder-strewn rim or who view it from the air. This now-famous terrestrial scar is known today as Meteor Crater, a name first formally proposed in 1906 by geologist Herman L. Fairchild of the University of Rochester [N.Y.],[4] but not officially adopted by the U.S. Interior Department's Board of Geographical Names until 1946.[5] In the intervening years, a half dozen other names, including Crater Mound, its official designation from 1932 to 1946, appeared in published discussions of this remarkable pit.

The name Meteor Crater is not a misnomer though purists would have it Meteorite Crater—the term "meteorite" properly applying to extraterrestrial objects that not only enter the earth's atmosphere but penetrate to the earth itself. In this instance the word "meteor" had a geographical as well as an astronomical derivation and thus adhered to standard practice for geological nomenclature. Although Fairchild certainly intended to suggest the origin of the crater, the word also referred to "Meteor, Arizona," the designation of what was then the nearest post office, a facility established five miles north of the crater just as the controversies over Coon Mountain began. Fairchild noted in proposing the name that the geographical reference "gives official standing to the word."[6]

The matter of names is the least of the long and sometimes bitter controversies that have swirled around this strange hole in the ground since it first came to the attention of science in 1891. Basic to these controversies was the question of Coon Mountain's origin—whether formed by some familiar geological process such as volcanism or by an astronomical catastrophe, specifically by the collision of a large meteoroid with the earth. Such a cosmic cataclysm was unknown to history at the turn of the century and indeed hardly a hundred years had passed since the first recognition that meteors and meteorites originated in interplanetary space. The fact that some solar system body massive enough to blast out a large crater could strike the earth remained beyond human experience until well into the twentieth century.

The violent explosion in the remote Tunguska River region of central Siberia on June 30, 1908, marks a turning point here. It flattened hundreds of square kilometers of forest and set off atmospheric reverberations that were recorded as far away as England. Although the exact nature of the Tunguska event remains somewhat uncertain and the subject of sometimes frivolous conjecture,[7] it was for many years assumed to be the result of the impact of one huge meteorite or a cluster of meteorites and this assumption was reinforced by the destructive but much better documented meteoritic fall in 1947 in the Sikhote-Alin region of northeastern Siberia.[8] Far from being incredible today, the idea of the impact of large meteorites or small asteroids has been invoked in recent years as an explanation of the mass extinctions of biological lifeforms on the earth, such as that of the dinosaurs some sixty-five million years ago, that show up in the paleontological record in the earth's rocks.[9]

Something of the significance of the Coon Mountain controversies is reflected in the fact that Coon Mountain's crater inspired one of America's most eminent geologists to make an early attempt to explain the origin of the craters on the moon by impact,[10] an idea no less incredible at the time and which only many years later successfully challenged the long-held belief that these crateriform structures were analogous to terrestrial volcanoes. That meteoritic impact rather than volcanism has been the dominant process in bringing the surface of the moon to its present state is now widely acknowledged both in and out of science. With the recent revelations through spacecraft imagery that the other terrestrial planets and even the satellites of Mars, Jupiter, and Saturn are pockmarked with crateriform features, the idea of meteoritic impact has become a fundamental concept in modern theories of the origin and evolution of the solar system.

Something of the irony that marked these controversies is reflected in the fact that the same eminent geologist who early urged an impact origin for the lunar craters rejected at the same time such an origin for Coon Mountain's crater which is now considered to be their best terrestrial analog.[11] Many years of sometimes acrimonious debate were required before scientists quietly conceded that Grove Karl Gilbert was right about the moon but wrong about Coon Mountain.

Others have also been wrong about Coon Mountain. Even Barringer who first marshaled and published the evidence for its meteoritic origin in 1906 was wrong in believing that the crater represented a potentially fabulous mining bonanza that would return profits of up to $250 million from the nickel and platinoid metals to be recovered from the main mass of the meteorite buried beneath its floor. Despite the hundreds of thousands of dollars that Barringer and his associates and successors poured into exploration of the crater, no such mass has ever been found.

The history of the scientific investigation of Coon Mountain bears directly on the development of meteoritic impact theory as it relates to the craters on the moon. It is perhaps difficult more than a quarter of a century into the space age to realize how thoroughly the belief that the lunar craters were ancient volcanoes once permeated scientific and popular thinking about the moon. The volcanic analogy held sway for nearly three centuries, and despite sporadic challenges in the late 1800s and early 1900s has yielded its sovereignty over selenological thought only in the past fifty years.

Historically, one of the major objections to an impact origin of the lunar craters was the apparent absence of such features on the earth. The recognition that Coon Mountain's crater was such a feature and the subsequent identification of other meteorite craters in other parts of the world from criteria derived in studies at Coon Mountain eliminated this objection.

The discovery of terrestrial impact craters has been largely the work of geologists and others in related fields. In the second decade of the twentieth century, however, the controversies over Coon Mountain stirred new interest in the origin of the lunar craters on the part of astronomers who soon found they could eliminate another objection as well. In particular, they proposed an answer to the question of why almost all of the tens of thousands of craters on the moon are of a roughly uniform circular configuration when the random infall of meteoric bodies implied that some impacts must be at oblique angles and must thus form elliptical-shaped scars. This answer—that impacts involve explosions—was not immediately accepted, perhaps because of its simplicity or because it went beyond human experience. It had to be advanced independently a number of times before being admitted to serious consideration. It is not accepted by all scientists even today. From an historical standpoint, however, there can be no doubt that the idea that the impacts of large meteorites produce explosive forces has been a key factor in the development and eventual acceptance of impact as a cosmic process.

This book, then, has two separate but related themes. The first concerns the long-sustained efforts by Daniel Barringer and his associates to establish Coon Mountain's meteoritic origin and to exploit its crater commercially. These efforts climaxed in 1929 in a previously unreported "great debate," involving a number of prominent scientists, over the probable dimensions and disposition of the meteorite that produced the crater. This, in turn, is set in the matrix of the far longer but parallel development of meteoritic impact theory in general as it relates to the crateriform features on the moon and the other solar system bodies and thus to the genesis of the solar system itself.

More broadly, these two themes encompass the progress of a significant revolution in scientific thought, a revolution that was many years in the making and is the product of many minds. Its origins go back to the nineteenth century and to the dissatisfactions of a very few with the volcanic analogy as an explanation for the craters on the moon. It acquired momentum steadily through the first half of the twentieth century, with the recognition first of Coon Mountain's impact origin and later of the impact origin of other terrestrial features and with the progressive development of a theoretical basis for meteoritic impact. It became a fait accompli only after the dawn of the space age in 1957, and after man had achieved the technological capability of reaching the moon and other solar system bodies. Its triumph is evident in the fact that since that time thousands of scientists in a wide variety of disciplines have been concerned directly or indirectly in their research with the implications and ramifications of meteoritic impact—an idea which was once dismissed by scientists as fanciful.

While what follows is primarily history, it may also provide some insights into the sociology of science. The Coon Mountain controversies revolved around such issues as authoritarianism within disciplines, communication barriers between disciplines, professionalism versus dilettantism, questions of priority, the roles of government and commerce in science, and the processes by which the conventional wisdom of science is changed.

Some of the best-known and most respected scientists of the late-nineteenth and early-twentieth centuries were involved in one or another aspect of these controversies—geologists Gilbert, Fairchild, and George Perkins Merrill; astronomers Henry Norris Russell, William Wallace Campbell, and Forest Ray Moulton; and physicists Elihu Thomson and Percy Williams Bridgman, to name only a few.

Yet the key figure in these controversies—Barringer—was not a scientist at all in the professional sense, but a lawyer, mining engineer, and entrepreneur whose milieu was the world of business, industry, and high finance.

Barringer's dream of a mining bonanza at Coon Mountain became a consuming interest in the final twenty-five years of his life, and he counted heavily on science to help him "prove his claim." It is surely ironic and somewhat tragic that at the end science irreparably tore the fabric of his dream. At least one of his sons has suggested that this may have contributed to his death.[12]

ONE

THE CRATERS OF THE MOON

> Thus, when we see, on the surface of the moon, a number of elevations, from half a mile to a mile and a half in height, we are strictly entitled to call them mountains; but when we attend to their particular shape, in which many of them resemble the craters of our volcanoes, and then argue that they owe their origin to the same cause which has modelled many of these, we may be said to see by analogy, or with the eye of reason. Now, in this latter case, though it may be convenient in speaking of the phenomena to use expressions that can only be justified by reasoning from the facts themselves, it will certainly be the safest way not to neglect a full description of them that it may appear to others how far we have been authorized to use the mental eye.
>
> WILLIAM HERSCHEL, 1787

For nearly three hundred years and well into the twentieth century it was widely believed both in and out of science that the so-called "craters" on the moon had originated in forces internal to the moon itself, forces analogous to those which created the great volcanic craters of the earth. That this belief was held only with reservations in some quarters is perhaps evident from William Herschel's caveat. Yet despite some skepticism it persisted and by the end of the nineteenth century had become firmly entrenched in scientific thought. Throughout this period the word "crater" was defined with specific reference to the bowl-shaped depressions at the summits of terrestrial volcanoes and thus, when applied to the lunar features, was itself an expression of the analogy.[1]

Herschel himself reinforced the analogy when he described three "volcanoes" he had observed on the moon on two consecutive nights in April 1787, one of which he declared "shews an actual eruption of fire." This eruption, which he estimated to involve an area about three miles in diameter, "exactly resembled a small piece of burning charcoal, when it is covered by a very thin coat of white ashes," and was similar to one he had seen on the moon in 1783 which, "though much brighter than that which is now burning, was not nearly so large in the dimensions of its eruption." On at least eleven earlier occasions dating back to 1587, other observers had reported "bright spots" or "ruddy" areas on the surface of the moon, but Herschel seems to have been the first to describe such phenomena in terms of the volcanic analogy.[2]

EARLY SELENOGRAPHERS

Direct analogies between the physiographic features of the earth and the moon began with Galileo who in 1609 was the first to view the lunar surface with the aid of the newly invented "optic tube." Before this time, earth-moon analogies based solely on naked-eye observations of the moon's more obvious albedo features were largely anthropomorphic or zoomorphic, as Galileo noted in reporting his lunar observations in the *Siderius Nuncius* in 1610.[3] "Of the moon's appearance I find no more exact description than that some say it resembles a human face; others that it is like the muzzle of a lion; still others that it is Cain with a bundle of thorns on his back."

Through his crude telescope Galileo observed "prominences" and "cavities" on the moon's surface which he compared to mundane mountains and valleys. One cavity "larger than the rest, and perfectly round in shape" he likened to a specific region of Bohemia "if that were enclosed on all sides by mountains arranged exactly in a circle," as indeed it seemed to be. Such comparisons proved to be daring for the time. They challenged long-held Aristotelian doctrines regarding the perfection of the celestial spheres and the uniqueness of the earth, and they eventually got him into trouble with the Peripatetic authorities of the Church of Rome.

For Galileo, what soon came to be called the "craters" on the moon were simply "spots" which he distinguished by their smaller size and greater number from "large" or "ancient" spots which "are plain to everyone and have been seen throughout the ages." These smaller spots were "so numerous as to occur all over the lunar surface" and gave the brighter areas of the moon especially a broken and uneven appearance. One such area, "where it is spotted as the tail of a peacock sprinkled with azure eyes," he wrote, "resembles those glass vases which have been plunged while still hot into cold water and have acquired a crackled and wavy appearance from which they receive their common name of 'ice cups.'"

Subsequently, Galileo provided a more detailed, although less colorful, description of these smaller spots in his *Dialogue Concerning the Two Chief World Systems* in 1632.[4] "But what occur most frequently there," he declared, "are certain ridges, somewhat raised, that surround and enclose plains of different sizes and various shapes, but for the most part circular. In the middle of many of these there is a mountain in sharp relief, and some few are filled with a rather dark substance similar to that of the large spots that are seen with the naked eye; these are the larger ones, and there are a very great number of smaller ones, almost all of them circular."

In *Siderius Nuncius,* Galileo published five rough sketches of the moon, each of which shows the unevenness of the terminator line between the dark and sunlit portions of the lunar sphere, the "large" or "ancient" spots, and a number of smaller, circular features, none of which correlates exactly with features on modern charts of the lunar surface.

But more faithful drawings and maps of the moon were soon forthcoming, and much of the subsequent observational history of the moon is concerned with gathering data for ever more accurate lunar maps and atlases. Only three of these early maps are important here—those published by Michael Florent van Langren and Johannes Helvelius in 1645 and by Giovanni Riccoli in 1651—for together they established the nomenclature which is still used today and which enhanced terrestrial analogies.[5] Langren and Helvelius, for example, labeled the "large" or "ancient" spots "maria," or "seas" in Latin, and the bright regions "terra" or "earth," an analogy that Galileo himself had earlier suggested, citing the ancient Greek philosopher Pythagoras as his authority. Riccoli began the practice of naming other prominent lunar features such as the larger craters after persons renowned in philosophy and the sciences, particularly astronomy.

The early selenographers were interested primarily in recording the detailed topography of the moon as seen through the relatively crude telescopes of the time. Explaining the origin of this topography, however, was first considered by Robert Hooke, the erudite secretary of the Royal Society of London and one of the most penetrating and versatile minds of the seventeenth century.

In *Micrographia* in 1665,[6] Hooke reported observing a vat of boiling alabaster in which, "presently ceasing to boil, the whole surface will appear all over covered with small pits, exactly shaped like those of the Moon." From this he concluded that "these pits in the Moon seem to have been generated much after the same manner that the holes in Alabaster, and the *Vulcans* of the Earth are made." It was not improbable that the substance of the moon must be "very much like that of the Earth" and "being agitated, undermin'd, or heav'd up, by eruptions of vapour, may naturally be thrown up into the same kind of figured holes, as the small dust, or powder, of the body of Alabaster." Nor was it improbable "that there may be generated, within the body of the Moon, divers such kind of internal fires and heats, as may produce such Exhalations; for since we can plainly enough discover with a *Telescope,* that there are multitudes of such kind of eruptions in the body of the Sun itself, which is accounted the most noble Aetherial body, certainly we need not be much scandalized at such kind of alterations, or corruptions, in the body of this lower and less considerable part of the universe, the Moon. . . ."

Hooke also experimented by dropping musket balls and other solid objects into a "well-tempered mixture of Tobacco-pipe clay and water," producing crater-like indentations. In Hooke's day, however, it was not possible to explain lunar craters by a simple extrapolation from pits or depressions made by the impact of musket balls, wads or pellets of mud or clay, or even raindrops, on a smooth, plastic surface. For interplanetary space as then conceived was a void where no material bodies existed to impact on the moon; thus Hooke could not imagine from "whence those bodies should come."

Hooke's conclusion that the forces responsible for the lunar craters were internal and analogous to the "Vulcans" of the earth was not seriously challenged for more than two hundred years. During this period no evidence of lunar volcanism was adduced, but neither was any evidence found to confute the idea which was accepted as a common-sense proposition in the absence of a better explanation. There were difficulties, of course, in explaining both the great size of some of the lunar craters and their great number; but these were not insurmountable and with some imagination could be explained within the range of human experience.

The post-Galilean history of the moon is marked by relatively brief periods of considerable interest, interspersed with long periods when the problems of the moon were almost entirely ignored, except by a handful of observers. From Hooke's time and through most of the eighteenth century, in fact, astronomers paid very little attention to the moon at all, although mathematicians now took up its study, particularly during the 1740s when Alexis Claude Clairault, Jean le Rond D'Alembert, and Léonard Euler worked out the moon's motions in space in what proved to be a triumphal vindication of Isaac Newton's law of gravitation and Newtonian mechanics in general.[7] Even the early interest in lunar mapping evaporated. The last of the early moon maps was published in 1680 by Gian Domenico Cassini, and another did not appear until 1775, when a chart prepared by Tobias Mayer in 1750, notable for its use of a coordinate system for reference, was published posthumously. Between 1780 and 1783, a lunar map by J. H. Lambert appeared; and, of course, Herschel's observations of the moon came during the 1780s, early in his astronomical career.

A whole new era in selenography was ushered in in 1791, when German observer Johann Heironymous Schröter published the first of the two volumes of his *Selenotopographische Fragmente* and, for the first time, put the study of the physical conditions of the moon on a comparative basis.[8] In 1795, the English portrait painter John Russell produced a magnificent pastel drawing of the moon from his own telescopic observations and followed this two years later with a lunar globe, or "Selenographia," specially mounted to depict the moon's appearance at any age and in any position relative to the earth. In 1805–06, Russell also engraved two maps, known as "lunar Planispheres," one showing the full moon as it is seen from the earth, the other as it would appear with sunlight falling obliquely on its surface.[9]

That a volcanic explanation of the craters on the moon was held to be fully credible at this time is not particularly surprising in view of the general credulity in all matters lunar that seems to have then prevailed. In the late 1770s and early 1800s, for example, it was widely believed that the moon was habitable and indeed inhabited by "lunarians," "lunites," "selenites," or what have you, who lived lives much like those of the earth's inhabitants. This view was popularized in part through the sermons and writings of various pluralistic theologians such as the Reverend Timothy Dwight, president of Yale College from 1795 to 1817; Thomas Chalmers, leader of an evangelistic movement in Scotland; and the Reverend Thomas Dick, also a Scot.[10] But it was also espoused by a number of more or less eminent astronomers, including Schröter and Herschel. Herschel, in 1780, had declared it to be "almost an absolute certainty" that the moon was inhabited; and, although he begged not to be called a "lunatic" for the belief, he added that if given a choice between the earth and the moon, "I should not hesitate a moment to fix upon the moon for my habitation."[11] Schröter subsequently reported green fields on the surface of the moon and what he took to be a city.[12]

In the 1820s, the German selenographer Franz von Paula Gruithuisen went further, claiming that he had actually seen men and animals on the moon, that he could distinguish between lunar meadows and cultivated fields, and that he had observed a large fortress on its surface and, nearby, a star-shaped temple from which he deduced that the "lunites" worshipped the stars.[13]

The denouement of this overly imaginative extension of the earth-moon analogy seems to have begun in August 1835 with the celebrated "moon hoax" perpetrated in the *New York Sun*.[14] This involved a series of articles purportedly describing lunar observations made with a telescope constructed on "an entirely new principle" by Herschel's son, Sir John F. W. Herschel, who was then observing the southern skies from the Cape of Good Hope in South Africa. The articles allegedly were written by Herschel's assistant, a Dr. Andrew Grant, and reprinted from a recent issue of the *Edinburgh Journal of Science*. Actually, they were written by Richard Adams Locke, a $12-a-week *Sun* reporter with a considerable knowledge of astronomy and an imagination surpassing even Gruithuisen's.

Often in vivid detail the articles reported that Sir John had seen sheep, pygmy zebra, and animals similar to the bison, the giraffe, and the fabled unicorn roaming lunar grasslands, as well as bipedal winged creatures four feet tall and covered with "short and glossy copper-colored hair" to which Herschel had assigned the name of "Verspertilio-homo," or manbat. Like Gruithuisen, he supposedly had seen a temple. Perhaps the most credible part of Locke's fascinating but fabricated account was his reference to Sir John's observation of a volcano, a "flaming mountain."

The articles caused a great furor, for while they were probably intended as a satire on the whole idea of a populated moon,[15] they were taken literally by many of the *Sun's* readers and by others who learned of their content secondhand. Reprints were quickly run off and sold. A few days after the series ended, however, Locke admitted his authorship to a reporter from the New York *Journal of Commerce* which promptly exposed the deception. Whether the *Sun's* editor, Benjamin J. Day, knew of their spurious nature prior to their publication is problematical, but in any case he could console himself with the knowledge that they had increased his newspaper's circulation to then-record levels.

Many people undoubtedly felt foolish, angry, or both, for having been thus flummoxed, but Herschel himself, once word of the hoax reached him in South Africa, seems to have enjoyed the whole affair and to have considered it a good joke. He was in fact quite tolerant on the subject of lunar life, believing, literally, that there might be two sides to the question, as there were two sides to the moon. This view he based on a suggestion, made earlier by the mathematical astronomer Peter Andreas Hansen, that the moon's shape was elongated toward the earth as a result of the earth's gravitational attraction, and thus any air or water on its surface might be concentrated on its far side and closer to the moon's center of gravity. "It by no means follows, then, from the absence of water or air on this side of the moon," Herschel later wrote, "that the other side is equally destitute of them, and equally unfitted for maintaining animal or vegetable life."[16]

Two other events occurred at about this same time that provided the *coup de grâce* to any idea of a populated moon. The first of these was the publication by German selenographers of two new lunar maps, one between 1824 and 1835 by Wilhelm Gotthelf Lohrmann, and another in 1834 to 1836 by Wilhelm Beer and Johann Heinrich Mädler. The latter was accompanied the following year by a book, *Der Mond,* which contained a wealth of descriptive data, including micrometric measurements of hundreds of lunar features.[17]

Beer and Mädler's work, in particular, remained the basis for selenography for more than forty years. As the erudite historian of astronomy Agnes M. Clerke would later note, it "gave form to a reaction against the sanguine views entertained by Helvelius, Schröter, Herschel, and Gruithuisen as to the possibilities of agreeable residence on the moon, and relegated the 'Selenites' to the shadowy land of the Ivory Gate. All examples of change in lunar formations were, moreover, dismissed as illusory."[18]

Even more important in this regard was the advent of lunar photography,[19] starting in 1839 when Louis-Jacques-Mandé Daguerre revealed his invention of the first practical photographic process, and shortly thereafter tried to photograph the moon through a telescope. He failed, but the following year John W. Draper succeeded in obtaining a small image of the moon which showed some surface detail. Significant photographs of the moon, however, came only after 1850 and the work of William Cranch Bond and Lewis M. Rutherfurd in America and Warren de la Rue in England. Although not of the quality that lunar photographs would eventually achieve, these early images provided strong reinforcement to the idea, urged by Beer and Mädler, that the moon not only was a forbidding place for life in any form, but that it was in fact geologically a dead and unchanging world.

By the early 1860s this view of the moon had become so well established that subsequent reports of changes on the lunar surface caused brief sensations within astronomy. In 1866, for example, selenographer J. F. Julius Schmidt, the director of the Athens Observatory, made the startling announcement that the crater Linné, a deep crater more than six miles in diameter on both Lohrmann's and Beer and Mädler's maps, had disappeared, apparently obliterated by some sort of igneous flow, and was now a shallow whitish spot with a small pit in its center.[20] The idea that a change had occurred was strengthened by the facts that Schröter earlier had seen the area much as Schmidt reported it and that Linné was not recorded on Russell's lunar globe and maps around the turn of the century. Again, in 1878, German observer Herman J. Klein reported the appearance of what he considered to be a newly formed crater on the moon, near the known crater Hyginus, which he called "Hyginus N."[21] Both of these reports stirred considerable discussion but were later dismissed as accidents of observation, i.e., as the effects of subtle changes in the angle of incident light falling on the lunar surface relative to the observer in the first case, and the discernment of a small, but previously unnoticed surface feature in the second.

It is significant that such reports had not always been received as anomalous. In the interim between the elder Herschel's 1787 report of lunar "volcanoes" and Schmidt's 1866 report on Linné, more than fifty observations of bright spots, reddish or purplish areas, patches of vapor, "lightning," or "volcanoes" on the moon had appeared in the literature without attracting any particular attention.[22] Clearly by the 1860s, attitudes had changed, and the limit of credulity in matters relating to the moon had been drastically lowered.

That the moon was now seen to be a lifeless and unchanging body, however, did not alter the premises on which the volcanic analogy was based beyond setting the presumed occurrence of lunar volcanism back to some earlier period in the moon's evolutionary history. The analogy, in fact, had been elevated to the status of virtual orthodoxy some years before, and by no less an authority than Sir John Herschel himself. In 1849, in his *Outlines of Astronomy,* Sir John declared that the craters of the moon "offer . . . in its highest perfection, the true *volcanic* character, as it may be seen in the crater of Veseuvius, and in a map of the volcanic districts of the Campi Phlegraei [i.e., the "burning fields" near Naples, Italy], or the Puy de Dôme [a region of ancient volcanoes in central France]."[23] Sir John's pronouncement remained unmodified through the many editions of his book over the next fifty years and became a standard citation in subsequent books and writings by others dealing with the physical condition of the moon, although a few authors preferred to draw their own comparisons.[24]

GEOLOGISTS AND LUNAR CRATERS

The interest in the origin and nature of lunar craters, moreover, was not confined entirely to astronomy; and during the first half of the nineteenth century the young but rapidly developing science of geology gave further credence to lunar volcanism through studies of specific volcanoes on the earth and of volcanic processes in general.[25] As early as 1831, French geologist J.B.A.L.L. Élie de Beaumont compared certain terrestrial craters with those on the moon and suggested that the latter originated in forces thrusting upward from beneath the lunar surface, with subsequent collapse forming crateriform features called "calderas." This was, in fact, an extrapolation from an hypothesis, known as the "craters of elevation" theory, advanced some years before by the German geologist Leopold von Buch to explain terrestrial craters, and involved essentially the same processes that Hooke had observed in his vat of boiling alabaster.

In 1840 the American geologist James Dwight Dana, in the course of four years of traveling to various remote parts of the world, stopped briefly at the Sandwich Islands and subsequently gave the volcanic analogy a new twist.[26] While there, he visited the great shield volcano of Kiluaea on the island of

Hawaii; and within its summit caldera or "pit," as he preferred to call it, he found "pools of boiling lava in active ebullition" in which "the circular or slightly elliptical form of the moon's crater's is exemplified to perfection." For Dana, these boiling lakes of lava also explained the great diversity in size, as well as the uniformity of shape, of the lunar craters. "If the fluidity of lavas, then, is sufficient for this active ebullition, we may have boiling going on over an area of an indefinite extent; for the size of the lake can have no limits except as may arise from a deficiency of heat. The size of the lunar craters is therefore no mystery. Neither is their circular form difficult of explanation; for a boiling pool necessarily, by its own action, extends itself circularly around its center. The combination of many circles, and the large sea-like areas [on the moon] are as readily understood."

In the 1870s, three major works devoted entirely to the moon appeared in book form which together indicated the directions of most future selenographical or selenological studies. One of these, Edmund Neison's *The Moon and the Condition and Configuration of Its Surface,* published in 1876, need be mentioned only in passing here.[27] It continued the rigorous descriptive tradition that had been established by Beer and Mädler and would be followed by a number of eminent, if mostly amateur, selenographers, especially in England. Neison carefully eschewed speculation in his book, and while he clearly favored a volcanic explanation for the lunar features, he managed to avoid committing himself unequivocally on the point.

The two other books, however, proved to be more important in regard to attempts to explain the genesis of the moon's topography. In one, *The Moon: Considered as a Planet, a World, and a Satellite,* selenographers James Nasmyth and James Carpenter explored the volcanic analogy in great detail, adding refinements of their own.[28] In the other, *The Moon: Her Motions, Aspect, Scenery, and Physical Condition,* astronomer Richard Anthony Proctor briefly but effectively outlined an alternate explanation—the origin of the craters on the moon through meteoritic impact.[29]

Nasmyth and Carpenter's book, published in 1874, was a monumental work notable, among other things, for its illustrations—simple graphic drawings explaining the processes of lunar crater formation under discussion in the text, interspersed with photographs of a remarkable series of clay sculptures depicting various lunar landscapes as seen in the telescope, and for comparison, in one instance, the Campi Phlegraei on earth. Nasmyth and Carpenter, however, were concerned with more than merely describing the moon's topography in great detail. They sought to specify precisely the nature of the processes that had produced crateriform features on the moon and, further, to identify the source of the forces that had powered these processes.

Their basic premise in this was the volcanic analogy. "We can scarcely doubt," they wrote, "that where a lunar crater bears a general resemblance to a terrestrial crater, the process of formation has been nearly the same. . . . Where variations present themselves they may be reasonably ascribed to the differences of conditions pertaining to the two spheres."

Geologists, they pointed out, were not all in agreement on the processes involved in terrestrial volcanism, and most of them had rejected hypotheses based on von Buch's old "craters of elevation" theory. Consequently, they noted, "we are led to regard a volcano as a pyramid of ejected matter, thrown out and around an orifice," and they then postulated the formation of the lunar craters by a process analogous to a fountain, with ejected material spewing from a relatively small vent in the moon's crust and "being deposited around the hollow and forming an embryo circular mountain." Such fountainlike action, moreover, could account for the central cones or peaks observed in some of the lunar craters, for as the violence of the eruption subsided, ejected material would fall back on or near the vent itself and accumulate there. To explain craters without central cones or peaks, they invoked "the outflowing and welling forth of fluid lava" as observed in terrestrial volcanism.

The fountain idea, they argued, also explained the "most striking difference" that existed between lunar and terrestrial craters. "The greatest dissimilarity is in the point of dimensions; the projection of materials to 20 or more miles distance from a volcanic vent appears almost incredible." But the moon's lesser gravity, only one-sixth that of the earth, and its "proportionately larger surface in proportion to its mass" were pertinent here. "The rate at which it parted with its cosmical heat must have been much more rapid than in the case of the earth, especially when enhanced by the absence of the heat-conserving power of an atmosphere of air or water vapour. . . . We thus find in combination conditions most favorable to the display of volcanic action in the highest degree of violence. . . . the ejected material is left free . . . to deposit itself at distances far greater than those which we have on the earth." Incidentally, they assigned the formation of those craters with seemingly concentric ring walls, such as Copernicus, to "landslip phenomena," i.e., the slumping of material from the precipitous inner wall of the primary crater rim.

Nasmyth and Carpenter also explored possible sources of the eruptive energy on the moon. "Steam is the agent to which geologists have been accustomed to look for the explanation of terrestrial volcanoes," they pointed out, "but we are debarred from referring to steam as an element of lunar geology, by reason of the absence of water from the lunar globe. . . . We might suppose that a small proportion of water once existed; but a small proportion would not account for the immense display of volcanic action which the surface exhibits." Dana, they noted, had suggested sulphur because "of the part which it appears to play in the volcanic igneous operations of our globe, and on account of its presence in cosmical meteors that have come within the range of our analysis." However, they reasoned, "Any matter sublimated by heat in the substrata of the moon would be condensed upon reaching the cold surrounding space. . . . It does not appear clear how expansive vapours could have lain dormant till the moon assumed a solid crust, as all such would doubtless make their escape before any shell was formed, and at an epoch when there was ample facility for their expansion."

Rather, they preferred an explanation of their own based on what they called the principle of "expansion upon solidification" of molten material beneath the moon's cooling and contracting crust. "Molten matter of a volcanic nature," they declared, "when about passing to the solid state, increases its bulk to a considerable degree." To illustrate the principle they cited an experiment by the English physicist John Tyndall with molten bismuth in a stoppered "stout iron bottle" which, as the bismuth cooled and solidified, had been "rent open." When this principle was applied to the moon at some time in its past, they suggested, the molten material under its surface, "in approaching its solid condition, expanded and burst open or rent its confining crust." Thus, they concluded, the "principle of pre-solidifying expansion" provided "the key to the mystery of volcanic action on the moon."

Not all of the crateriform features on the moon were formed in this manner however, they conceded, for they found the origin of "the great ring formations not manifestly volcanic," i.e., the very largest craters and the lunar maria, to be beyond explanation, adding that others who had offered explanations "have been no more happy in their conjectures." Hooke's analogy to bubbles rising and bursting they rejected, pointing out that "if it were reasonable to suppose the great rings to be the foundations of such vast volcanic domes, we must conclude these to have broken when they could no longer sustain themselves, and in that case the surface beneath should be strewed with *debris,* of which, however, we can find no trace." They also dismissed a suggestion, advanced by one M. Rozet in 1846,[30] that "circular, vortical movements" formed the raised ring walls by throwing scoria from the center to the circumference, noting that "we are at a loss to account for the origination of such vorticose movements, and M. Rozet is silent on the point."

Interestingly, Nasmyth and Carpenter also explored briefly the idea that sublunarean explosions might have been responsible for these structures, and particularly their generally circular shape. "The true circularity of these structures appears at first view a remarkable feature," they declared. "But it ceases to be so if we suppose them to have been produced by some very concentrated force of an upheaving nature, and only if we admit the homogeneity of the moon's crust." Given this latter condition, "then *any* upheaving force, deeply seated beneath it, will exert itself *with equal effects at equal distances from the source:* the lines of equal effect will obviously be radii of a sphere with the source of the disturbance for its centre, and they will meet a surface over the source in a circle. . . . Thus we see that an intense but extremely confined explosion, for instance, beneath the moon's crust, must disturb a circular area of its surface. . . ." [emphasis in original]. But Nasmyth and Carpenter confessed difficulty in accounting "for such a very local generation of a deep-seated force; and granting its occurrence, we are unprepared with a satisfactory theory to explain the resultant effect of such a force in producing a raised ring at the limit of the circular disturbance."

Thus, of all hypotheses considered, the two selenographers ended by cautiously opting for the "boiling lava lake" or "ebullition" theory advanced by Dana as the explanation of these larger features. "As regards the smallest craters with cones, we believe few geologists will refuse their compliance with the supposition that they were formed as our crater-bearing volcanoes were formed," they declared. "But when we come to ring-mountains having no cones, and of such enormous size that we are compelled to hesitate in ascribing them to ejective action, we are obliged to face the possibility of some other causation. . . . Professor Dana's appears the most rational, since it is based upon a parallel found on the earth. In citing it, however, we do not necessarily endorse it."

Proctor's book *The Moon,* which appeared in 1873 and just prior to Nasmyth and Carpenter's landmark work, was written in a casual and popular vein, and as a consequence its often penetrating and innovative ideas seem to have been taken less seriously by astronomers and selenographers of that day.

Proctor was a knowledgeable astronomer and an extraordinarily prolific writer on astronomical subjects, publishing no less than fifty-seven books as well as numerous articles in popular magazines and professional journals before his tragic and untimely death at age fifty-one.[31] Born on March 23, 1837, he was educated at King's College in London and at St. John's College at Cambridge, where he studied mathematics and emerged with a baccalaureate degree and as twenty-third wrangler in the mathematical tripos, the university's top honors examination. In 1866, he was elected as a Fellow of the Royal Astronomical Society and later became an honorary secretary and editor of its *Monthly Notices.* This same year, he was forced, because of financial losses related to the failure of a bank, to turn to popular writing to support himself and his family. In 1875, two years after the first publication of *The Moon,* he emigrated to America where he settled in Florida and continued his astronomical writing. In 1888, enroute back to England, he was taken ill in New York City; doctors there, fearing that he had contracted yellow fever in Florida, ordered him out of his hotel and into a raw, windy, rainy night. He died twelve hours later, on September 12, 1888.

The ideas that Proctor advanced in *The Moon*, while not all entirely new, could hardly have been formulated a hundred years earlier when, as noted above, interplanetary space was widely believed to be a void, *le vide planètaire* as the French called it, which contained nothing more tangible than some vaguely defined ethereal fluid. Moreover, through the eighteenth century, reports of "shooting stars," "fire stones," "thunderstones," and the like were still explained as terrestrial phenomena caused by atmospheric perturbations, "exhalations" of sulphur, lightning, or, such was the lingering superstition of the times, as manifestations of supernatural forces. The emotions that such objects stirred as they streaked across the sky, emitting brilliant flashes of light, sparks and smoke, and loud rumbling, hissing and explosive noises, were those of fear and awe rather than curiosity.[32]

Reports of "stones" falling from the sky were usually received with incredulity, and more often than not were dismissed as absurd. In 1777, to cite but one example, a trio of savants from the *Acadèmie Royale des Sciences* in France, which included the great French chemist Antoine Lavoisier, studied a so-called "thunderstone" which had been brought to the *Acadèmie* some years earlier and, in their report, ridiculed the contention of the finders that it had fallen from the sky, concluding that it was simply an ordinary stone that had been struck by lightning.[33] Certainly one reason for this skepticism on the part of the contemporary scientific community was the fact that all such reports stemmed from peasants or other relatively unsophisticated observers; meteoriticists even now are faced with much the same problem, for meteoric or meteoritic phenomena are largely events of an unpredictable instant and simply do not occur at the convenience of their science.

There were, of course, a few exceptions to these beliefs. Leaving aside ancient speculations by such philosphers as Anaxagoras and Seneca, it can be noted that astronomer Johannes Kepler, Galileo's great contemporary, in the early 1600s had suggested that shooting stars or meteors, and fallen stones or meteorites, were related and that both were of extraterrestrial origin.[34] "Falling stars," he wrote, "are composed of inflammatory viscous materials. Some of them disappear during their fall, while others indeed fall to the earth, drawn by their own weight. Nor, indeed, is it improbable that they have been formed into globes from feculent materials mixed with the ethereal air itself, and thrown from the ethereal region in a straight line through the air like very small comets. . . ." Early in the 1700s, England's Edmund Halley also had concluded that meteors and meteorites were extraterrestrial and were caused by material disseminated through interplanetary space falling into the sun and encountering the earth in its sunward passage.[35]

THE MODERN FOUNDATION OF METEORITICS

But it was not until the end of the eighteenth century that *le vide planètaire* began to be filled, so to speak, with material bodies; and to Ernst Florens Friedrich Chladni, a German physicist best known for his studies in acoustics, goes the principal credit for placing meteorites and meteoritic theory on their modern foundations. In 1794 Chladni published a memoir in which he postulated that the planets had been formed by the coming together of "particles of matter, which were before dispersed throughout infinite space," or by the fragments of much larger planets "that have been broken to pieces, either perhaps by some external shock, or by an internal explosion."[36] By either hypothesis, however, "it is not improbable, or at least contrary to nature, if we suppose that a large quantity of such material particles, either on account of their too great distance, or because prevented by a stronger movement in the other direction, may not have united themselves to the larger accumulating mass of a new world; but have remained insulated, and, impelled by some shock, have continued their course through infinite space, until they approach so near to some planet as to be within the sphere of its attraction, and then by falling down to occasion the phenomena. . . ." It was partly as a result of Chladni's 1794 memoir that two students at the University of Göttingen named Brandes and Benzenberg observed more than four hundred meteors over a two-month period late in 1798 and, from sightings of the same meteors from different locations, established their appearance at far greater height than had hitherto been thought, and their probable origin in outer space.[37] Chladni, in a second memoir in 1819, extended his earlier work, concluding that meteors, meteorites, and even comets were related phenomena and were all manifestations of the fact that space was indeed filled with matter in discrete forms.[38]

Moreover, the belief that "stones" simply could not fall from the heavens eroded further after 1803 when, on April 26, more than three thousand such stones fell in an elliptical area near the French village of L'Aigle, a fall witnessed by hundreds of people. This event was investigated by Jean-Baptiste Biot, a highly respected scientist and member of the French National Institute, whose report left no doubt as to the reality of the fall.[39]

One other event at the beginning of the nineteenth century is of significance here. This was the discovery in 1801, by Giuseppe Piazzi of Palermo, of the first of what proved to be numberless small bodies circling the sun between the orbits of Jupiter and Mars. In 1802, after the German observer Wilhelm Heinrich Matthias Olbers found a second such body, William Herschel pronounced them an entirely new class of solar system objects, for which he proposed the name of "asteroids."[40]

What is important here about these developments is that, once it was realized that interplanetary space contained such bits and pieces of matter whirling around the sun and occasionally colliding with the earth, it became possible to postulate an impact origin for the craters on the moon. Such an origin was in fact proposed as early as 1802 by the von Bieberstein brothers in Germany,[41] and again in 1815 by K. E. von Moll.[42] But their proposals received only limited circulation and indeed were too far ahead of their time to have any effect whatever on the development of selenographical or selenological thought.

In 1829 Gruithuisen also suggested an impact theory of sorts which he developed somewhat later by outlining his concept of the mechanics of the impact process. His theory was intended to account not only for the lunar craters, but certain terrestrial features that Élie de Beaumont had described as "lunar ring mountains . . . without volcanism." Gruithuisen envisioned concentrically layered spherical bodies "sinking" into the "soft" plastic surface of the earth or moon largely under their own weight. At impact, the sides of the sphere would be sheared off, and these, penetrating only partly into the surface, formed the visible circular walls. The remaining cylindrical portion continued to penetrate beneath the surface, the crushed and fused rock ahead of it flowing outward and up its side to bolster the upraised rim. But Gruithuisen's extreme views on lunar life precluded a serious consideration of his suggestion, as selenographer J. F. Julius Schmidt would later note. As a result of Gruithuisen's "fantasy," he wrote in 1856, "Today, one directs to astronomers questions about the moon only in jest."[43] Actually, it was not until the 1870s, when Proctor advanced it with considerable diffidence but in considerable detail, that the idea of the impact origin of the craters on the moon appeared in significant form.

Importantly, in *The Moon*, Proctor rejected all terrestrial analogies to explain the moon's topography, declaring that "it is impossible to recognize a real resemblance between any terrestrial feature and the crateriferous surface of the moon." Moreover, he took as his basic premise the idea that the solar system itself had been formed at some remote time through a process of accretion of "meteoric" bodies, rather than from a contracting, rotating gaseous mass, or nebula, as had been proposed by Pierre Simon Laplace in the 1790s in his *Exposition du Système du Monde*. Proctor declared that the solar system "had its birth, and long maintained its fires, under the impact and collisions of bodies gathered in from outer space." Furthermore: "According to this view, the moon, formed at a comparatively distant epoch in the history of the solar system, would not merely have had its heat originally generated for the most part by meteoric impact, but while still plastic would have been exposed to meteoric downfalls, compared with which all that we know, in the present day, of meteor-showers, aërolitic masses, and so on, must be regarded as altogether insignificant."

To Proctor, at least, the idea of meteoritic impact as the origin of the lunar craters was entirely new, and as he anticipated that it might be the subject of criticism, he prefaced his exposition with a disclaimer. "It may seem, indeed, at a first view, too wild and fanciful to suggest that the multitudinous craters on the moon . . . have been caused by the plash of meteoric rain—and I should certainly not care to maintain that as the true theory of their origin; yet it must be remembered that no plausible theory has yet been urged respecting this remarkable feature of the moon's surface."

The volcanic analogy, he argued, would not suffice. "As blowholes, so many openings cannot at any time have been necessary, whatever opinion we may form as to the condition of the moon's interior and its reaction upon the crust." Moreover, he pointed out, it was generally regarded that water, which was absent from the moon, was absolutely essential to volcanic processes. Hooke's explanation that the craters on the moon were analogous to the breaking of bubbles or blisters rising in a vat of boiling alabaster, Proctor also dismissed as "inadequate" as "it is impossible for such bubbles to be formed on the scale of the lunar craters."

For the smaller lunar craters in particular, Proctor found "nothing incredible" in the supposition that they were formed by "meteoric rain" falling on the moon when its surface was in a plastic condition. "Indeed, it is somewhat remarkable how strikingly certain parts of the moon resemble a surface which has been rained upon while sufficiently plastic to receive the impressions, but not too soft to retain them."

Impact, furthermore, could even explain their circularity, he maintained. "Nor is it any valid objection to this supposition, that the rings left by meteoric downfall would only be circular when the falling matter chanced to strike the moon's surface squarely; for it is far more probable that even when the surface was struck very obliquely and the opening first formed by the meteoric mass or cloud of bodies was therefore markedly elliptic, the plastic surface would close around the place of impact until the impression formed had assumed a nearly circular shape."

The larger crateriform features, Proctor conceded, were a different problem, for they "under close telescopic scrutiny by no means correspond in appearance to what we should expect if *they* were formed by the downfall of great masses from without." Yet, he suggested, it was at least possible that the tremendous heat generated by such a downfall would liquify a vast circular region of the lunar surface "and that in rapidly solidifying while still traversed by the ring-waves resulting from the downfall, something like the present condition would result." Such a massive impact, he added, might even cause an upwelling of material from the lunar interior. "But such ideas as these require to be supported by much stronger evidence than we possess before they can be regarded as acceptable." The theory that these larger features were once great lakes Proctor also rejected for it "seems open to difficulties at least as grave as the one I have just considered, and to this further objection, that it affords no explanation of the circular shape of these lunar regions."

Proctor's "meteoric" impact hypothesis stirred neither criticism nor controversy and, in fact, it was largely ignored by his contemporaries in astronomy, who continued to favor the volcanic analogy in general. Nasmyth and Carpenter's massive work, which appeared within a year, quickly overshadowed his self-styled "wild and fanciful" ideas; and indeed, after 1874, those who wrote at all about the physical condition of the moon usually adopted Nasmyth and Carpenter's work, illustrations and all.[44]

Proctor, it has since been pointed out, apparently had so little confidence in his impact hypothesis himself that he soon abandoned it, and indeed it is missing from the second edition of *The Moon*, which appeared in 1878, and from subsequent editions. Moreover, in what he planned as his magnum opus, *Old and New Astronomy,* the manuscript of which was completed after his death by A. Cowper Ranyard and published in full only in 1895, he opted for the volcanic analogy, adopting a somewhat modified version of Nasmyth and Carpenter's views, including their conditional espousal of Dana's "lava lakes" analogy.[45] As published, the book also contains a number of Nasmyth and Carpenter's illustrations which, however, may have been selected by Ranyard.

The situation, however, is more complicated than this. In a brief preface to the 1878 edition of *The Moon,* Proctor explained that certain material relating to lunar theory, or to the motions of the moon, had been deleted as he had "found reasons to believe that portions of the original were too difficult for the general reader," and that his final chapter on the physical condition of the moon had been "changed considerably," although he did not specify what changes had been made, or why.

But in this regard, it may be noted that sometime between 1873 and 1876 Proctor found that his "meteoric" impact hypothesis was not entirely original, although he still seems to have been unaware of the speculations of the von Biebersteins, von Moll, and Gruithuisen. In an essay on "The Past and Future of Our Earth" in his 1876 book, *Our Place Among the Infinities*,[46] he wrote that on the primordial earth, "the solid crust, contracting at first more rapidly than the partially liquid mass within, portions of this liquid matter would force their way through and form glowing oceans outside the crust. Geology tells us of regions which, unless so formed, must have been produced in the much more startling manner, conceived by Meyer, who attributed them to great meteoric downfalls."

Proctor did not further identify Meyer or his works, but in a lengthy footnote to this passage he opined that there "is very little new under the sun" and noted that he, too, had suggested "meteoric downfall" for the origin of the craters on the moon in 1873. "I find that Meyer had far earlier advanced a similar idea in explanation of those extensive regions of our earth which present signs of having been in a state of igneous fluidity," he added.

Even more interesting, however, is the fact that in 1878, the same year that his impact hypothesis disappeared from *The Moon*, Proctor restated it elsewhere with greater emphasis and in greater detail, adding some significant embellishments.[47] "When I first advanced this theory (in 1873)," he explained in a long essay entitled "The Moon's Myriad Small Craters," "I had not yet fully recognised the evidence both *a priori* and *a posteriori* in its favour.... I now view the subject differently. The evidence in favour of the meteoric theory of the small craters is much stronger than I at first supposed, the difficulty of forming any other plausible theory much greater. I may even go so far as to say that it would be a problem of extreme difficulty to show how a body formed like the moon, exposed to similar conditions, and for the same enormous time-intervals, could fail to show such markings as actually exist on the moon. A theory of which this can be said stands on a somewhat strong basis."

Proctor began by again arguing solar system origins by accretion against Laplace's nebular theory, and with a bow to the now established "uniformitarian" school of geologists which had sprung up toward the end of the eighteenth century. "Let it suffice to mention that the theory of planetary and solar growth by the gathering in, during past ages, of immense quantities of meteoric and cometary matter, is one which has this immense advantage over the nebular theory, that it assumes the former action of a process which is going on at the present time."

Then, using calculations by others of the length of time required for the molten moon to cool to a solid state, he concluded that in its plastic stage, the moon's surface "would receive and retain any impressions which it might receive from without, much as the surface of a nearly dried pool receives and retains the impressions of raindrops. Or rather, as such a surface, of stones thrown upon it, allows the stones to pass through, and shows thereafter a shallow depression where the stone has fallen, so if any large mass fell upon the moon's surface while in the plastic state, the mass would pass below the surface, and a circular, saucer-shaped depression only would show where the mass had fallen."

For the purpose of his argument, Proctor limited the period of this plastic stage to three million years and, extrapolating from estimates of the rate of meteoric infall on the earth, he concluded that some ten million "large" meteorites must have fallen on the moon during this time. "Among these ten million meteorites ten only in a thousand might be very large, so as to leave where they fell circular depressions from a quarter of a mile to a mile in diameter," he noted. "For the diameter of the aerolites themselves, of course, would not be nearly so large as that of the circular depressions left where they had fallen." Moreover, among these meteorites, "probably some—it might be only one in a thousand—would be excessively large, from a quarter to a half a mile perhaps in diameter."

Proctor here noted, and then argued away an objection to meteoritic impact theory that would come up many times in future years: that no such massive meteorites were known to have struck the earth in historic times, or that no large meteoritic masses had been recognized by geologists on or within the earth's crust. But, he declared, "so many instances are on record of the passage of masses apparently as large as 100 yards in diameter through our air . . . that we cannot but believe in the existence even to this day of many enormous meteorites, and in the probability that at long intervals they fall upon our earth's atmospheric shield."

Other potentially significant ideas appear here and there in Proctor's discussion, most of which he left undeveloped. He declared without elaboration, for example, his belief that "the substance of the moon once formed a ring around the earth," and that "almost every mass which . . . strikes the moon must be vaporized by the intense heat excited as it impinges on the moon's surface." In an apparent reference to the linear configurations of small craters observed on the moon, which had been cited in support of the volcanic analogy, Proctor suggested that "in the arrangement of the smaller lunar craters, peculiarities might sometimes be recognised indicating the occasional fall of flights or strings of meteors." As to the great number of lunar craters, Proctor postulated a far greater density of "meteoric distribution in the solar domain" during the moon's plastic stage than in later epochs. And finally, he recognized that the effects of this intensified "plash of meteoric rain" in a remoter time, while they would remain on the surface on an airless and waterless moon, would no longer be apparent on earth itself. "It is certain," he wrote, "that the surface of our own earth must once have been in a similar way pitted with the marks of meteoric downfalls. . . . Yet we know that of those impressions which the earth then received no traces now remain. Again and again has the surface of our earth been changed since then. By the denudation of continents, by the deposit of strata under the seas, and by the repeated interchange of seas and continents, every trace of the primeval surface of our globe has long since been either removed or concealed."

ALTERNATIVE HYPOTHESES

Proctor's 1878 essay, which he reprinted in his book *The Poetry of Astronomy* in 1881,[48] also attracted little attention. But, in fact, interest in the moon, as reflected in the contemporary astronomical literature, was now again at a low ebb, although the lingering debate over the disappearance of Linné and the appearance of Klein's Hyginus N still surfaced sporadically in some journals. Astronomers, awakening to the rich possibilities provided by such rapidly developing techniques as photography and spectroscopy, were turning to more distant and more intriguing objects than the earth's now prosaic satellite. Except in matters relating to the mathematical theory of the moon's motions, professional astronomers in general ignored the moon, and the work of selenography was left largely to a small group of dedicated lunar observers, most of them English or German and most of whom would now be considered as amateurs. This group, furthermore, was far more interested in describing in great detail the minutest features they could discern on the moon in their small telescopes than in interpreting what they had seen or speculating on origins. Thus in 1882, Klein, in writing "On Some Volcanic Formations on the Moon," could declare that "the search after detail is, properly speaking, only now beginning," and that selenographers "have generally had in their eyes only the more universal and larger features of the lunar surface. . . . It is greatly to be desired, on the part of geologists," he added, "that the lunar features should be submitted to thorough study founded on eye-scrutiny; the result would be remarkable for the better knowledge we should gain of lunar and also terrestrial formations."[49]

That the volcanic analogy was widely if more often than not tacitly accepted at this time is also clear, though its adherents conceded problems. The English selenographer W. R. Birt, for example, wrote in 1881 that it "appears pretty evident that the manifestation of volcanic action in the production of salient lunar features has not been disputed; but are we certain that this action is still continued? and of the manner of its operation we are still quite ignorant. There is some reason to believe that it is still in operation; but the evidence we possess is not sufficiently conclusive, inasmuch as it is out of our power to point to a particular part of the moon and to say 'here volcanic action is in operation.'"[50]

Birt, parenthetically, urged his fellow selenographers to try to establish a relative chronology for the various types of lunar topography, noting that J. Chacornac in 1869 had referred "the great continental formations to an epoch anterior to that of the production of the great plains, this again being anterior to the period of explosive energy. . . . It would greatly assist our study if some attempt, however rough, were made to arrange a list of formations according to their supposed ages."

In the years immediately following Proctor's enunciation of his "meteoric" impact hypothesis, at least two other investigators, both of them outside astronomy, advanced impact explanations for the lunar craters. In 1879, architect August Thiersch and his father, Heinrich, a theologian, writing under the pseudonym of "Asterios," suggested such an origin in the course of strongly criticizing Nasmyth and Carpenter's fountain analogy.[51] And in 1877 and again in 1882, another architect, A. Meydenbauer at the University of Marburg, proposed on the basis of experiments reminiscent of Hooke's that the moon's surface was covered with cosmic dust and that the fall of masses of such dust had fashioned its topography. Meydenbauer declared that his experiments with highly pulverized materials showed "not similarity merely, but almost complete identity with the smallest details of the moon's surface, and that not in regard to the so-called craters alone." The largest craters and maria, however, he felt were formed by the impact of solid bodies of sulphur or phosphorus which, moreover, produced melting and fusion on impact.[52]

Hooke's boiling alabaster analogy also was now briefly revived. Jules Bergeron in France attempted to create somewhat larger and stronger crateriform structures than Hooke's vat had yielded by blowing streams of hot air through a mass of molten metal, in this case Wood's alloy, and succeeded in producing a "perfect model" of a lunar crater.[53] Moreover, he found that when the process was interrupted, by turning off the stream of air and then turning it on again, a second inner ring was formed that reminded him of the concentric appearance of such lunar craters as Copernicus. His experiments convinced at least one Fellow of the Royal Astronomical Society, George F. Chambers, who cited them in the 1889 edition of his three-volume work on astronomy to support the statement that the lunar craters were "probably due to the escape of gas from the interior of the moon when that body was in a semi-fluid state, as it is conceived once to have been."[54]

During the 1880s, two other hypotheses attempting to explain the moon's topography were advanced. One of these, proposed in 1886 notably by John Ericsson in England[55] but also by S. E. Peal in India,[56] was based on the assumption that water had once existed on the lunar surface and postulated that the craters on the moon were actually "annular glaciers." The lunar craters, so this reasoning ran, had once been pools of water that had been vaporized by heat from the moon's interior. The rising columns of vapor therefrom, condensing at their outer limits in the cold of space, then fell back as rain in the center of the pools and as snow at their edges. This process, continued over an extended period, gradually built up annular rings of snow and ice around the perimeters of the pools. As to the presence of water on the moon, Ericsson declared flatly that the "supposition that this neighboring body is devoid of water, dried up and sunburnt, will assuredly prove one of the greatest mistakes ever committed by physicists."

Ericsson specifically disputed the volcanic analogy, focusing his argument on the circular wall of the crater Tycho, which he calculated contained 1,382 cubic miles of material. "The supposed transfer of this enormous mass, in a molten state, a distance of 25 miles from the central vent imagined by Nasmyth, and its exact circular distribution at the stated distance, besides its vertical height of nearly 3 miles, involve, I need not point out, numerous physical impossibilities. . . . A rigid application of the physical and mechanical principles to the solution of the problem proves conclusively that water subjected successively to the action of heat and cold has produced the circular walls of Tycho. The supposition that these stupendous mounds consist of volcanic materials must accordingly be rejected, and the assumption admitted that they are inert glaciers which have become as permanent as granite mountains by the action of perpetual intense cold."

Ericsson did not banish the volcanic analogy entirely, however, for he found it convenient to account for the central peaks or cones of the lunar craters. The central part of the heated pools, he suggested, would soon acquire a high temperature, "admitting the formation of a vent for the expulsion of lava, called for as the moon, whose entire dry surface is radiating against space, shrinks rapidly under the forced refrigeration attending glacier-formation. Lava cones similar to those of terrestrial volcanoes, and central to the circular walls, may thus be formed, the process being favoured by the feebleness of the moon's attraction."

The idea that water might have once existed on the moon was not, of course, new; and Ericsson, for example, could and did cite Sir John Herschel to the point. Nor was the idea of lunar glaciation new, although it had not been previously applied specifically to the moon's craters. Some twenty years before, the English chemist Edward Frankland, whose theory of valency is one of the foundations of modern chemistry, had suggested on the basis of his own study of the lunar surface that at least a few of its features had been formed by glacier action. Frankland thought it probable that the moon, like the earth, had passed through glacial eras [ice ages] and that the valleys, rills, and streaks on its surface were evidences of this glaciation. He suggested specifically that two arcuate ridges, one of them double, associated with a broad, bright streak or ray radiating from the crater Tycho, were actually glacial terminal moraines. At the time, the noted English observer, the Reverend T. E. Webb, had pointed out that under Frankland's "ingenious theory" traces of any glacial action should be far more numerous than observation had revealed.[57] Subsequently, Proctor discussed Frankland's suggestions at some length, and also dismissed them.[58]

A second hypothesis, which gained somewhat more credence in light of the contemporaneous researches of George H. Darwin into lunar tides and tidal friction, posited the origin of the moon's topography in tidal forces. It was first proposed by the French astronomer and theoretician Hervé Faye in 1881,[59] and later advocated and developed by H. Ebert in Germany in 1890,[60] and J. B. Hannay in England in 1892.[61] It envisioned an era before the moon became locked in a synchronous rotation with the earth, and when its molten interior was racked by great periodic tides raised by the earth's gravitational attraction which cracked and tore its thin crust. With each rising tide, molten material was extruded onto the surface through these cracks, with some flowing back into the moon's interior after the tide had ebbed. Some of the material, however, and particularly at the outer limits of the flow, would congeal, and eventually a solid circular ridge would be formed. Ebert tested the efficacy of this process in the laboratory and found that he could quite closely duplicate the appearance of the lunar craters. Moreover, the tidal process could explain the presence of central peaks or cones in some lunar craters as manifestations of the final weak extrusions of progressively congealing material.

The tidal hypothesis also found a supporter in the English observer Nathaniel E. Green, who gave it a somewhat different slant.[62] Green felt that when the moon was still in a sufficiently fluid state to be strongly influenced by tidal forces, tidal friction had then brought it quite rapidly into synchronous rotation with the earth, its visible face turned always toward its primary. "This agreement once set up would expose the side turned towards the earth to a steady strain on the still liquid interior, inducing it to pour out through any openings in the crust and this tendency would be increased by the contraction of the whole mass in cooling."

None of these suggested alternatives to the volcanic analogy stimulated more than a casual, passing interest within science, and certainly none of them constituted a serious challenge to the prevalent belief that the craters on the moon could best be accounted for by volcanic processes of some sort. But they were nevertheless symptomatic of a gradually growing suspicion on the part of at least some people that the volcanic analogy was seriously flawed and that the differences between the lunar craters and terrestrial volcanoes were indeed far greater than the similarities.

By this time, however, the volcanic analogy had become deeply entrenched not only in astronomy and selenography, but in geology as well, as a number of works by eminent geologists attest. In the 1880s, for example, geologist John W. Judd, a Fellow of the Royal Society, declared in the several editions of his book *Volcanoes* that the moon "exhibits on its surface sufficiently striking evidences of the action of volcanic forces. Indeed the dimensions of the craters and fissures which cover the whole visible surface are such that we cannot but infer volcanic activity to have been far more violent on the moon than it is at the present day upon the earth."[63] And Dana, in his widely used *Manual of Geology*, stated flatly that the moon "presents to the telescope a surface covered with the craters of volcanoes, having forms well-illustrated by the earth's volcanoes, although of immense size. The principles exemplified on the earth are but repeated in her satellite."[64]

The then unchallenged assumption behind such statements, of course, was that all naturally formed terrestrial craters were volcanic in origin. But this assumption now would be brought into question through inquiries by a renowned geologist into the origin and nature of a peculiar crateriform structure on the earth itself, previously unknown to science. This curious feature, on an uninviting high desert plain in the northern part of what was then the Arizona Territory, was known locally by several names, but more often than not it was called "Coon Mountain."

TWO

IRON AND DIAMONDS

> Just as in the domain of matter nothing is created from nothing, just as in the domain of life there is no spontaneous generation, so in the domain of mind there are no ideas which do not owe their existence to antecedent ideas which stand in the relation of parent to child. It is only because our mental processes are largely conducted outside the field of consciousness that the lineage of ideas is difficult to trace.
>
> GROVE KARL GILBERT, 1896

On August 20, 1891, an eminent mineralogist, Arthur E. Foote, reported to a meeting of the American Association for the Advancement of Science in Washington, D.C., that a large number of meteoritic irons had been found at the base of a "circular elevation . . . occupied by a cavity" a few miles southeast of the Canyon Diablo trading post in northern Arizona. One of the irons, he added, had been found to contain tiny diamonds.[1]

Foote's announcement of the discovery of diamonds alone was sufficient to capture the interest of his audience, which included leading members of the various scientific societies in the Washington area. But at least one of his listeners, Grove Karl Gilbert, chief geologist for the U.S. Geological Survey, was more intrigued by the association of meteoritic irons with what Foote had described as a large crateriform structure not obviously volcanic in origin. In the discussion that followed Foote's presentation, Gilbert remarked that "this so-called crater"—Foote had used the word sparingly and with care—"was like the depressions on the surface of the moon produced by the impact of an enormous meteoric mass."[2]

Gilbert's suggestion was not without its antecedents, and he himself later recalled that at the time he had in mind the idea that the earth and the other solar system bodies might have been formed by the accretion of meteoric material, and the possible effects such a process might have on terrestrial landforms. "Speculating on this line, I had asked myself what would result if another small star should be added to the earth, and one of the consequences that occurred to me was the formation of a crater." This idea had come to him from "the many familiar instances of craters being formed by collision," and here he cited raindrops falling in soft ooze, pebbles thrown into pasty mud, and steel projectiles firing into armor plate—familiar examples indeed. "Analogy," he added, "easily bridged the interval from the cannon ball to the asteriod. So when Dr. Foote described a limestone crater in association with iron masses from outer space, it at once occurred to me that the theme of my speculations might here find its realization."[3]

The events which led Foote to make what was the first scientific investigation of Coon Mountain and its crater, however, had nothing to do with such speculations. They were inspired entirely by a more mundane concern for pecuniary profit, a concern which pervades much of the history of this unusual feature.

DISCOVERY OF THE IRONS

The irons were first discovered late in 1886 by one Mathias Armijo, a herder for a sheep owner in the area named Harry Melbourne.[4] In all Armijo found about ten pieces, the largest weighing 154 pounds. From their weight and their metallic luster, he thought they were lumps of silver, not an uncommon assumption on the part of finders of meteoritic iron in the nineteenth century,[5] and consequently he said nothing about them for several years. Early in 1891, however, he showed his finds to the sheep camp's cook, an erstwhile prospector named Craft, who recognized them as iron and, seeing "a possibility of profit," as Gilbert later remarked, quickly staked out mining claims at Coon Mountain.[6] In March Craft submitted samples of the iron to the mining firm of N. B. Booth & Company in Albuquerque, New Mexico, claiming that he had found a "vein" of "pure metallic iron" forty yards wide, two miles long, and "finally disappearing into a mountain," from which a "carload" of the metal could be picked up from the surface with but little effort. Craft wanted the Booth firm to interest a railroad company in buying his claim.[7]

The Booth firm immediately sent one of the pieces to an assayer for analysis and then arranged for Craft to show the property. These negotiations were not concluded, however, because in Gilbert's words, "Mr. Craft, having borrowed money on the strength of his great expectations, mysteriously disappeared."[8] Instead, a man named Hollingsworth, who had grubstaked Craft, appeared at the crater and began staking out claims of his own.[9]

In April, the Booth firm received the assay report which showed that the sample was composed of "75.7 percent iron, 1.8 percent lead, ½ oz. of silver, and a trace of gold." This was a highly unusual finding, and as Foote pointed out, the presence of lead, silver, and gold in the iron were "probably the results of the materials used in making the assay." Nonetheless, the Booth firm now had a forty-pound iron broken up with a trip hammer and sent one of the pieces to General James A. Williamson, land commissioner for the Atlantic and Pacific Railroad,* whose mainline tracks to California ran some five miles north of Coon Mountain. Williamson, in turn, sent the piece to Foote at his mineralogical establishment in Philadelphia along with a description of Craft's claims and a request for Foote's opinion as to the value of the purported "mine."[10] Thus did Coon Mountain come to the attention of science.

Foote immediately recognized the iron as meteoritic from its "peculiar pitted appearance" and from the "remarkable crystalline structure of the fractured portion," concluding that "the stories of the immense quantity were such as usually accompany the discovery of so-called native iron mines or even meteoric stones." In June, Foote journeyed to Coon Mountain and confirmed that the quantity of iron had been "greatly exaggerated," but he conceded that the area displayed "some remarkable mineralogical and geological features which, together with the character of the iron itself, would allow a good deal of self-deception in a man who wanted to sell a mine."

In describing the "circular elevation" near which the irons had been found, Foote erroneously identified it from a U.S. Geological Survey map as "Sunset Knoll," a somewhat higher volcanic feature nine miles to the southeast of Coon Mountain. The "cavity" in its center, he reported, was about three-quarters of a mile in diameter and from 50 to 100 feet deep, "the sides of which are so steep that animals that have descended into it have been unable to escape and have left their bleached bones at the bottom." Foote did not see any evidence that might suggest Coon Mountain's origin in what he briefly surveyed, although the possibility of volcanism did occur to him. "The rocks which make up the rim of the so-called 'crater' are sandstones and limestones and are uplifted on all sides at an almost uniform angle of from thirty-five to forty degrees," he noted. "A careful search, however, failed to reveal any lava, obsidian, or other volcanic products. I am therefore unable to explain the cause of this remarkable geological phenomenon."

As a mineralogist, Foote quite naturally devoted the bulk of his report to the meteoritic irons. In a "thorough examination of many miles" of the surrounding plain he and a crew of five men turned up only three irons of any appreciable size, the largest weighing 201 pounds and showing "extraor-

---

*Later the Atchison, Topeka, and Santa Fe Railway.

dinarily deep and large pits, three of which pass entirely through the iron," and a total of 131 smaller masses ranging in weight from one-sixteenth of an ounce to just over six pounds. But following Foote's departure, local interest stimulated by his visit led to the discovery of another ton of meteoritic material, including three large irons weighing 625, 506, and 145 pounds, along with many smaller pieces of up to seven pounds, some of which were found buried in the ground. Traders in the area, notably Fred W. Volz, proprietor of the Canyon Diablo trading post, developed a lucrative business selling meteorites, which became known as "Canyon Diablo irons" to museums and other collectors throughout the world until, in a few years, the market was glutted.[11]

Foote and his crew had also collected several hundred pounds of a black "oxidized and sulphuretted" material scattered among the irons at the base of Coon Mountain that appeared to be identical to a substance incrusting some of the irons and partially filling some of their pits.[12] Also, he noted a brownish-white incrustation on a few of the irons which he thought was "probably aragonite," a form of calcium carbonate ($CaCO_3$) often associated with iron ore deposits. Such an incrustation, he pointed out, "has rarely been found" on meteoritic irons and was "especially interesting as showing that the meteoritic irons must have been imbedded a long time, as the formation of aragonite would be exceedingly slow in this dry climate."

As Foote had other business in the West, he was not able to return directly to Philadelphia at the conclusion of his brief investigation, and consequently he shipped one of the larger irons to Professor G. A. Koenig of the University of Pennsylvania for immediate metallurgical examination. Koenig found the iron to be of such "extraordinary hardness" that it required a day and a half to cut a section for study, with several chisels being "destroyed" in the process. The section revealed cavities containing a "pulverulent" iron carbide as well as the minerals troilite and daubreelite, both previously known to occur in meteorites. Moreover, in an attempt to polish the exposed surface to bring out the so-called Widmanstätten figures characteristic of meteoritic iron, an emery wheel was ruined; and on closer examination, Koenig found that the cavities also contained "diamonds which cut through polished corundum as easily as a knife through gypsum."

To Foote, as to most of his listeners, this first discovery of diamonds in meteoritic iron was the most important result of his investigation. "The diamonds exposed were small, black and, of course, of little commercial value," he declared. Actually, he added, one of the cavities in the iron, when treated with acid, had yielded a single white diamond 0.5 millimeter in diameter which, unfortunately, had been lost "in manipulation." The iron itself contained about three percent nickel, Koenig reported, and about 1.5 percent cobalt, both figures which later proved to be off the mark; and as the black oxidized material found on, in, and around the irons seemed to contain these elements in about the same proportion, he concluded that it had "therefore presumably derived by a process of oxidation and hydration of the latter." Foote himself suggested that the "remarkable quantity" of this oxidized material "would seem to indicate that an extraordinarily large mass of probably 500 to 600 pounds . . . had become oxidized while passing through the air and was so weakened in its internal structure that it had burst into pieces long before reaching the earth."

For Gilbert at least, Foote's report effectively disposed of the mineralogy at Coon Mountain, although, as will be seen, there was more mineralogy there than had met Foote's eye. "For the crater of non-volcanic rock he offered no explanation," Gilbert later noted, "but the iron he pronounced of celestial origin—a shower of fallen meteors. It has long been known that many of the bodies that reach the earth from outer space are composed of iron, and that such iron is of a peculiar character. . . . So Doctor Foote, in characterizing the iron as meteoric, merely referred it to a well-established class. His explanation was not tentative, but final, and has not been called into question by any subsequent investigator."[13]

Gilbert seems to have been the only one in Foote's audience to grasp the broader implications of Foote's report and to recognize that there was much more to be explained at Coon Mountain than the nature and composition of a few hundred pounds of iron and iron oxide. As a geomorphologist whose primary concern was with terrestrial landforms and the processes which created them, he was fascinated by the problem presented at once by Coon Mountain's crater and the occurrence of meteoritic iron in its immediate vicinity.

GROVE KARL GILBERT

In August 1891 when Foote spoke, Gilbert, at age forty-nine, was already renowned, even revered as a scientist, and well on his way to eventual canonization by his colleagues in geology in general and the members of the U.S. Geological Survey in particular. There are a number of more or less excellent memorials and biographies of him extant,[14] but it is necessary here to review briefly his life and work, since his conclusions at Coon Mountain, and more important the authority that his great prestige gave to these conclusions, are factors in the subsequent Coon Mountain controversies.

A native of Rochester, New York, Gilbert had studied mathematics and the classics at the University of Rochester, graduating in 1862 and receiving a master's degree there in 1872. Following his graduation, he had tried teaching, but soon took a clerical position in that remarkable Rochester institution of science and commerce, Ward's Natural History Establishment, where, despite the tediousness of his work, he developed a casual interest in geology into a commitment to a career. After six years of sorting and classifying specimens and occasional forays into the field to collect fossils for Ward's inventory, Gilbert in 1869 joined the Ohio State Geological Survey. There, he served a two-year geological apprenticeship under Dr. John Strong Newberry, the surgeon and geologist who had accompanied Lt. Joseph Christmas Ives of the U.S. Army Topographic Engineers up the Colorado River by steamboat in 1857 and had made the first geological studies of its awesome Grand Canyon.[15] In 1871 Gilbert became a member of the Army's Geographical Survey West of the 100th Meridian, headed by Lt. George Montague Wheeler,[16] and in the course of this work made extensive studies in Utah, Arizona, and New Mexico which led, among other things, to his recognition and definition of the distinctive landforms of the West's vast Basin and Range Province.

From time to time during his travels for the Wheeler Survey, Gilbert encountered Major John Wesley Powell, the energetic, one-armed Civil War veteran who had become famous for his daring explorations of the Colorado River and its great gorges in 1869 and 1871. Gilbert was sufficiently impressed by the ebullient major's grasp of scientific problems and his forcefulness and imagination in pursuing them, that Powell in 1874 had no trouble persuading him to join his own Geological and Geographical Survey of the Rocky Mountain Region, the youngest and smallest of the western surveys. Although of widely differing backgrounds and personalities, both men had many interests in common and quickly became fast friends, with Gilbert serving as Powell's close confidant for the rest of Powell's life and as his executor and his first and most sympathetic biographer after his death in 1902.

As a member of the Powell Survey, Gilbert spent three extended field seasons in the West studying landforms, primarily in Utah, where his work helped to build the foundations of modern geomorphology. In 1879, when political bickering among the various competing and often overlapping surveys forced their consolidation,[17] Gilbert became one of six senior geologists in the new U.S. Geological Survey and remained associated with it until his own death in 1918. In 1881 Powell himself took over the directorship of the new agency and increasingly over the ensuing years, as his varied other interests occupied more and more of his time, Powell turned over its administrative responsibilities to Gilbert. Gilbert, although he disliked such work and not alone because it severely curtailed his own research activities, accepted it gracefully and performed it well. In 1888 Powell appointed him chief geologist, a post which further reduced his research opportunities.

During his forty-nine year career, Gilbert made many important contributions to geology, particularly in geomorphology, but also in such areas as glaciology, hydrology, and even selenology. In all he published more than 150 papers, reports, and monographs, the majority of them after 1892, some of which remain classics of the geological literature. He was a lucid, if somewhat stilted and pedantic writer and an excellent illustrator, as well as an articulate speaker who sometimes reported his researches, including his investigation of Coon Mountain, in the form of carefully prepared addresses often keyed to philosophical themes, which he delivered before one or another of the many scientific organizations in which he was active. He is most renowned, however, for two monographs, his *Report on the Geology of the Henry Mountains*, published in 1877,[18] and his *Lake Bonneville* which, although based largely on his early work with the Wheeler and Powell surveys, did not appear until 1890.[19] In the first of these, Gilbert developed the concept of dynamic equilibrium, the idea that landforms reflect a state of balance between the processes acting upon them and their structure and composition, and described an entirely new landform which he called a "laccolite,"* domed mountains of eroded sedimentary strata uplifted by the intrusion of molten magmatic material, which he found not only in the Henrys but in other isolated ranges of southeastern Utah as well. In his *Bonneville* monograph, which he considered his most important work, he described the vast Pleistocene pluvial lake that had once covered much of the northern part of Utah, of which the Great Salt Lake is today but a dwindling remnant, and the hydrostatic forces that had shaped its vast basin as its thousand-foot deep waters gradually evaporated.

Gilbert, his biographers agree, was a scholarly man with a reflective and philosophical bent of mind and a dry sense of humor. He was gentle, friendly, tolerant, even-tempered, and uncomplaining, whether enduring the intellectual tedium of bureaucratic work in Washington or physical hardship in the field, officious politicians on the banks of the Potomac or recalcitrant pack mules on a precipitous mountain trail. His sense of fairness, his willingness to share his knowledge and experience, and his hesitancy to criticize others were such that his advice and counsel were widely sought and highly prized, and he became a sort of father-advisor to his colleagues in and out of the Survey. He acquired enormous prestige and authority as a scientist, and the word "beloved" has been used to describe him as a man. He was, perhaps, the closest equivalent to a saint that American science has yet produced.

*Later changed to "laccolith" to avoid confusion with mineralogical nomenclature.

In his science, Gilbert was pragmatic and methodical, combining careful, systematic observation with an uncomplicated and uncritical form of analogic reasoning. Simple analogy, in fact, loomed large in his thinking, and just as he believed that all ideas had their antecedents in other ideas, so he believed that hypotheses, which were no more than mere "scientific guesses," were "always suggested through analogy." An unexplained phenomenon "resembles in some of its features another phenomenon of which the explanation is known," he noted. "Analogic reasoning suggests that the desired explanation is similar in character to the known, and this suggestion constitutes the production of a hypothesis."[20] A thorough knowledge of the "consequential relationships of nature" was helpful, he thought, "and he who has the greatest number has the broadest base for the analogic suggestion of hypotheses."[21]

Thus for Gilbert analogy was the mother of invention. The researcher's first task, then, was to use analogies to "invent" hypotheses, and the more the merrier for any given problem, as multiple hypotheses tended to eliminate any personal bias that the investigator might bring to a single, self-invented hypothesis. "The great investigator is primarily and preeminently the man who is rich in hypotheses," he declared. "The man who can produce but one cherishes and champions that one as his own, and is blind to its faults. With such men, the testing of alternative hypotheses is accomplished only through controversy. Crucial observations are warped by prejudice, and the triumph of truth is delayed."

Once invented, the multiple hypotheses must, in turn, be subjected to "crucial tests," also to be invented by the investigator, by which they could be measured against the physical evidence of the field or the laboratory to determine which among them best accounted for the phenomenon under consideration. "If the phenomenon was really produced in the hypothetic manner," he noted, "then it should possess, in addition to the features already observed, certain other specific features, and the discovery of these will serve to verify the hypothesis. . . . If they are found the theory is supported; and in case the features thus predicted and discovered are numerous and varied, the theory is accepted as satisfactory. But if the reexamination reveals features inconsistent with the tentative theory, the theory is thereby discredited, and the investigator proceeds to frame and test a new one."[22]

The failure of an hypothesis to survive such tests, however, was neither to be deplored or ignored. "Rejected hypotheses," he pointed out, "have a positive value in the domain of the subject in which they belong, and he who makes them public gives to his fellow-workers in the special field the fullest advantage of his material. Some steps of his progress, which did not prove suggestive to him, will find fertile ground in the mind of another and bear fruit. This consideration," he added, "places the progress of knowledge before the glory of the individual. . . ."[23]

This quite straightforward method, which Gilbert noted had its own antecedents and thus was not original with him,[24] he applied at Coon Mountain where he apparently felt it served so well that he subsequently cited his investigation there as an illustration of the method itself. But the results thus obtained, he emphasized then, "are ever subject to the limitations imposed by imperfect observation. However grand, however widely accepted, however useful its conclusion, none is so sure that it can not be called into question by a newly discovered fact. In the domain of the world's knowledge," he declared grandiloquently, "there is no infallibility."[25]

On these latter points, however, it must be noted that for Gilbert, controversy of any kind, but especially scientific controversy, was extremely distasteful; and even when he was strongly provoked, as he must have been during the Coon Mountain controversies, he contrived to avoid personal involvement. When his own ideas were criticized, which was seldom enough, his response was to ignore both the critic and the criticism, a response that some of his biographers suggest reflected his belief that controversies too often involve conflicts of personalities rather than of ideas or issues, as well as his strong desire to eliminate the personal factor from science, certainly a noble goal.[26] But silence, of course, can be itself a form of argument and, when used by someone of Gilbert's stature, can carry great weight, suggesting as it so often does that the unanswered criticism is not worthy of comment. Moreover, silence is an extremely difficult argument to refute.

As it turns out, Gilbert's conclusion regarding Coon Mountain's origin was simply wrong, a fact which in itself is of no great importance, for error in science is common enough and can be corrected in time and with further work, as Gilbert surely understood. A more grievous error, certainly, in view of his towering prestige and, indeed, his own stated position on such matters, was his silence on the problem of Coon Mountain's origin once his own conclusion regarding it was "called into question" by new and significant evidence presented by others. The effect of his silence, which he maintained to his death, was compounded by the refusal of his colleagues and successors, particularly in the Survey, to publicly question his great authority, to place "the progress of knowledge before the glory of the individual" as it were, even when they privately conceded that he had been wrong. As a result, a fuller understanding of the processes involved in Coon Mountain's origin, and indeed the development of meteoritic impact theory in general, undoubtedly was delayed for many years.

Foote's report had strongly aroused Gilbert's curiosity. "The thought of examining the scar produced on the earth by the collision of a star was so attractive that I desired to visit the crater," he later recalled.[27] His duties in Washington, however, made this immediately impractical, and consequently he dispatched a colleague, Willard D. Johnson, to make a reconnaissance of Coon Mountain and report back on what he found there. Johnson, a civil engineer who had served as Gilbert's topographer in the Powell Survey and on later field excursions and who was knowledgeable and experienced in geological work, spent several days at the crater in the early autumn of 1891, studying its rock formations, sketching a map, and reporting his findings by letter to Gilbert, who incorporated them in his later summary of his own investigation. At first Johnson thought the crater might represent the wreck of a laccolite, reasoning from the upturned, outward dipping strata he found as he climbed to the top of its low rim. But after descending into the crater and examining its walls and floor, he rejected this hypothesis because he could find no trace of the igneous rocks required for the formation of such structures; and, moreover, the hypothesis could not account for the vast crater itself. He also thought that the crater was somewhat oval, with the longer diameter running east and west, and that the east rim contained more material than the west rim, and briefly speculated that it might represent the scar left by a ricocheting meteorite.

His final conclusion, however, was that the crater had been formed by a subterranean explosion and that, as Gilbert later paraphrased his report, "in some way, probably by volcanic heat, a body of steam was produced at a depth of hundreds or thousands of feet, and the explosion of steam produced the crater." On this hypothesis, Gilbert noted, the presence of the meteorites was incidental, and their association with the crater accidental.[28]

Johnson's description of Coon Mountain was more complete than Foote's had been, "but instead of satisfying my curiosity," Gilbert later recalled, "tended to whet it, and I availed myself of the first opportunity to make a personal visit." Foote had effectively disposed of any idea that the irons might be terrestrial and native to the area, and Johnson had eliminated the idea that the crater might have been originally a laccolite. For Gilbert, only two viable hypotheses concerning the crater's origin remained to be considered—Johnson's volcanic steam explosion idea and his own suggestion of meteoritic impact. "If my visit was to aid in the determination of the problem of cause," Gilbert later wrote, "it must gather the data which would discriminate between these two theories, and an attempt was accordingly made to devise crucial tests."[29]

GILBERT VISITS METEOR CRATER

In devising these tests, Gilbert made an important assumption, one which would be bitterly debated nearly forty years hence when the Coon Mountain controversies came to a head. "If the crater was produced by the collision and penetration of a stellar body, that body now lay beneath the bowl," he reasoned, "but not so if the crater resulted from explosion. Any observation that would determine the presence or absence of a buried star might therefore serve as a crucial test. Direct exploration by means of a shaft or drill hole could not be undertaken on account of the expense," he added, "but two indirect methods seemed feasible."[30]

The first of these was strictly quantitative. "If the crater were produced by explosion, the material contained in the rim, being identical to that removed from the hollow, is of equal amount," he noted, "but if the star entered the hole the hole was partly filled thereby and the remaining hollow must be less in volume than the rim."[31] Gilbert thus simply proposed to compute the two volumes and compare them.

His second test was scientifically somewhat more sophisticated and involved an additional assumption that if the crater concealed a "buried star," the star itself should be similar in composition to the meteoritic irons scattered over the surrounding plain, and therefore its presence or absence might be detected with a magnetic needle. "If it were absent," he explained, "the compass would point in the same direction, whatever its position in reference to the crater, whether within or without, on one side or the other, nearby or miles away; but if a mass of iron large enough to produce the crater lay beneath it, its attraction would pull the needle one way or the other, producing local variations."[32] This assumption, as Gilbert would discover, also posed some unforeseen problems.

Late in the afternoon of October 22, 1891, Gilbert boarded a train at Washington for the four-and-a-half-day trip to Flagstaff, Arizona Territory, a small lumbering and ranching community at the foot of a 12,670-foot extinct volcano known as the San Francisco Peaks some forty-five miles west-northwest of Coon Mountain.[33] He was accompanied by Marcus Baker, another Survey colleague who was experienced in measuring terrestrial magnetism; and, along with the paraphernalia required for routine geological field work, they carried with them a full set of instruments for making magnetic surveys.

That Gilbert expected his investigation to show that Coon Mountain and its crater had originated in a meteoritic impact is clear enough. "At this time," he later wrote, "it seemed to me that the presumption was in favor of the theory ascribing the crater to a falling star, because the theory explained, while its rival did not, the close association of the crater with the shower of celestial iron. . . . So when Mr. Baker and I started for the crater it seemed rather probable than otherwise that we should find a local deflection of the magnetic needle, and that we should find the material of the rim more than sufficient to fill the hollow it surrounds."[34]

Gilbert and Baker arrived in Flagstaff late in the afternoon of October 26. The next morning, Gilbert rode the train eastward to the Canyon Diablo station to learn what he could about Coon Mountain and the Canyon Diablo irons from Volz, the trader there, and his employees. Gilbert's notes contain no entries for Wednesday, October 28, beyond the fact that he was in Flagstaff and that he dispensed $1.50 for what he recorded as "antics." Surely, however, he spent part of the day assembling men, equipment, and supplies, and arranging for horses and a wagon for the expedition to Coon Mountain. The following morning Gilbert and Baker, with teamster Gaines Cameron, cook George Haldy, and striker John W. Tinker, began a three-day trek to the crater, riding southeast from Flagstaff through a forest of ponderosa pines to Mormon Lake and Mormon Mountain, then east across "Ashurst" Mesa,* where the tall pines yielded to scrubby junipers, and finally down onto the treeless western reaches of the broad, arid valley of the Little Colorado River. Gilbert geologized as he rode, recording abundant evidence of former volcanic activity in the area during the first two days of the trip. On the final day, however, as they descended the eastern rim of the mesa, the ancient lava rocks vanished and the underlying sedimentary stratum, then known as the Aubrey limestone,† appeared beneath the sparse, sandy soil of the barren, windswept plain. The party reached Coon Mountain on Saturday evening.[35]

For the next seventeen days Gilbert occupied himself with topographic studies of what he now referred to in his notes as "the amphitheatre," making a plane table survey of the entire area, examining the structure and stratigraphy of the rocks, and taking a series of photographs of the crater and its environs. Baker assisted in this work, but his principal concern was to survey the area for magnetic anomalies. To this end, he established two lines of magnetic stations at right angles to each other across the crater, one of which was on the magnetic meridian and extended for more than three miles out onto the plain. At each station, the horizontal direction of the compass needle, its vertical dip, and the magnetic intensity were measured.

---

*Now Anderson Mesa. "Ashurst" was probably a local reference to the pioneer family that had operated a ranch stop atop the mesa for several years up until 1890.
†Now called Kaibab limestone.

Other members of the party were assigned the task of scouring the outer slopes of the crater rim and the surrounding plain for meteorites, volcanic rocks, or anything else that might look interesting. Along with a large number of irons, this search turned up only five small pieces of volcanic rock, two of which were worn fragments of vesicular basalt which Gilbert concluded were "undoubtedly industrial and imported" and all of which he thought probably had been brought to the crater by Indians. Gilbert himself found a walnut-sized "piece of black lapilli" and wondered whether it had been "imported by accident—or by the wind?"[36]

The crater's dimensions proved to be somewhat different from Foote's estimates. While it was indeed about three-quarters of a mile in diameter, Gilbert found its floor to be some 400 feet below the surface of the surrounding plain, or more than 300 feet deeper than Foote had thought, and its rim to be only 150 to 200 feet above the plain, less than half the height Foote had surmised. Nor was the crater somewhat oval, as Johnson had thought, for Gilbert's measures showed only minor differences in its various diameters.[37]

Close inspection of the crater's inner walls showed that it had in fact been formed in nonvolcanic rocks. Descending stratigraphic sections at various points revealed remnants of a layer of red Moenkopi sandstone up to twenty feet thick, then up to 250 feet of the cream-colored Aubrey limestone underlain to an unknown depth by a yellowish-gray Aubrey sandstone* that was largely hidden from view by talus sloping down to the sandy, rock-strewn floor of the crater. From the talus and the condition of the floor, Gilbert inferred "erosion amounting to several tens of feet since the building." Between the limestone and the underlying sandstone, there was also a zone of mixed shattered and brecciated rock. The exposed strata of the walls were sharply upturned and dipped outward at angles ranging from 15° to 80°, and he noted evidences of upward faulting, shearing, and crushing of the rock in the crater's walls. No meteoritic material was found within the crater itself, however.[38]

The same rock he found inside the crater he found on its rim. There, he noted a great jumble of limestone and sandstone boulders and blocks (he estimated that one of the larger ones weighed perhaps 500 tons) intermixed with smaller fragments grading down to grains, along with some meteoritic iron and "a plentiful black mineral" which was associated with the scaly incrustations on some of the meteorites and with what he called "slaggy masses." This, presumably, was the same black "oxidized and sulphuretted" material that Foote had found in such profusion around the crater. The debris atop the

---

*Now called Coconino sandstone.

rim extended radially outward to form a continuous mantle over its outer slopes and beyond, the fragments of limestone and black scale gradually diminishing in size and quantity until, at a radius of about three and a half miles from the rim's crest, they disappeared. On the eastern and southern slopes, he found a "pumice" noticed earlier also by Volz and Johnson and which he thought resembled in structure "the amorphous or coralline tufa of Lake Bonneville," along with deposits of a white sand, and a distinctive "soft white earth," all of which he considered to be part of the ejecta from the crater.[39]

At first glance, Gilbert was struck by the resemblance between the topography of the crater's rim and a glacial moraine, and specifically the great terminal moraine of the Laurentian region of the northeastern United States with which he was familiar. But the absence of glacial phenomena in the neighboring mountains and the yellow sandstone in place hundreds of feet below indicated to him that it was not a moraine. Yet "certainly the age of the deposit is of the same magnitude with the great moraine." The glacial analogy interested him enough, in fact, that just before closing out his notebook on his Coon Mountain investigation, he discussed it again. "How closely the description of the ejecta matches the moraine of the great glacier! Irregular mounds, grouped sometimes in indefinite lines, originally enclosing basins, studded with occasional or numerous boulders and large blocks which are not water worn, and which are derived from the subjacent bed rock or from still lower strata." But the rocks were not striated, as would be expected in glacial action, and moreover: "Tracts several hundred feet in extent consist of one kind of rock and contiguous tracts of another. In these tracts, the orientation of the original dip of the fragments shows that they were originally joined in blocks of immense size (100–300 ft.) and have fallen apart where they lie. . . ."[40]

Gilbert's days at the crater were spent mainly in routine topographic work and in studying the configuration of the surrounding plain to learn its precrater contours. He spent his evenings in his tent at the camp on the crater's west rim, writing notes on his observations and recording his thoughts on the day's work. There is only a single entry in his notebook that does not relate directly to this work. "Today," he wrote on Friday, November 6, "I have seen two hawks riding the wind as it rises against the west face of the rim." The routine was broken briefly, too, when a Washington colleague, geologist George Perkins Merrill of the U.S. National Museum, visited the crater, arriving Sunday evening, November 8, and leaving the following Tuesday morning.[41] Merrill, who was in the area collecting specimens of volcanic rock for exhibition at the forthcoming Columbian Exposition,[42] would visit the crater again some fifteen years hence, and eventually contribute to the understanding of the processes that had created it. But his thoughts on this first visit are not known; at least Gilbert did not record them.

Scattered among the observational data in Gilbert's notes are comments and speculations about Coon Mountain and the progress of his investigation there. After only four days at the crater, he summed up his first impressions, reviewed his two principal hypotheses, and indicated that all was not going well with Baker's magnetic tests. "It is evident that the disturbance was violent," he wrote. "Rocks of great size are moved to great distances and all must have been moved upward. Two general processes seem competent: (1) the fall of a star, (2) the explosion of some substance beneath. The fall of a star is supported—was suggested—by the meteoric iron adjacent and associated. It lacks the important support of magnetic variation."[43]

Baker's magnetic measurements, in fact, were yielding negative results, and Gilbert now questioned the sensitivity of his instruments. Over the next few days, he performed a series of experiments in his tent with a small bar magnet and a compass to determine by extrapolation "the depths apparently necessary to have the supposed star out of range of observation." His computed results indicated that any appreciable mass of iron should be detectable at a reasonable depth below the crater floor. Then, postulating a hypothetical meteorite a half-billion cubic feet in volume, 220 billion pounds in weight, and at a depth of "probably not over 2,000 feet," he calculated that if such a mass had no influence on Baker's instruments, then a 635-pound meteorite found near the crater should have no effect on the compass needle at a distance of only one and a third inches. This, experiment showed, was "highly absurd" and, he concluded, "the observations have gone far enough to show that a large meteor does not lie near the surface. Either the hole was not made by a meteor, or the meteor went through the crust. What is the relation between the weight, size, velocity and penetration of a projectile?" he wondered. "For the same materials and velocity does a large projectile penetrate farther than a small? I think it will."[44]

The absence of any magnetic anomalies continued to concern Gilbert, and consequently he soon devised a second method for detecting magnetically the presence or absence of a buried meteorite. "I suspect that the needle does not afford the means to determine the depth of the star, nor its presence," he wrote on November 12, pointing out that even small meteoritic irons seemed to attract both ends of the needle more or less equally and that at a distance the difference would be too small to produce observable results. "But the attraction should change the weight," he reasoned. "A magnet should weigh more over the star than with the star away, and it should weigh more at the bottom of the crater than on the rim if the star is in there. The delicate way to measure differences in weight is by the pendulum, a magnetic pendulum should swing quicker over an attractive body." A series of further experiments with the magnet and various small meteorites persuaded him that such a pendulum "should be sensitive to the star."[45]

Over the next two days Gilbert and Baker conducted tests with the small magnet suspended as a pendulum at the bottom of the crater and more than 500 feet above the floor on the rim, finding that it "swings just as fast at the top of the rim as at the bottom of the hole. It follows, then, from my postulates," Gilbert concluded, "that the meteor is not in the hole. It is still worth while to test the postulate by observations near a great mass of iron, but chief attention should be given to other explanations of the crater."[46]

Gilbert now turned to his alternative hypothesis—that the crater had been formed by a volcanic steam explosion. "Can any explosive action be conceived as possible?" he had written in his notebook four days after arriving at Coon Mountain. "Whatever burst Krakatoa is sufficiently powerful. It was probably steam. If steam lifted this [the crater], it lifted 1,000 feet or more of rock and its pressure was 1,000 or more pounds to the inch. To exert this pressure, its temperature must be high, but probably not higher than liquid basalt. The country abounds in basaltic eruptions, and it is conceivable that an injection occurred beneath this. If steam thus blew the top off, why did not lava follow up?"[47]

Following the failure of the pendulum tests, Gilbert returned to this line of speculation in his notes. "If a missile did not enter the crater, it was made by some cause operating from below," he opined. "A gas accumulation with competent energy could hardly have been attained gradually. The covering rocks are too porous." There must, then, have been a rapid conversion of solid or liquid material into gas, he reasoned, and the only such "natural conversion" known to geology was that from water into steam. "In volcanism that is the great explosive. To explode water in a leaky vessel, the sudden application of great heat is necessary. This can be accounted for by the intrusion of lava." To his earlier question as to why the lava did not "follow up" after the explosion, he now answered that it "must be anhydrous not to froth over." The entire process, he pointed out, would be facilitated if the water were preheated, adding that it "must of course be superheated just before the explosion."[48]

Gilbert also concluded that the "locus of the explosion is manifestly at the base of the Aubrey sandstone," and that it had generated enough energy to lift between 1,000 and 1,500 feet of rock from the crater. Here he pointed out that the competency of steam to produce this amount of energy could be tested quantitatively by making a "crude estimate" of the energy required to excavate the crater in terms of foot pounds of work performed, and comparing this with the energy in the form of steam produced by the heat of a unit of molten lava. This, indeed, he would later do and, although he conceded his result was a "rough approximation," he concluded that "it served to show that the assumed cause was of the same order of magnitude as the result accomplished."[49]

Gilbert spent his final four days at Coon Mountain completing his topographic survey and making additional photographs. On Wednesday morning, November 18, the expedition broke camp and rode to the railroad at Canyon Diablo, where some instruments and equipment, specimens of rock and meteoritic iron, and photographic negatives were shipped back to Washington.[50] From Canyon Diablo the party made its way to McMillan's ranch, north of Flagstaff, from which, on November 20, Gilbert and Baker embarked on nine more days of field work involving "comparative studies" of the great volcanic field around the San Francisco Peaks with its multitude of lava flows and cratered cinder cones.[51] There, according to a subsequent report in Flagstaff's weekly newspaper, the *Coconino Sun*, "a search was made among the craters for the purpose of finding one similar to Coon Mountain but none was discovered."[52] Actually, Gilbert found at least three structures that brought Coon Mountain to mind, including one in particular, known as "Moon Crater," whose walls "resemble those of Coon Crater" and whose floor, like those of Coon Mountain's crater and the craters on the moon, was lower than the level of the surrounding plain. Other evidences of a more gradual volcanic process, however, dissuaded him from classifying Moon Crater as "an explosion crater of the Coon type."[53]

The *Sun*, incidentally, summed up Gilbert's Arizona expedition with admirable detachment in a report, entitled "Meteor or Volcano—Which Was It That Made the Coon Mountain Crater," that appeared a few days after Gilbert's departure for Washington. "While there is no evidence of volcanic action," the anonymous author wrote, "there has evidently been an eruption there as rock has been blown from the bottom to the top, and enough of it to make a mountain 150 feet high." The article then briefly described Gilbert's two principal hypotheses of the crater's origin and predicted that his final report "will be interesting to scientists as well as the public generally."[54] Elsewhere, however, press reports of Gilbert's investigation, although apparently derived from the *Sun* report, dealt solely with his meteoritic impact hypothesis and in such a way that they seem to have created an impression that Gilbert had assigned this origin to the crater. Lick Observatory's director, Edward S. Holden, for instance, drew this conclusion from such an item in the *San Francisco Examiner,* headlined "A Meteoric Crater," which he found sufficiently interesting to reprint in the *Publications of the Astronomical Society of the Pacific.*[55]

Gilbert, however, was not yet ready to put his conclusions about Coon Mountain on the record, for there were a number of matters relating to his investigation to be followed up back in Washington. These included routine petrographic and chemical examinations of the rock and iron specimens he had collected and the preparation of a detailed topographic map and a chart of Coon Mountain's stratigraphy. In addition, tests must be made to further assess the sensitivity of Baker's instruments, and experiments carried out, by firing clay balls at a clay target, to determine the relationship between the dimensions and velocities of impacting projectiles and the diameters of the crateriform structures which they produced. The deduced values thus obtained, Gilbert thought, might be pertinent to his computations of the volume of the crater vis-à-vis the volume of the ejecta on the rim and on the surrounding plain, but in his final analysis they proved to be superfluous. He also made an extensive survey of the literature on volcanoes to determine whether there might be other craters on the earth with characteristics similar to those of Coon Mountain's crater. Finally, to pursue to its logical conclusion the initial analogy which had led him to Coon Mountain, he embarked on a brief but intense investigation of the origin of the craters on the moon. This corollary investigation, completed from postulate to publication in only a few months during the late summer and fall of 1892, was conducted as a "private study" while he was on a leave of absence from the Survey.[56] It proved to have no significance whatsoever for Gilbert as far as Coon Mountain was concerned nor, as will be seen, did his Coon Mountain investigation influence his conclusions about the origin of the lunar craters.

GILBERT'S CONCLUSION—VOLCANISM

Actually, the details of Gilbert's Coon Mountain investigation and the conclusions to which his "crucial tests" there had led him did not appear in print for more than five years. The political storms that battered the Survey, and Powell in particular, during much of this period seem to have been principally responsible for this delay. Hostility and criticism within Congress against Powell and the Survey had been building for several years, largely as a result of Powell's extra-scientific activities, especially in the areas of western irrigation and land use policy; in 1892 these antagonisms came to a head. Congress slashed the Survey's budget in half; and in the consequent reduction in force, Gilbert's post of chief geologist was only one of many that were abolished. Gilbert, who managed to avoid any personal involvement in the political situation, survived, drawing the unenviable task of notifying many of his colleagues of their termination; he continued to serve for a time as Powell's chief aide on an unofficial basis and at Powell's request. In 1893, Powell sent him to the high plains just east of the Rocky Mountains to study the hydrology of artesian wells, an aspect of irrigation that Powell's critics had accused him of neglecting; and following Powell's departure as Survey director in 1894, Gilbert continued this work through the 1894 and 1895 field seasons under Powell's successor, Charles D. Walcott.

During these years, Gilbert referred only rarely to his investigation of Coon Mountain. His Arizona expedition is mentioned briefly in his annual reports for the Survey for 1891 and 1892, but only to restate the problem and to outline the two rival hypotheses proposed to resolve it without, however, indicating his preference for one or the other of them.[57] In his 1892 report, perhaps to further allay criticism emanating from Congress, he also emphasized the potential significance of the problem both to science and to commerce. "In one case," he explained, "it represents a factor of volcanism not elsewhere known to be isolated, and therefore instructive in its contribution to the physical history of volcanoes; in the other it is of important practical value as indicating the presence beneath its hollow of many thousands of tons of nickeliferous iron, and it is of scientific importance as illustrating a method of planetary aggregation by the falling together of similar masses. . . . A report of conclusions on the structure and origin of Coon butte has been partially prepared," he added, "but has not yet received final form for publication."[58] In August 1892, however, Gilbert apparently gave some indication of what these conclusions might be in discussing "Coon Butte and the Theories of Its Origin" at a joint meeting of the American Association for the Advancement of Science and the National Academy of Sciences held at Rochester. This lecture was not reported at the time, nor was it subsequently published; but many years later the geologist who had invited him to speak recalled that Gilbert favored the volcanic theory.[59]

Gilbert's promised report never materialized; and, instead, his first and indeed his only formal statement of his conclusions came in the form of an address on "The Origin of Hypotheses, Illustrated by the Discussion of a Topographic Problem," the topographic problem, of course, being Coon Mountain. This he delivered as president of the Geological Society of Washington on December 11, 1895, before the Scientific Societies of Washington, and it was subsequently published in full by the journal *Science*.[60]

The philosophical part of this address dealt with what for Gilbert, and probably for most of his listeners, was simply the conventional wisdom of science as it related to scientific method. But now he emphasized two points in particular, both of which he felt had been strongly reinforced by his Coon Mountain experience. The first of these underscored the importance of inventing multiple hypotheses and then subjecting them to rigorous tests to determine, by a process of elimination, which best explained the phenomenon in question; the second involved the danger inherent in the fact that the hypothesis that initially appears to explain most plausibly a given set of phenomena does not always prove to be the correct one in the light of such tests. He had gone to Coon Mountain, he told his audience, expecting to find the scar made by the impact of a great iron meteorite. But the conclusion that he had reached there, indeed the conclusion mandated by his methodology, was that Coon Mountain and its crater were created by a volcanic steam explosion.

To illustrate these points, Gilbert described his Coon Mountain investigation in considerable detail and then explained the observations and reasoning that had led him to this conclusion. Of Baker's magnetic measurements, he reported, "the magnetism was found to be constant in direction and intensity at all stations, the deviations from uniformity being not greater than the unavoidable errors of observation. So if the crater contains a mass of iron its attraction is too feeble to be detected by the instruments employed." To further evaluate this result, he added, "the delicacy of the instruments was afterward tested at the Washington Navy Yard, by observing their behavior when placed in certain definite relations to a group of iron cannon whose weight was known, and the following conclusions were reached: If a mass of iron equivalent to a sphere 1500 feet in diameter is buried beneath the crater, it must lie at least 50 miles below the surface; if a mass 500 feet in diameter lies there its depth is not less than 10 miles. So the theory of a great iron meteor is negatived by the magnetic results, unless we may suppose either the meteor was quite small as compared to the diameter of the crater, or that it penetrated to a very great depth."

Gilbert's second "crucial test," which relied on the data from his topographic survey, yielded a similar result. "During its progress," he explained, "the configuration of the surrounding country was carefully studied, and its general plan was found to be so simple and regular that the original contours before the creation of the crater could be restored without great liability of error." Such a restoration was made and Gilbert then computed the volumes of the rim and the crater relative to the original level of the surrounding plain. "The magnitude of the hollow was found to be 82 million cubic yards," he reported, "and the magnitude of the rim was also found to be 82 million yards. It, therefore, appears that if the rim could be dug away down to the level of the ancient plain, and the material tightly packed within the hollow, it would suffice to precisely fill the hollow and restore the ancient plain. The excess of material required by the theory of a buried star was not found. Thus each of the two experiments whose testimony had been invoked declared against the theory of a colliding meteor; and the expectation founded on the high improbability of fortuitous coincidence nevertheless failed of realization."

His meteoritic impact hypothesis thus disposed of, Gilbert now turned to "the only surviving theory," that of a volcanic steam explosion, declaring that "all the various features discovered in the local study were considered with reference to it," but adding that to "describe and discuss them on this occasion would lead too far from our subject, and they may be passed by with the remark that, while not all are as yet fully understood, they seem not to oppose the theory."

Instead, Gilbert argued this hypothesis simply by citing accounts in the literature of "other natural explosions where steam was the agent," including the historical eruptions of Vesuvius and Krakatoa, and particularly the 1888 explosion of the Japanese mountain Kobandai, where "the agency of steam distinctly appeared." Kobandai, he pointed out, had blown "a cloud of rock" 4,000 feet into the air, and the resulting crater, although less regular in form than Coon Mountain's crater, was nineteen times as capacious and had continued to emit "fierce jets of steam" for months after the explosion. "The competency of volcanic steam for the production of a crater is thus shown by a parallel instance," he declared, "and the only difference between the Japanese case and the Arizonian lies in the fact that in the one the disrupted rock was volcanic and in the other it was not. This difference seems unessential, for in neither case was there an eruption of liquid rock; the ancient lavas of Kobandai had been cold for ages, and their relation to the catastrophe was wholly passive. Moreover," he added, "the manifestation of volcanic energy is no more exceptional on the Arizona plateau than in the Bandai district. The little limestone crater is in the midst of a great volcanic district. The nearest volcanic crater is but ten miles distant, and within a radius of fifty miles are hundreds of vents from which lava has issued during the later geologic periods."

Nor did volcanic steam explosions occur only at the tops of volcanic mountains, he argued; and here he cited various crateriform structures in Europe, and particularly the Rhine River Valley, that were believed to be volcanic and, as most of them contained lakes, were called "maars." These, like Coon Mountain, had floors depressed below the surface of the surrounding plain, were "hollowed chiefly from non-volcanic rocks, limestone, sandstone and slate," and in some cases had low rims "composed of fragments of similar rocks." Still another example was Lonar Lake in India, a circular crater in volcanic rock some 300 to 400 feet deep and with a partial rim 40 to 100 feet high composed of blocks of the same basalt exposed in the outward dipping rocks of the crater's inner walls. Gilbert here quoted a geologist who had studied Lonar Lake and who had concluded that it was "impossible to ascribe the hollow to any other cause than a volcanic explosion."

Gilbert also reported that he had considered and rejected two other hypotheses of Coon Mountain's origin, one that it was a limestone sink, and the other, founded on the moraine-like appearance of its rim, that ice had somehow been involved in its formation. "Each," he noted, "was based on a single feature of the crater but failed to find verification in any other features." Still another idea that he discarded, advanced in 1894 by a man named Warren Upham, combined the meteoritic and volcanic explanations. This postulated that the underlying rocks had become heated by some volcanic process "so that conditions were ripe for an explosion, and that the mine was actually fired by a falling star, whose collision ruptured a barrier between water and the hot rock, or in some other way touched the volcanic button." Such an explanation, he pointed out, "demands a coincidence of what may be called the second order, for the colliding star is supposed not only to have chanced upon the prepared locality, but to have arrived opportunely at the critical epoch."

Yet despite his confidence in his "crucial tests" of the meteoritic hypothesis and his consequent advocacy of a volcanic explanation, Gilbert concluded his summary of his Coon Mountain investigation on a note of uncertainty. A colleague, Edwin E. Howell, he reported, had raised questions concerning the basic assumptions on which his tests rested. Howell had examined many of the meteoritic irons, concluding that each was a "complete individual" and that, while all were found together in a relatively small area, apparently none had been broken off from a larger iron mass. "Reasoning by analogy from the characters of other meteoric bodies, he infers that the irons were all included in a large mass of some different material, either crystalline rock such as constitute the class of meteorites called 'stony,' or else a compound of iron and sulphur, similar to certain nodules discovered inside the iron masses when sawn in two. . . . If it be true that the iron masses were thus imbedded, like plums in an astral pudding, the hypothetical buried star might have great size and yet only small power to attract the magnetic needle," he pointed out. "Mr. Howell," he added, "also proposed a qualification of the test for volumes, suggesting that some of the rocks beneath the buried star might have been condensed by the shock so as to occupy less space."

These questions, Gilbert conceded, were "eminently pertinent to the study of the crater and will find appropriate place in any comprehensive discussion of its origin." Yet despite "their ability to unsettle a conclusion that was beginning to feel itself secure," he chose not to discuss them, then or later, concluding his address with a brief homily in which he compared a knowledge of nature to an account in the bank.

Indeed, Gilbert never discussed Coon Mountain publicly again.

THREE

MOONLETS AND OTHER MATTERS

> But selenographers are not the only students of the
> moon's face. There are also selenologists, who use the
> telescope comparatively little, but cogitate much, and
> who have evolved theories of great ingenuity and variety.
> Far be it from me to say aught to their disparagement,
> for this evening I join myself to their ranks.
>
> GROVE KARL GILBERT, 1892

Grove Karl Gilbert's brief excursion into the realm of selenography and selenology in 1892 was the direct result of his investigation at Coon Mountain. From his speculations there, he later wrote, "I was led to give attention to the crateriform hollows of the moon, which have been ascribed by some writers to the impact of meteoric masses falling to its surface."[1] Yet beyond this, there is no further connection between the two investigations, and in reporting one Gilbert did not even refer to the other. Certainly, in his study of the moon, he no longer drew any analogies between its craters and "the little limestone crater" in Arizona.[2]

Gilbert's study of the moon and the impact hypothesis that emerged from it, as will be seen, had little or no influence on the subsequent development of impact theory in general. His hypothesis was rejected in the main by those few of his contemporaries in astronomy and geology who considered it in detail at all, and then was largely ignored for nearly sixty years while impact theory was being developed along different lines by others. Even Gilbert himself, after his initial advocacy, ignored it, deserting selenology as abruptly as he had espoused it, and confining his researches and his hypothesizing to terrestrial problems for the remaining twenty-four years of his life.[3]

In its general premise, of course, Gilbert's hypothesis was not new, and its polite but skeptical reception surely reflects the fact that the idea of an impact origin for the lunar features vis-à-vis the volcanic analogy was no less "wild and fanciful" in the intellectual milieu of 1892 than it had been in 1873 when Proctor so diffidently had advanced it. But in addition, Gilbert had embellished his hypothesis with a set of special conditions which were hardly more credible at the time than the idea of impact itself. It is in these special conditions, however, that his hypothesis is unique, and consequently that his venture into speculative selenology is of interest.

GILBERT AND THE MOON

Gilbert's hypothesis, astronomer Harold Jacoby noted after hearing it outlined before a meeting of the New York Academy of Sciences in February 1893, "agrees with the meteoric theories of Proctor, Meydenbauer and others in that it ascribes the craters to the impact of bodies colliding with the moon. It differs," he added, "in the previous history of the incident bodies."[4]

For where Proctor had envisioned meteors traveling at planetary velocities through interplanetary space and Meydenbauer had proposed that these meteors consisted of cosmic dust, Gilbert was obliged to banish meteors as such entirely from his thinking and to restrict his "incident bodies" to a Saturn-like ring of what he called "moonlets" orbiting the earth at some primordial epoch in its history. Pointing out that astronomers believed Saturn's rings to consist of "an infinitely larger number of very small bodies revolving about the planet in parallel orbits," Gilbert proposed that "a similar ring of minute satellites once encircled the Earth, and that these gradually became aggregated into a smaller number of larger satellites, and eventually into a single satellite—the Moon. The craters mark the spots where the last of the small bodies collided with the surface when they finally lost their independence and joined the larger body."[5]

Thus, in seeking to account for the moon's topography, and indeed specifically the circularity of the lunar craters, Gilbert incidentally postulated a process for the origin of the moon itself, a point which was neither emphasized by Gilbert himself nor mentioned by subsequent critics. As this postulate was based on assumptions not yet known to be valid but not thereby invalid, its time may yet come.* Unfortunately, however, Gilbert did not specify the de-

---

*Saturn, of course, is no longer unique as a ringed planet. A Uranian ring system was discovered in 1977 from earth-based photometry, and subsequently Voyager spacecraft imagery in 1980–1981 revealed a single ring around Jupiter.

tails of ring and satellite formation, and thus his hypothesis does not provide any insights for modern selenologists and planetary astronomers.

Gilbert carried on his investigation into the nature and origin of the moon's topography from the summer of 1892 through to the spring of 1893. He first outlined his hypothesis before a meeting of the National Academy of Sciences in Baltimore on November 1, 1892, and then made it the subject of his address as retiring president of the Philosophical Society of Washington on December 10, 1892. Subsequently, he restated it at a meeting of the New York Academy of Sciences on February 16, 1893. His full paper was finally published by the Philosophical Society of Washington in April 1893.[6]

In the course of his study, Gilbert observed the moon telescopically on eighteen nights through two lunations from August to October 1892, using the 26-inch refractor at the U.S. Naval Observatory at Washington, with a magnifying power of 400 which he found to be "most serviceable" to his eye. These direct observations he supplemented by studying lunar photographs made at the Lick Observatory. In addition, he investigated the mechanics of crater formation with what by then were tried and true methods, i.e., by firing balls of clay and Wood's alloy at targets of the same materials, in a physical laboratory at Columbia College in New York City. Finally, he read extensively in the selenographical and selenological literature, where he found the works of Beer and Mädler, Dana, Proctor, Neison, Meydenbauer, and Ebert, among others, all of whom he either subsequently cited or quoted.

His exposition of his "moonlet" hypothesis is prefaced by a brief description of the lunar topography and a summary of the problem of its origin and the various hypotheses that had been advanced to solve it. "It is hardly profitable to discuss the suggestion that the greater walls were formed about vortices of a primeval liquid moon, nor the suggestion, albeit advanced independently by several authors, that the vast encircling cliffs of the moon are remnants of Cyclopean bubbles that have burst," he declared. But the tidal theory of Faye and Ebert merited some attention, he felt, both because of the extensive researches of George Howard Darwin into the effect of tides on the past history of the earth-moon system[7] and because of Ebert's experimental demonstration of the tidal process in the laboratory.[8] He concluded, however, that the crust of the moon as postulated by this hypothesis would not be thick enough or strong enough to support the weight of the rims of the larger craters, and he wondered whether, "if the crust was cracked and fissured by lunar tides as required by the theory, these fissures would not merely "gape, instead of forcing out the liquid through apertures here and there?" Finally, there were many smaller craters on the rims and outer slopes of the larger craters, and "the initiation of these by tidal processes seems impossible."[9]

The glacial, or "snow" theory as he called it, posed greater difficulties, even beyond the question of the existence of water or ice on the moon. "If the rim were built up by the quiet fall of an infinitude of ice particles or snow flakes, its configuration should be smooth and regular instead of exhibiting the rugosity actually observed," he declared. Nor could the theory explain the central peaks in some craters and like the tidal theory, it failed to account for the small craters on the rims and slopes of larger ones.

Understandably, in view of prevailing belief, Gilbert discussed the volcanic analogy at somewhat greater length. "By a majority of writers the craters are assumed to be volcanic," he noted, "and as they differ in size, abundance, and form from terrestrial volcanoes, it is thought that they represent some special type of volcanism determined by physical conditions peculiar to the moon."

From the standpoint of abundance, moreover, he found little difficulty with this interpretation. Using Faye's count of up to 30,000 visible lunar craters, Gilbert estimated from his own field experience and from his reading in the literature that there were perhaps 3,000 volcanic craters on the North American continent, an area roughly equivalent to that of the moon's visible face. This estimate, he pointed out, included only craters that the processes of erosion had not demolished. Had older terrestrial craters been exempt from erosion, like those of the moon, they might now be equally abundant.

The size of the lunar craters, however, was another matter, particularly in their horizontal dimensions. Using ratios between the largest, and the average of the ten largest, craters on the moon and on the earth as then known, he concluded that, even allowing for a lunar gravity one-sixth that of the earth's, the diameters of the largest craters on the moon were four to nine times greater than those of their counterparts on the earth. In citing these ratios, "considered as obstacles to the acceptance of the volcanic theory," Gilbert added three comments. First, he noted, the diameters of the ten largest terrestrial craters were closely grouped around a maximum of about 15 miles, while the diameters of the ten largest lunar craters were "widely scattered . . . like the distances of aberrant shots from the bull's-eye." Thus, he reasoned, "we should predict that the complete exploration of the earth will bring to light other craters about as large as those now known, but will discover none larger; but we could not make a similar prediction as to the maximum crater on the opposite side of the moon." Second, he noted that conditions affecting volcanic action in earlier geologic periods "were doubtless different from those determining the size of the craters we can examine;" and third, that the material of the moon might differ from that of the earth "in such a way as to affect the size of volcanic craters." In the vertical dimensions, he found "no important discrepancy," the depths of the lunar craters, again allowing for the lesser lunar gravity, being roughly proportionate to those of terrestrial craters.

But the differences in form between the lunar and terrestrial craters were the greatest factor and, for Gilbert, were fatal to the volcanic analogy. With craters of the ordinary, or Vesuvian type, which included 95 percent of all terrestrial volcanoes, "the lunar craters have little in common," he declared.

> Ninety nine times out of a hundred the bottom of a lunar crater lies lower than the outer plain; ninety nine times out of a hundred the bottom of a Vesuvian crater lies higher than the outer plain. Ordinarily the inner height of the lunar crater rim is more than double its outer height; ordinarily the outer height of the Vesuvian crater rim is more than double its inner height. The lunar crater is sunk in the lunar plain; the Vesuvian is perched on a mountain top. The rim of the Vesuvian crater is not developed, like the lunar, into a complex wreath, but slopes outward and inward from a simple crest-line. If the Vesuvian crater has a central hill, that hill bears a crater at its summit and is a miniature reproduction of the outer cone; the central hill of the lunar craters is entire, and is distinct in topographic character from the circling rim. . . . The smooth inner plain characteristic of so many lunar craters is either rare or unknown in craters of the Vesuvian type.

Thus, "through the expression of every feature the lunar crater emphatically denies kinship with the ordinary volcanoes of the earth. If it was once nourished by a vital fluid, that fluid was not the steam-gorged lava of Vesuvius or Etna."

Volcanic craters of the Hawaiian type, Gilbert noted, were "somewhat rare, but their rarity does not affect their value as interpreters of extra-telluric phenomena," for as Dana had pointed out, "they resemble the moon's craters much more closely than do those of ordinary volcanoes." But while Hawaiian volcanoes agreed with lunar craters in having inner plains and to some extent in the terracing of their inner walls, they differed in the fact that they were on mountain tops, in the absence of central peaks, and in the presence of level terraces. "In my judgement," he concluded, "the differences far outweigh the resemblances, and I have not succeeded in imagining such peculiarities of local condition as might account for the divergence in form."

Volcanic maars were even rarer than Hawaiian-type volcanoes and represented an "antithetic phase of volcanism" in that their formation "includes no eruption of lava, but merely an explosion of steam." Less than fifty were known on earth at the time, he noted, and these were small, the largest being less than two miles in diameter. "They resemble the craters of the moon, in that their bottoms are depressed below the general level, and in that the volumes of their rims are approximately equal to the capacities of their cavities. They lack the wreath, the inner terraces, the inner plain, and the central hill. Thus characterized, they differ widely from the lunar craters of medium and maximum size, but they resemble those of smaller size." If this resemblance was accepted as satisfactory, he added, half the lunar craters could be explained. But to "adapt the explosive hypothesis to the larger craters it is necessary not merely to think of a greater explosion, but to imagine some phase or accompaniment of explosive action which will furnish the rim with a system of concentric ridges and the cavity with a level bottom and a central eminence. If the attempt at adaptation fails, as I think it must, then the explanation can be accepted for the small craters only by divorcing them from the large...."

Summing up, Gilbert the geomorphologist declared that "The volcanic theory as a whole, is therefore rejected, but a limited use may be found for the maar phase of volcanic action in case no other theory proves broad enough for all the phenomena."

Meteoric impact theories,[10] Gilbert found, also offered problems. In introducing them, he first drew a familiar analogy between the lunar craters and the crateriform structures produced by dropping pebbles into mud, raindrops falling "on a slimy surface," and projectiles fired into plastic targets. "As the present study is primarily physiographic," he declared at the outset, "these similitudes have been considered with great care, and it is my belief that all features of the typical lunar crater and of its varieties may be explained as the result of impact."

In extant hypotheses however, Gilbert pointed out, the impacting bodies were considered to be meteoric and cosmic in origin. "Nevertheless," he opined, "it is incredible that even the largest meteors of which we have direct knowledge should produce scars comparable in magnitude with even the smallest of the visible lunar craters." To overcome this difficulty, he noted, earlier writers had assumed that meteors had once been larger in size and greater in number, "and as no evidence has been found that the earth was subjected to a similar attack, there is assigned to the lunar bombardment an epoch more remote than all the periods of geologic history, any similar scars produced on the earth having been obliterated by the processes which continually reconstruct and remodel its surface."

Earlier hypothesizers also had found it necessary to imagine "a condition of the lunar surface which should admit at the same time of plastic molding and the preservation of the resulting forms. But this "does not really escape the difficulty, for it will not do to postulate a degree of softness incompatible with the survival of lofty cliffs." To Gilbert the geologist, however, this difficulty was "only imaginary and not real. Rigidity and plasticity are not absolute terms but relative," he explained, "and all solids are in fact both rigid and plastic. When great masses and great forces are involved, as, for example, in the making of continents or mountain chains, the distinction loses value."

If the lunar features were created by impact, he reasoned, the masses of matter involved had been greater than those of terrestrial mountain ranges and the concentration of energy at the point of impact would be correspondingly great. "Moreover a portion of this energy may have been converted into heat, with the result that the parts affected were rendered less rigid or even molten, and it even appears necessary to assume a result of this sort to account for the level surfaces of the inner plain of the craters." Here, he cited computations by a colleague, R. S. Woodward of the U.S. Coast and Geodetic Survey, showing that a body falling into the moon drawn only by the force of lunar gravity would have a velocity of impact of 1.5 miles per second, and that its equivalent energy on impact, if composed of "ordinary volcanic rock," would be sufficient to raise its temperature 3,500°F. "In other words, the quantity of heat developed would be greater by one-half that that necessary to fuse the body. The average velocity of shooting stars is estimated at 45 miles per second, and it is easy to understand that the heat developed by the sudden arrest of a fragment of rock travelling with such speed might serve not only to melt the fragment itself, but also to liquify a considerable tract of the rock mass by which its motion was arrested."

Still another difficulty with meteoric theories, for Gilbert at least, involved an assumption he had made earlier at Coon Mountain, i.e., that if the craters were produced by impact, "we should naturally expect to find in the rim the entire volume of the matter displaced in the formation of the hollow plus the volume of the moonlet. . . ." But this relation did not appear to hold for lunar craters, as indicated both by his own observations and by computations based on the measurement of shadows cast by lunar crater rims made by Ebert.[11] Ebert, he noted, had compared the volumes of the rims and cavities of ninety-two craters from eight to one hundred miles in diameter and had found the rim volume greater in only twenty-eight of them. "Though the imperfection of the data gives a large probable error to the determinations," he added, "there can be no question of the general fact that in many instances the rims of large craters are quite inadequate to fill the cavities they surround." This conclusion in itself did not necessarily invalidate impact theories, however. For in his experiments, Gilbert noted that on occasion the material of the rims of his clay craters, when pared away, did not fill the cavities and, with further experimentation, "the cause of this result was discovered." In those instances where the general mass of the target was softer than its surface, he found, "the uplift consequent on the production of the hollow was only partly localized about its periphery, the remaining part being widely distributed through flow of the softer material below."

THE PROBLEM OF CIRCULARITY

But for Gilbert the "most formidable difficulty" was "the circular contours of the craters." Except for minor details of form, he noted, "some of the lunar craters are as nearly circular as can be determined by measurement; others are slightly elliptic; a few only are elongate." Indeed his own measurements of 120 craters shown on the Lick Observatory photographs indicated that 75 percent had ellipticities of less than 0.1, almost 92 percent had ellipticities of less than 0.2, and only about 3 percent had ellipticities greater than 0.3. "It is inferred that the predominant direction of the incident bodies supposed to have formed them was vertical to the lunar surface, or nearly so," he pointed out, "but it can be shown from simple geometric considerations that the predominant angle of incidence of swift-moving meteoric bodies approaching from all directions would be 45 degrees, and the scars produced by such collisions would be predominantly oval instead of predominantly circular."

Gilbert knew of only one earlier attempt to obviate this problem. "It was suggested by Proctor that immediately after the shock of collision that there might be an elastic return to a circular form," he noted, but this idea required "a high tensile elasticity, such as we do not know in rocks, but only in certain substances of organic origin, and it thus fails to receive support from the phenomena of our terrestrial experience."

Gilbert introduced his "moonlet" hypothesis by noting that along with "the nomadic and apparently individual meteors of space," there was at least one group of small bodies that were "symmetrically arranged and moving in a systematic and orderly way"—the "meteors" in Saturn's rings. "It is my hypothesis that before our moon came into existence the earth was surrounded by a ring similar to the Saturnian ring; that the small bodies constituting this ring afterward gradually coalesced, gathering first around a large number of nuclei, and finally all uniting in a single sphere, the moon. Under this hypothesis, the lunar craters are the scars produced by the collision of those minor aggregations, or moonlets, which at last surrendered their individuality."

This postulate of a Saturnian ring around the earth, he pointed out, yielded a "material difference" in the conditions affecting the infall of bodies onto the moon in that "all the minor bodies colliding with the moon have initial orbits lying approximately in the same plane," and were moving, moreover, at relatively slow velocities and in the same direction. Gilbert here worked out the geometry of the two cases in some detail and with diagrams to show that while the greatest number of impacts by random infalling meteors would be at 45°, the maximum number of impacts of his "moonlets" would occur very near the vertical and, indeed, that half of all the moonlets would strike the lunar surface at angles of less than 30° from the vertical. Thus, he wrote, "the law of incidence angle for ring-derived moonlets agrees with the law suggested by the roundness of the impact scars in that it indicates a predominant approximation to verticality, and it therefore accords better with the phenomena than does the law of incidence angle derived from the theory of cosmic meteors. The introduction of the hypothesis of a Saturnian ring thus accomplishes much toward the reconciliation of the impact theory with the circular outline of the lunar craters."

Having thus eliminated the third dimension, as it were, Gilbert now turned to the laboratory to find corroborative evidence. Using a slingshot device by which a ball of clay was made to strike a clay target of varied viscosity at various measured angles and velocities, he found that the ellipticity of the resulting impact scars was a function not only of the incident angle, but of the vicosity of the clay and, inversely, the velocity of impact. "No attempt was made to discover the precise character of this complex relation," he noted, "because it was immediately evident that the experiment could not be made to deal with velocities and strengths of material comparable to those associated with the production of the lunar craters." Nonetheless, he concluded that the ellipticity increased only gradually with an increase in the angle of incidence up to 30° or 40°, and more rapidly at higher angles, and this seemed to agree with the evidence of the lunar craters themselves.

To further strengthen this argument he investigated the nature of the orbits of his postulated moonlets, relying again on the computational skills of his colleague Woodward. This investigation, in which "the influence of the earth's attraction essential to a rigorous discussion, was ignored," yielded a simple expression, $n = \sqrt{\sin i}$, n being the relative number of moonlets with an incident angle of less than i.[12] Applied to his impacting moonlets, this expression indicated that 58 percent would deviate less than 20° from the vertical, 70 percent less than 30°, and 80 percent less than 40°. "The theoretic distribution obtained by this partial treatment," he declared, "accords so well with the phenomena under discussion that greater refinement seems not to be required."

His final point regarding the circularity of the lunar craters related to the tangential component of the motion of an impacting moonlet which, if its velocity were greater or less than the rotational velocity of the lunar surface, would either accelerate or retard the moon's spin. But the aggregate result of all collisions, he argued, "would be such a rotation of the moon that its surface speed would equal the average of the tangential components of the velocities of the moonlet impact. It is evident that if the tangential component of a moonlet's motion coincided exactly with the motion of the moon's surface, the impact phenomena would be the same as if the moonlet fell vertically on a motionless surface; and the harmonious adjustment of moon rotation to the motions of a system of moonlets would reduce to a minimum the ellipticity of the craters."

But tangentiality did not always produce harmonious adjustment.

If, as I have assumed, the moonlets approached the moon approximately in the plane of the equator, the fact is not attested by the grouping of the craters in a medial zone, and so it is necessary to assume further that the axis of rotation was not constant. This assumption need occasion no difficulty, for unless the approaching moonlets moved *precisely* in the plane of the moon's equator, their collisions would disturb its axis of rotation, and there is no reason to suppose that these disturbances would be compensatory rather than cumulative. Under the successive impulses thus given the moon's equator may have occupied successively all parts of its surface, without ever departing widely from the plane of the moon's orbit [emphasis Gilbert's].

Summarizing his hypothesis, then, Gilbert declared that, "In fine, the hypothesis of the Saturnian ring, by restricting the colliding bodies to a single plane, by substituting a low initial velocity and thus rendering the moon's attraction the dominant influence, and by introducing a system of directions controlling, and therefore adjusted to, the moon's rotation, relieves the meteoric theory of its most formidable difficulty. It also explains in a simple way the abundance of colliding bodies of a different order of magnitude from ordinary meteorites and aerolites." Moreover, "the idea that the ring, although possessed of sufficient stability to assume a definite form, nevertheless suffered some disturbance or underwent some process of evolution by which its stability was destroyed, is likewise familiar to celestial mechanics, and it does not appear necessary in this connection to speculate as to the precise manner in which the integration of its discrete elements was effected. . . ."

Gilbert also discussed the mechanics of impact cratering which, for him, involved the production of heat. The greater the size of the impacting body, the greater the temperature generated, and to this he ascribed the restriction of the inner plains, and less directly, of central peaks, to the larger craters. In the case of small craters, the impacting bodies "either were crushed or were subjected to plastic flow, and in either case were molded into cups in a manner readily illustrated by laboratory experiments with plastic materials." The material displaced in the formation of a crater was built into the rim, partly by overflow at the edges, but chiefly by outward mass movement in all directions. This outward and upward movement was accentuated, "possibly through the agency of heat, about the immediate edge of the cup, occasioning the special elevation called the wreath." Larger bodies produced larger cups, and at the same time fused a portion of the lunar material and "softened" other portions. The walls of these larger cups, however, were so lofty that they could not sustain their own weight and, further weakened by the effects of heat, "settled downward and their lower portions flowed toward the center," producing the inner plains and terraces. In addition, this settling resulted in an upward movement at the center of the crater, perhaps enhanced by the "elastic recoil" of the impacted material, producing central peaks which, in some cases, might also contain the unfused remnant of the impacting body. "The impact theory as thus developed," he concluded, "appears competent to explain the origin of all typical features of lunar craters."

Gilbert's lengthy exposition did not deal entirely with theoretical matters, however. He also reported what he called "sculpture" on the moon, noting that certain crater rims and level tracts were scored by grooves or furrows exhibiting parallelism of direction and that groups of small hills, which reminded him of glacial drumlins, had parallel axes as well. "Tracing out these sculptured areas and plotting the trend lines on a chart of the moon, I was soon able to recognize a system in their arrangement, and this led to the detection of fainter evidences of sculpture in yet other tracts." The trend lines, he found, converged at a point near the center of the Mare Imbrium, a vast, partially ring-walled plain more than 700 miles in diameter. "Associated with the sculpture lines is a peculiar softening of the minute surface configuration," he added, "as though a layer of semi-liquid matter had been overspread, and such I believe to be the fact; the deposit had obliterated the smaller craters and partially filled some of the larger. These and allied facts, taken together, indicate that a collision of exceptional importance occurred in the Mare Imbrium, and that one of the results was the violent dispersion in all directions of a deluge of material—solid, pasty, and liquid."

The great forces involved, he noted, were particularly well illustrated by a series of "gigantic" and "remarkably straight" furrows that scored the lunar surface at great distances from Mare Imbrium itself, the best-known of which were the Rheita and Alpine valleys of the moon. Through the telescope, these furrows reminded him of "the rude grooves sometimes seen on glaciated surfaces where the corner of a hard boulder, dragged forward by the ice, has plowed its way through a brittle rock. . . . but the graving tool in this case, instead of being slowly pushed forward by a matrix of ice, moved with high velocity and was controlled only by its own inertia." More than half these furrows trended toward Mare Imbrium, he found, "and thus it appears possible, if not probable, that they were produced by the Imbrium deluge, and the implication of power is thereby rendered even more impressive. What must have been the violence of a collision whose scattered fragments, after a trajectory of more than a thousand miles, scored valleys comparable in magnitude to the Grand Canyon of the Colorado!"

The effects of this Imbrium deluge also suggested a rough relative chronology for the lunar features.[13] "Through the entire region lying between the Mare Imbrium and the Maria Serenitatis and Tranquilitatis, sculpture and associated veneering have so modified the surface that there is no difficulty in discriminating the craters of later date from those of earlier. The whole topography may be classified as antediluvial and postdiluvial."

Other features on the moon's surface also supported the impact hypothesis, Gilbert noted, and especially the white streaks, or rays, radiating from some of the larger, brighter craters. Here he cited an unpublished letter from one William Würdemann to Benjamin Apthorp Gould, the founding editor of the *Astronomical Journal*. Würdemann's explanation of the streaks or rays was "that a meteorite, striking the moon with great force, spattered some whitish matter in various directions. Since gravitation is much feebler on the moon than with us, and atmospheric obstruction of consequence does not exist, the great distance to which the matter flew is easily accounted for." This explanation appealed strongly to Gilbert, who wondered "why the idea has not sooner found its way into the moon's literature."

A final point: Gilbert also noted that his earth-orbiting moonlets would impact on the earth as well. "While the moon was growing the relations of orbits and attractions were such that any moonlet which narrowly escaped collision with the moon was enormously perturbed, acquiring an entirely different orbit about the earth. Many must have been so directed as to collide with the earth, and the traces of their collisions, if ever discovered, will tie together at a new point the chronologies of satellite and planet."

CRITIQUES AND COUNTERTHEORIES

Given Gilbert's great stature in science, it is hardly surprising that his presentations of this hypothesis before various scientific groups were reported briefly in abstract in several leading journals,[14] despite the fact that his exposition does not lend itself well to abstraction. As the director of Lick Observatory, Edward S. Holden, remarked in reprinting such an abstract, Gilbert was "the first geologist of high standing to give his authority to the hypothesis that the lunar craters have, *in general*, been formed by the bombardment of the lunar surface by meteorites [emphasis Holden's]." He added, however, that "my own studies of the moon photographs do not lead me to the same conclusion."[15] Holden was, in fact, an adherent of the volcanic analogy.[16]

But if Gilbert's stature assured that his hypothesis would be noted, it did not assure that other scientists would be persuaded by his arguments, and indeed none were. The few reviews that appeared following the April 1893 publication of his full paper were either noncommittal or critical of his reasoning. The *American Naturalist*[17] and *Science*[18], for example, merely summarized his thesis at somewhat greater length than had the earlier abstracts, although the *Science* reviewer supplied some additional historical background on the problem itself. Specific criticisms, however, were presented in *The Nation*[19] by the eminent Harvard geologist William Morris Davis, who thirty years later became Gilbert's principal biographer;[20] and in the English journal *Knowledge*,[21] by selenographer A. Cowper Ranyard, who, as noted earlier, was Proctor's literary executor.[22]

Ranyard, a volcanist, found not only that Gilbert's hypothesis "does not commend itself at all to my mind," but rejected Proctor's impact hypothesis as well, along with Hooke's "blister or bubble" analogy and the glacial or snow theory, which he discussed in the version advanced by Peal. Citing Roche's limit to the effect that a ring formed more than 2.44 radii from the center of a planet "would break up," Ranyard argued that a moon "formed just outside such a ring would have an ellipticity greater than that of an ordinary hen's egg; and as tidal action carried the moon away from its primary it would gradually approximate to a spherical form. One can hardly conceive that such a change of shape could take place without obliterating scars on its surface."

Still another objection, "which, to my mind, is even more conclusive," involved "the many lines or strings of small craterlets which fall evidently into a line with one another. If we are forced to treat them as scars upon a target, we must regard their alineation as the result of mere chance distribution; but the number of such strings precludes any such assumption; there must therefore be a physical reason for the alineation, and the most obvious explanation seems to be that the craterlets mark out a line of weakness on the crust of the moon and lie along a volcanic fissure or fault. . . . We therefore seem driven back to the volcanic hypothesis," Ranyard concluded, "and have to explain why upon the moon, which is so much smaller than the earth, the volcanic outbreaks have been on so colossal a scale."

Davis prefaced his more detailed critique of Gilbert's hypothesis by assessing the current status of the problem itself. "For a number of years there has been in astronomical literature a slender undercurrent of doubt as to the volcanic origin of the 'craters' of the moon. In spite of their question-begging name, their analogy with terrestrial volcanoes is very incomplete." Some observers, he noted, conceded the differences between typical terrestrial volcanoes and the lunar features, and increasingly the analogy was being restricted "to those special and somewhat exceptional terrestrial volcanic forms to which the name 'caldera' is coming to be applied. These are commonly explained as resulting from the destruction of former cones by violent explosive action, and it has therefore come to be the fashion to explain lunar craters in the same way, thus placing them in a class somewhat apart from ordinary volcanoes. . . ."

What was new in Gilbert's hypothesis, Davis pointed out, was not his suggestion of an impact origin for the lunar features, but his postulate of a terrestrial ring of moonlets, "yet this alone, if undeveloped, would be a contribution of small value." And if "the ingenious and logical discussion of the angles at which impact would most frequently occur is an admirable example of scientific deduction," still the "weakest point" in Gilbert's hypothesis was the assumption on which this discussion was based, i.e., that his moonlets all moved in the same plane. Davis had difficulty imagining that these "numerous little bodies" could have always moved in such an orderly fashion. The wide departure of the asteroids from a common orbital plane was well known, "and even if only a moderate departure be admitted for the moonlets, their average angle of incidence on the lunar surface might be decidedly altered from the favorable values determined by the assumed conditions of simpler movement."

A second "weak point" was the "essential absence" of very elliptical craters on the moon "such as would certainly sometimes be formed by oblique impacts." Under Gilbert's postulates, such impacts would be "relatively rare, but the very elliptical craters are still rarer. The centre of the visible face of the moon is just where they might be expected to occur, and yet there they are conspicuously absent. The point needs further elucidation."

This seems to have been the extent of the immediately published comment on Gilbert's hypothesis. But in 1896 it came in for criticism on still another point from one Ephraim Miller, a professor of mathematics and astronomy at the University of Kansas, in the course of proposing a bizarre hypothesis of his own.[23] Miller declared that Gilbert's hypothesis "contains inherent and radical difficulties" and argued that collisions between Gilbert's larger, growing moonlets of approximately the same size "would break both bodies into a multitude of smaller bodies, and hurl them outward in space in a thousand different directions. This process would be carried on just as fast as the meteoritic masses of the old ring should be drawn toward the center. A swarm of asteroids would therefore, under such conditions, have taken the place now occupied by the Moon."

Miller proposed instead that the moon's craters were simply solidified remnants of ancient "moonspots," which he considered to be analogous to sunspots, which would also become permanently fixed in the form of craters like those on the moon. This suggestion too was soberly criticized in the astronomical literature by an anonymous reviewer who did not consider it a "tenable theory."[24]

Despite such theorizing, the volcanic analogy continued to dominate selenography and selenology through the 1890s and well into the twentieth century. It was, as Davis noted, "ingrained" in the astronomical literature, and he might have added that it was ensconsed in the geological literature as well in those few instances where geologists concerned themselves at all with the problems of lunar topography. In 1895, a well-known European geologist, Edward Suess, opted for the volcanic explanation in a paper before the Imperial Academy of Sciences in Vienna,[25] praising Dana's comparison between lunar craters and Hawaiian volcanoes and arguing that the physical conditions of the moon "seem to have favored the intensity of volcanic action, and to have produced the bizarre forms of craters on the Moon."

But very few scientists in any discipline at this time considered the matter of the origin of the lunar features, and the volcanic analogy seems to have been accepted quite casually and uncritically, as if the problem, for all practical purposes, had long been resolved except for a few minor details. W. W. Payne, editor of the newly established journal *Popular Astronomy,* surely expressed the prevailing view in 1894 when, in a general discussion of the moon, he declared that its features were "occasioned, probably, by volcanic forces of surprising magnitude in some very remote period in past lunar history."[26]

Harvard College astronomer William H. Pickering, a leading American selenographer, was not even sure that lunar volcanism need be consigned to an earlier era, inquiring in print, "Are there at present active volcanoes on the Moon?" and wondering "whether volcanic activity upon the Moon is really dead?"[27] Pickering, in a few years, would adopt the tidal variation of lunar volcanism as a result of experiments inspired by England's J. B. Hannay, who in 1892 had reported observing small craters forming in the slag floating on the undulating surface of cooling molten iron.[28] Pickering, in turn, found he could produce such lunarlike forms in containers of hot paraffin, a material "more readily handled," by using a small piston pump to simulate both volcanic and explosive volcanic action.[29]

When Pickering described these experiments subsequently in his book *The Moon,*[30] reviewer R. G. Aitken of the Lick Observatory pronounced them "very interesting" and added that they "will leave little doubt in the minds of

1 The earth's moon.

2 Franz von Paula Gruithuisen

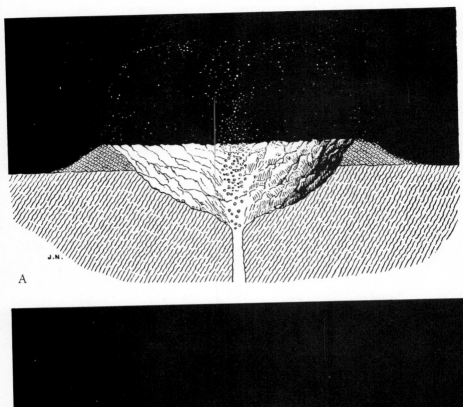
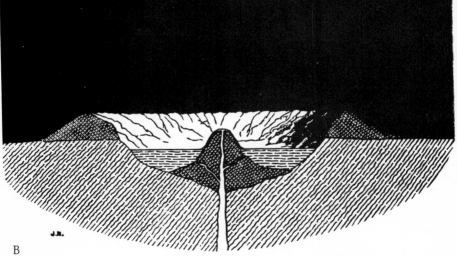

3 Nasmyth and Carpenter's (a) "Fountain" and (b) the resulting crater.

4 R. A. Proctor

5 A. E. Foote

6 Willard D. Johnson at plane table.

7 Coon Mountain (Meteor Crater) from the south.

8 A wide-angle view of Meteor Crater.

9 Grove Karl Gilbert

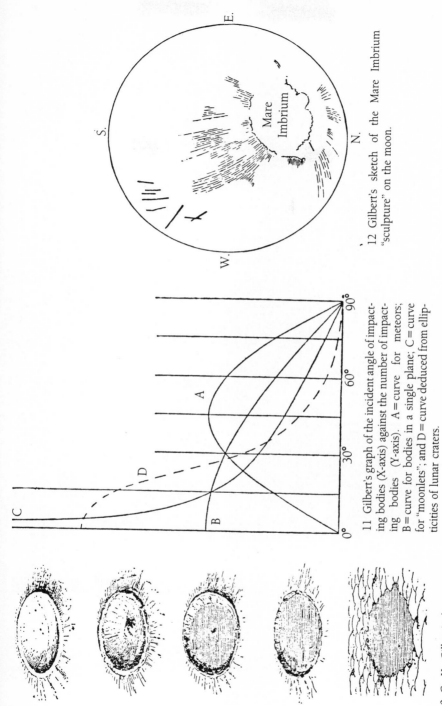

12 Gilbert's sketch of the Mare Imbrium "sculpture" on the moon.

11 Gilbert's graph of the incident angle of impacting bodies (X-axis) against the number of impacting bodies (Y-axis). A = curve for meteors; B = curve for bodies in a single plane; C = curve for "moonlets"; and D = curve deduced from ellipticities of lunar craters.

10 G. K. Gilbert's lunar crater forms as related to size, ranging from the smallest at the top to the largest at the bottom.

13 W. H. Pickering

most readers that the source of the lunar craters was volcanic activity."[31] A reviewer in England, S. A. Saunder, added that "Whatever may be thought of Prof. Pickering's well-known opinions as to the existence of ice and vegetation upon the moon, or as to the frequency of volcanic eruptions, there can be no doubt . . . that his explanations are as consistent as any others that have been proposed to account for the appearances; if, in the opinion of some who have not had his advantages, his conclusions are not yet proved, he has at all events gone a long way towards rendering them probable."[32]

The idea of an impact origin for the moon's craters, moreover, came in for an attack in general terms in a sometimes lively and wide-ranging debate carried on sporadically in 1897 and 1898 in the letters to the editor columns of the *English Mechanic and World of Science*. This began when an anonymous correspondent signing himself "R. P." suggested an impact hypothesis strongly reminiscent of Proctor's and subsequently quoted Proctor to his purpose.[33] He drew only a single supporter, who also remained anonymous as "J. S. G.," but they were soon outnumbered two-to-one by critics who were joined later in the controversy by the noted astronomer E. M. Antoniadi, then at the Flammarion Observatory at Juvisy in France.[34] To make his point, Antoniadi cited what he called an "irresistable [sic] onslaught" against the meteoric theory made by selenographer William Noble at the November 1897 meeting of the British Astronomical Association.[35] "With his genuine common-sense," he wrote, "Capt. Noble said that 'seeing that a shot made a hole its own size in a target,' he would ask, 'How did a meteor four miles in diameter manage to make a hole 20 miles in diameter in the moon?' In light, then, of this argument, only a bullet 142 miles in diameter would give rise to a depression like Clavius. . . . Meteors therefore of 300, 420 and 700 miles would be required to produce the Maria Crisium, Serenitatis and Imbrium respectively; but to use Capt. Noble's own words, our moon would long ago have been 'blotted out of the face of the sky' were she to have ever come to grief with such imaginary roving masses."

Such arguments, simplistic as they may be in light of later knowledge, were not untypical of the time. Even at this late date, for example, the idea that meteors and meteorites were, in effect, volcanic "bombs" hurled to earth by lunar volcanoes was being seriously discussed by some scientists.[36] This suggestion, advanced by Laplace and others nearly a century earlier, was explored at length by Sir Robert S. Ball, Astronomer Royal for Ireland and later Lowndean professor of astronomy and geometry at Cambridge University, in the several editions of his book, *The Story of the Heavens,* published between 1886 and 1905.[37] For Ball, the question was not whether meteorites were of volcanic origin, for he considered that this had been conclusively shown by an Austrian mineralogist named Tschermak, but the identity of the solar system body whence they came. It was not the moon, he declared. "The journey of a meteorite from the moon to the earth is only a matter of days, and therefore, as meteorites are still falling, it would follow that they still must be constantly ejected from the moon." But as the lunar volcanoes were no longer active, it was "utterly out of the question . . . that . . . she should continue to launch forth meteorites." Looking elsewhere, Ball concluded that the source of these volcanic meteorites was "most probably" the earth itself. "No one supposes that the volcanoes at present on the earth eject fragments which are to form future meteorites," he explained, "but it seems possible that the earth may be slowly gathering back, in these quiet times, the fragments she ejected in an early stage in her history."

Gilbert's hypothesis was considered again in 1903 by another Harvard geologist, Nathaniel Southgate Shaler, who, however, discussed it simply as a "meteoric" impact theory and in general terms without reference to Gilbert's postulate of a terrestrial ring of moonlets.[38] Shaler's studies of the moon involved one hundred nights of direct observation with the 15-inch Merz refractor at the Harvard College Observatory from 1867 to 1872, and thereafter the inspection of lunar photographs "in a desultory manner, from time to time, for a third of a century."

Certain aspects of Shaler's thinking regarding the origin of the lunar topography were insightful, although they since have been largely ignored, and they will be discussed later in this study as they relate to the subsequent development of impact theory. But here it may be noted that he rejected impact as an explanation of the lunar craters, or "vulcanoids," as he preferred to call them, citing a long list of objections to the idea. Among them were the "linear order" of some craters and craterlets, pits within cavities "so centrally located that they cannot be explained by the chance in-falling of bodies," and the absence of radiating fractures on the lunar surface. He also pointed out that the craters "all have the axes of their pits at right angles to the surface," and if formed by meteorites, "it does not seem possible that they could have all come upon the sphere in a path normal to its surface."

Shaler conceded that the "apparent lack of order" in the distribution of the "vulcanoids" gave "some warrant for the hypothesis that they must owe their origin to other than volcanic origin," but he also declared that it "seems incredible that if the lunar vulcanoids were due to bolides they should not have fallen in somewhat greater numbers upon the earth because of its greater gravitational attraction." Indeed, "the fall of a bolide of even ten miles in diameter would, by the inevitable development of heat due to its arrest, have been sufficient to destroy the organic life of the earth, yet this life has evidently been continued without interruption since Cambrian time."[39]

There were, however, certain other features on the lunar surface "which may be explicable by the impact of large bodies," and Shaler declared himself "disposed to hold with Gilbert and other inquirers that the maria are the result of large masses falling upon the surface of the sphere. All the facts indicate that these vast sheets of lava did not come from the interior." Shaler here offered four arguments: (1) the "extensive melting" caused by the conversion of momentum to heat, (2) the "singularly uniform fluidity" of the maria material, (3) its darker color, and (4) the availability of sufficiently large bodies among the asteroids to produce such impacts. The objections to an impact origin for "vulcanoids" were "insuperable," he concluded, but for the maria, the idea of impact was "the only working hypothesis that I have been able to find which in any way serves to explain these remarkable lunar features." Nonetheless, Shaler rejected Gilbert's contention that a major impact event had formed Mare Imbrium because "in the passages between the connected maria there is no evidence of scouring action such as would have been brought about by the swift movement of great masses of lava."

Although Gilbert's hypothesis was cited occasionally in subsequent years, it acquired few adherents or advocates. In 1905, for example, geologists T. C. Chamberlin and R. D. Salisbury mentioned it in a footnote in their widely used textbook, *Geology*, but only as a "meteoric" impact theory and without reference to Gilbert's terrestrial ring or to his moonlets. In their text the authors discussed lunar craters in terms of the volcanic analogy, comparing them to volcanic explosion craters on the earth and indeed, citing Gilbert, to Coon Mountain's crater.[40] The following year, astronomer Forest Ray Moulton noted Gilbert's hypothesis in his popular textbook, *An Introduction to Astronomy*, but again only briefly as a "meteoric" theory in a neutral discussion of the origin of lunar craters.[41] In 1910, astronomer Thomas Jefferson Jackson See declared that while Gilbert's theory included many excellent features, "it was not without difficulties," and in a footnote opined that Gilbert's "supposed ring of moonlets about the earth is scarcely admissable, so that part of his reasoning seems to be vitiated."[42] See's own ideas, incidentally, subsequently led to a discussion of the origin of the lunar features among members of the British Astronomical Association who, however, referred to the impact theory in terms of Proctor's, Shaler's, and See's work, rather than Gilbert's. In 1917, a Donald Putnam Beard discussed Gilbert's lunar hypothesis favorably, but a few years later Beard also wrote favorably about a theory which proposed that the lunar crater Copernicus was a "vast coral atoll."[43]

Davis, in fact, remained Gilbert's severest critic. In his exhaustive biography of his long-time colleague and friend published in 1926, Davis again reviewed Gilbert's selenological "excursion" and pronounced it "manifestly ingenious, but its acceptance is hindered by three difficulties that Gilbert's essay does not overcome."[44]

First, Davis again cited the paucity of at least a few very elliptical craters on the moon "which even under the moonlet theory ought to make no insignificant fraction of the total." Second, there was the unsystematic distribution of craters, for Gilbert's hypothesis required that "the off-equator impacts were so far from compensatory that they tilted the moon into various positions, and this in spite of the earth's action in holding the supposedly prolate moon with its larger axis directed earthward." The last few thousand moonlets, Davis argued here, "would be unable to tilt the moon appreciably, and . . . ought therefore to have produced craters chiefly around an equatorial great circle, or . . . if they departed from the equator as widely as the few thousand last-formed craters do, the impacts of those falling in at high latitudes, north and south, would be so oblique that the resulting craters would be plainly oval. . . ." A final difficulty was that the postulated influence of the impacting moonlets on the moon's rotation "would be greatly weakened if not wholly overcome in the late period when the last few thousand impacts occurred, by the earth's action on the supposedly prolate moon. . . ."

Davis frankly admired "this lunar excursion of Gilbert's," and praised the ability of the "excursionist" to apply mathematical concepts to physical problems and to state these problems with "remarkable facility and clarity. . . . Nonetheless," he concluded, "it still seems that the necessity of tipping the moon this way and that after it had essentially reached its present size, in order to explain the systemless distribution of craters under the moonlet theory, is a serious embarrassment to a beautifully conceived process."

FOUR

STAKING A CLAIM

> The number of arguments which between us we have worked out, in support of the theory that this gigantic hole is an impact crater, will be set forth. . . . It must be remembered that while a great deal of the evidence collected by us is positively in favor of the theory, much of it is negatively so; that is to say it disproves the theory that this great hole is the crater of an ancient volcano, or was produced by an explosion of steam. . . .
>
> DANIEL MOREAU BARRINGER, 1906

Daniel Moreau Barringer learned of Coon Mountain, its crater, and its associated meteoritic irons during a "casual conversation" with a federal government employee named Samuel J. Holsinger "while we were smoking together one hot night on the porch of the old San Xavier Hotel in Tucson" in October 1902.[1]

The garrulous Holsinger had never actually visited Coon Mountain, but he had heard stories about it during his travels through the northern part of the Arizona Territory in the course of his official duties, and he found Barringer intensely interested in these stories. Undoubtedly he was encouraged in their telling by his awareness that Barringer was a highly successful mining entrepreneur and an owner of the rich Commonwealth silver mine at Pearce, Arizona Territory, along with a number of other productive or promising mining properties in the West and in Mexico. Holsinger was, indeed, ambitious to improve his situation.

Barringer later recalled his initial reaction: "I was naturally incredulous of the theory which, Mr. Holsinger informed me, was held by some of the people living in the neighborhood of Cañon Diablo, namely, that this great hole in the earth's surface had been produced by the impact of an iron body falling out of space, if for no other reason than that I realized that the crater must have been examined by members of the United States Geological Survey while making topographic maps of this region, and in their report they evidently did not accept this theory." Despite this skepticism, however, "an impression had been made upon what I am fond of calling my subconscious intelligence, and the persistence with which the Holsinger story kept recurring became almost annoying." On his return to his Philadelphia home, he remembered, he had informed his wife that "ridiculous as it might seem, I had made up my mind to inquire more fully into the matter."

Barringer at this time was forty-two years old and at the height of his considerable powers.[2] A man of strong personality and great energy and enthusiasm, he was physically robust and muscular, standing five feet nine inches tall and weighing over two hundred pounds. A sociable, outgoing man, he enjoyed good company, good food, good whiskey, and good cigars.

Barringer was born on May 25, 1860, in Raleigh, North Carolina, to former Congressman Daniel Moreau and Elizabeth Wethered Barringer; his mother died when he was seven, his father when he was thirteen, and he then became the ward of an older brother in Philadelphia. As child and youth, he attended a succession of schools—first "Mr. Ryan's School" in Raleigh, then "Bingham's Military School" in Mechanicsville, and finally "Colonel Malcolm Johnson's School" in Baltimore, Maryland. The military aspects of this early education undoubtedly fostered his life-long interest in guns and ballistics, an interest he would later find useful in the Coon Mountain controversies.

At fifteen, he entered Princeton University (then the College of New Jersey) as a sophomore, graduating in 1879 with a class which numbered among its members such future notables as Woodrow Wilson and Cyrus McCormick. He received an A.M. degree from Princeton in 1882, and that same year, an LL.B. degree from the University of Pennsylvania Law School, where he was president of his class. From 1882 to 1889 he practiced law as a partner in his brother's Philadelphia firm, but he seems to have found the milieu of the office and the courtroom uninspiring and overly confining. He much preferred travel and the outdoor life; and his penchant for big-game hunting, pursued vigorously throughout his life, led him during this period to invent and promote a rifle sight and to join with his friends and fellow hunting enthusiasts Theodore Roosevelt and Owen Wister in founding the Boone and Crockett Club, an organization still renowned among sportsmen.

Restless in the law, Barringer in 1889 enrolled in a special course in economic geology at Harvard, and the following year took courses in chemistry and mineralogy at the University of Virginia, studying under the distinguished metallurgist John W. Mallet, a native of Ireland who had come to America in 1853 and who, in addition to holding the Ph.D., M.D., and LL.D.

degrees, was a Fellow of the Royal Society of London.[3] In 1890, Barringer served a brief apprenticeship in the Arkansas State Geological Survey where his mentor was the noted geologist John Casper Branner, who would become president of Stanford University.[4] Both Mallet and Branner remained among Barringer's closest confidants throughout their lives, an he relied heavily on their experience and advice in the earlier years of the Coon Mountain controversies.

After 1889, Barringer considered himself a geologist and a mining engineer. In 1892 he formed a partnership with another Philadelphia geologist, Richard A. F. Penrose, and then spent the next two years examining mining prospects in Spain and in South America, and thereafter in the American West, but with little success. In 1897, however, his fortunes improved dramatically. That year he published two books, *Minerals of Commercial Value*, which he had begun under Mallet's tutelage, and *The Law of Mines and Mining in the United States*, which he wrote with John Stokes Adams and which for many years remained the authoritative work on the subject. Then, on October 20, he married Margaret Bennett of Phoenix, whom he had met under somewhat unusual circumstances four years before. "She was in danger of being left behind at La Junta, Colorado, when he pulled her aboard the already moving train," their eldest son Brandon has recorded. "La Junta means 'the meeting' in Spanish and was prophetic."[5] Finally, soon after his marriage, Barringer discovered the Commonwealth, a bonanza mine which would finance many of his subsequent mining ventures as well as his long and costly exploration of Coon Mountain's crater.

PROOF AND PROFIT

Barringer's interest in Coon Mountain was commercial as well as scientific. There can be no doubt that he was anxious to prove the validity of his newly acquired belief in its impact origin, but there can also be no doubt that he was motivated as well by the prospect of considerable financial gain. The two objectives, proofs and profits, were in fact compatible, at least up to a point, for both were founded on the same easy assumption that Gilbert had made more than ten years before, i.e., that a large mass of nickeliferous iron must lie somewhere beneath the crater's floor. Locating this mass, he was convinced, was simply a mining problem and called for the application of more or less standard mining techniques, with millions of tons of nickeliferous iron as the product and new scientific knowledge as a byproduct.

In January 1903, at a dinner of the Boone and Crockett Club in Washington, D.C., Barringer told his friend and fellow Philadelphian, Benjamin Chew Tilghman, what Holsinger had told him about Coon Mountain and of his own strong conviction of its impact origin. The scientifically minded Tilghman, a sometime associate of Barringer's in mining ventures, was persuaded in his turn and agreed to join in both the exploration and exploitation of the crater.[6]

With Tilghman's interest assured, Barringer now wrote Holsinger in Phoenix for more details about the crater and for "slivers" of the meteoritic iron, hinting that Holsinger himself might find the prospect of exploring and developing the crater profitable.[7] Holsinger replied promptly,[8] sending a sample of the iron and a brief analysis of its composition,[9] along with an inaccurate citation of Gilbert's "very exhaustive report" and a somewhat exaggerated description of the crater itself. He had also talked with Volz, the trader at Canyon Diablo, who had told him that over the years he had sold some 10,000 pounds of the irons, at $1.25 per pound, to museums and other collectors. Volz, he reported, "thought he had about supplied the market for the stuff," and "had prospected the country around the mountain with steel rods but could find no more."[10] As to Barringer's hint, he added, "I note what you say about considering me 'in' if there is anything promising in the proposition. I would certainly appreciate it for I assure you, I am quite anxious to try my hand at something. For five years I have kept a keen eye open to, if possible, hit upon something . . . through which I might better my position, but alas I have found nothing. . . . So if there should be any future in digging up meteorites, you can count me in."

Holsinger's most important information, however, was the fact that Coon Mountain's crater was situated on "unsurveyed unreserved public lands,"[11] and Barringer now resolved to claim these lands and patent them under federal mineral land laws. He realized that he would have to work through agents, for he was well enough known in Arizona mining circles from his association with the Commonwealth mine that his interest in the crater, should it become known, might attract undesired attention and perhaps even a few claim jumpers. That others might become curious about the crater was impressed upon him in mid-February, when the *Associated Press* reported that one of the tiny diamonds found by Koenig in the Canyon Diablo iron that Foote had sent him in 1891 had been put on public display at New York's American Museum of Natural History. Holsinger sent him a clipping of the report from a Phoenix newspaper, noting that "I doubt if this article will attract attention but I would feel better if it had not appeared. But it is not at all probable that any one will attempt to locate the ground. If I thought this I would go at once and locate it for you. I think, however, if you find anything prospective worth locating we will find Crater or Coon Mountain unappropriated."[12]

Barringer was not reassured, indeed he had already sent Holsinger detailed instructions to confer with his brothers-in-law in Phoenix, Thomas D. Bennett and E. J. Bennitt [sic] and then go to Coon Mountain as soon as possible to stake out mining claims there.[13] Holsinger, after talking with E. J. Bennitt, replied that he and Tom Bennett would go to the crater as soon as Bennett returned from a hunting trip.[14] "I am making out the locations today and will carry out the other details as speedily as possible," he advised Barringer. "If it should happen that Tom can not go I will go alone.... My presence there will create no suspicion, as I frequently make the rounds through that region...." A surveyor would not be needed, he noted, as "I am something of an engineer and have a complete surveying outfit. Mr. Bennett also understands surveying and we think we can do the work and thus avoid taking another party in on the secret. The surveyor would be hardest to manage," he added. "If we have only workmen employed we can make any kind of explanation for locating the land and it will go."

As it turned out, Bennett could go, and the two men arrived at Coon Mountain during the first week of March 1903 and began locating claims. At Winslow, a small railroad town some seventeen miles east of the crater, they talked with a Mexican who had found many of the irons and who confirmed Volz's earlier information that all the irons had been found scattered on the surrounding plain, rather than inside the crater. At Canyon Diablo, they also had "quite an interesting interview" with Volz himself. "He is of the opinion that it will not be difficult to locate the main body of the meteor," Holsinger reported. "Thinks it could not be over 200 feet from the surface of the basin and probably not over 100 feet. He says he has had various plans for discovering the meteorite but could never secure the financial backing to justify his undertaking the project.... The more I learn from persons who have had opportunity to observe conditions here the more I feel encouraged," he added. "I think with a comparatively small outlay the body of the meteorite may be discovered."[15]

Initially, four placer mining claims, named for the planets Venus, Mars, Jupiter, and Saturn and covering 640 acres or "all the ground we have any use for," were located and recorded in the names of Holsinger and Barringer's brothers-in-law. The two men also made inquiries about obtaining a crew of workmen to construct a trail into the crater and perform the other development work necessary to fulfill the requirements for patenting mining claims. In addition, they searched a wide area around the crater for possible mill sites and sources of water that would be needed for subsequent mining operations. Holsinger reported that they could get enough water to operate machinery by damming the intermittent flow of Canyon Diablo itself, a south-north trending gorge cut into the arid plain some two-and-a-half miles west of Coon Mountain.[16]

In Philadelphia, Barringer occupied himself in other preparations for the venture. On March 15 he chartered the prosaically named Standard Iron Company through which the crater was to be developed and to which the placer claims were assigned. Barringer held controlling interest with 448 of the 798 outstanding shares, while Tilghman took 250 shares and E. J. Bennitt 25. Holsinger, who was now "in," was assigned 75 nonassessable shares. The costs of mining for the meteorite were to be financed through proportionate assessments levied as needed against the stockholders until the mass was discovered and mining operations began to produce revenues which, it was confidently anticipated, would not only repay their investment but net a handsome profit.[17]

Barringer, parenthetically, maintained control over Standard Iron, and thus over Coon Mountain and its crater, until his death in 1929, when it passed to his children who, with their heirs, have retained it. Clearly, from the beginning, he realized that Coon Mountain was more than just another mining property to be exploited and then abandoned when it had been worked out and that, convinced of its impact origin, "it cannot fail to remain for all time one of the earth's greatest wonders."[18] Indeed, Barringer strengthened his control of the company early in 1910 when Tilghman, after a bitter dispute over operations at the crater, relinquished his shares, which Barringer then took over and held in trust under an arrangement whereby he would repay Tilghman's $45,000 investment with interest over time. Only Holsinger's seventy-five shares remained outside the direct control of the Barringer interests. Holsinger died in 1911, and in 1919 his shares were purchased from his estate for $5,000 by Barringer's friend and Princeton classmate, William Francis (Billy) Magie. Magie, a physicist who for many years was Princeton's dean of faculty, became one of Barringer's principal scientific allies in the later stages of the Coon Mountain controversies.[19]

Late in March 1903, construction began on a trail into the crater, and soon workmen were sinking shallow shafts into the crater's floor to meet the patent requirements. As Holsinger had predicted, the workmen posed no threat to the security of the venture. "We are regarded here as cranks," he reported, "who have 'no better place to put money than in that big hole.' I have given this opinion encouragement. . . . Even the workmen laugh at our supposed folly when our backs are turned. . . . Those who have quit us have helped disseminate this view of our operations."[20]

Holsinger's initial optimism, however, quickly faded, and in his reports on the development work, he offered rambling speculations that clearly reflected a growing skepticism over the eventual success of the venture. He advised Barringer, for example, that "if this basin was the work of a falling meteorite we will probably be disappointed in the size of it—when we discover it if we ever do. I would now place the extreme diameter at 300 or 400 feet and I should not be surprised if it was very much smaller. You cannot fail to reach the same conclusion when you speculate upon the effect which would probably be produced by a body 300 feet in diameter or even 100 feet and weighing 1 to 3 million tons falling upon the earth. . . . I sincerely hope it is 1000 feet in diameter but I am afraid it is not."[21] And again: "I am trying to combine the theory of explosion with that of the falling of a large meteoric body and I find it works well. Had that ever occurred to you, i.e., that after the projectile entered the earth the generation of steam and gas might have caused an explosion which threw out the fragments which have been found. . . ?"[22] Still later he wrote that "I must confess to you that the more I study the phenomenon of Coon Butte the further I am away from the meteorite theory. . . . I have, however, been forced to the conclusion that the excavation on Coon Butte could not possibly have been made by a meteor. It was made by some subterranean disturbance and we may as well concede this for we cannot explain the physical conditions in any other way. . . ."[23] At this point, however, Barringer called a halt to such gratuitous speculations, and Holsinger replied that he understood and that it was better not to discuss such matters through correspondence.[24]

Holsinger, in fact, was more acceptable to Barringer as an observer than as a theorizer. During one of his visits to Volz's Canyon Diablo trading post, he noticed Volz's stock of cedar stove wood and "remarked at the great age of the wood as shown by the annular rings." Volz, he informed Barringer, "told me that the wood was cut from Coon Butte and I then secured some of the sticks and ascertained that the trees were over 500 years old. From this I judge that there are trees growing out there which may reach an age limit of six or seven hundred years."[25] Holsinger thus produced the venture's first contribution to science by fixing a minimum age for the crater from the stunted junipers growing on its rim.

The question of expediting the application for patenting the claims also came up at this time; and Holsinger, wise in the ways of the federal bureaucracy, volunteered his experience to Barringer. "Of course," he wrote in mid-April, "you may deem it expedient to pay something in order to ease this application through, but as I look at it we are engaged in a perfectly legitimate business and we will ask for nothing we are not entitled to. I never took a bribe in my life and I will have to know that there is no other way to get out of it before I will consent to give one. . . ."[26] Two days later, Holsinger sent Barringer a letter he had just received from Bennitt, who was handling the patent matter in Phoenix. "Now it is quite evident," Holsinger wrote, "that he does not know that he has been among the Philistines. It was rather unfortunate that he should have conveyed to the Surveyor General the idea that you are desirous of securing an early patenting of our claims. The fact is that office will hold us up just in proportion, in amount, to the amount of anxiety displayed by us. Plainly speaking, they are a set of robbers and bleed every man they can. The chief clerk, Mr. Murphy, has made a small fortune in that business. Of course he divided up with the Surveyor General. . . ."[27]

The bureaucratic mill did, in fact, grind slowly. On May 6 Holsinger advised Barringer that the development work was completed, that the workmen and suppliers had been paid off at a cost of just over $1,300, and that "we are ready to apply for a patent."[28] The patent, however, would not be forthcoming for more than seven months.

In the interim, the enterprise attracted the attention of the Flagstaff newspaper, the *Coconino Sun,* which belatedly advised its readers in August that a "Philadelphia company is digging for a meteor southeast of Canyon Diablo which, found, is to constitute a bonanza iron mine." The reporter, however, proved to be a skeptic, with Gilbertian ideas: "It is a strain of the imagination to believe that a meteor, no matter what its velocity, under 3800 feet in diameter could have pierced 1000 feet—if we are to allow for surface erosion—into limestone and sandstone." "Conditions seem to indicate, rather, a blowout caused by exploding gases or confined steam. The pieces of meteoric iron are or have been encrusted with shale, showing that they fell as independent bodies rather than that they were fragments thrown off from one larger body as it struck the earth. By a simple coincidence a meteoric shower might have passed over our cone some time after its formation and thus their presence would be accounted for."[29]

Perhaps by coincidence too, Barringer two weeks later instructed Holsinger to begin digging trenches, sinking shafts, and running tunnels into the debris on the rim of the crater to "establish . . . whether fragments of the meteor are admixed with sandstone and limestone fragments from the present surface down."[30] This work in time produced evidence that Barringer and Tilghman would subsequently cite to show that the fall of the Canyon Diablo irons was not a "simple coincidence," but had been synchronous with the formation of Coon Mountain itself.

The patent on the claims—signed, incidentally, by Barringer's friend, President Theodore Roosevelt—was finally issued on December 24, 1903. Barringer, however, aware that winters in northern Arizona are usually severe, postponed his first visit to Coon Mountain for another three months, arriving there late in March 1904 "to see what kind of animal we have caught in our trap."[31]

EXPLORING THE CRATER

For the next two years, Coon Mountain was the scene of often frenzied activity. Barringer and Tilghman made numerous visits to the crater, supervising attempts to locate the mass of the meteorite in its depths and prospecting its rim and the surrounding plain for evidence for its impact origin and against Gilbert's volcanic steam explosion hypothesis.

In their exploration of the crater's environs carried out in the summer and autumn of 1904, Barringer and Tilghman picked up several thousand pieces of meteoritic iron, the largest weighing 225 pounds. They also collected up to a ton of the black iron oxide or iron "shale," as it was called locally, presumably the same material that had impressed Foote and that Gilbert had noted without comment during his later investigation. Most of this iron shale, which proved to be magnetic, occurred as small fragments or slightly curved flakes, with a laminated structure, strewn over the ground; but in some cases, it was found encrusting pieces of the meteoritic irons. Less frequently, it was in the form of large globular lumps that came to be called "shale balls." Some of these, when broken open, contained solid iron centers which, sectioned and etched with acid, displayed distorted but distinct Widmanstätten patterns characteristic of iron meteorities.[32] In time, it also was discovered that these iron centers exuded chlorides and that, after prolonged exposure to the air, they slowly disintegrated into iron oxide.[33]

In addition, they found what they could only estimate to be "many millions of tons" of finely pulverized silica, 98 to 99 percent pure silicon dioxide (SiO_2), on and in the bottom of the crater, underlying the entire rim and blanketing its outer slopes. This silica, or "rock flour" as it came to be known, apparently was the same material Gilbert had referred to briefly in his field notes as "white sand" and "soft white earth." Especially large deposits occurred on the crater's south rim, where they were cut by a dry wash more than ten feet deep and a hundred feet in length. Some of this silica, Barringer would later note, had the consistency of an "impalpable powder" and was so finely pulverized that "no grit can be noticed when it is placed between the teeth."[34]

Other evidence that the rocks in and around the crater had been subjected to a sudden, violent shock appeared in the exposed strata of the walls and in the boulders and blocks on the rim. There, they found what Barringer later referred to as "ghost sandstone," freestanding and seemingly solid rock, with its original stratification and bedding planes intact, that had nonetheless been shattered to the point where pieces could be broken off easily and crumbled with the fingers.

The crater itself, they also recognized, displayed a bilateral symmetry on a line running roughly north and south through its center. More of the larger boulders and blocks of limestone and sandstone were found on the east and west sides of the crater, while the high point of the rim was on the south and the low point on the north. More of the smaller irons and iron shale pieces were also found on the rim and on the plain to the north and northeast of the crater. Moreover, while the strata of the rim dipped quaquaversally, that is to say, downward and outward in all directions from the center, the strata on the southern rim obviously were more sharply upthrust, sections there having been nearly overturned. The rock on the south, too, had been more violently shattered, or "cominuted" as Barringer later noted, than at any other point on the rim. From this, they reasoned that the meteorite must have plowed into the earth at an angle slightly to the north of vertical.[35]

To explore the depths of the crater itself, Barringer and Tilghman planned to sink a deep shaft in the center of the crater at the junction of the four placer claims, a project which they confidently expected would not only quickly reveal the buried meteorite, but which could be carried out with only a few workmen using light timbering and a horse whim as a hoist. The work began almost immediately, and progress on the shaft progressed slowly but steadily as the workmen dug through the 80 to 90 feet of fossil-bearing lacustrine deposits and windblown debris that had accumulated on the crater's floor since its formation. Barely a month passed, however, before Holsinger reported a serious problem. "I have news from the crater which rather surprises me," Holsinger wrote Barringer in May. "They have actually struck water in the shaft." The workmen, he added, had reached a depth of 181 feet, where they had encountered a material that was "almost mud."[36] It was, in fact, quicksand, formed by water in the finely pulverized silica in the Coconino sandstone, the principal aquifer in the region. Repeated efforts to drive the shaft deeper proved futile, as the watery quicksand poured into the bottom of the shaft faster than the horse whim could haul it out.

Faced with this impasse, Tilghman now urged that a series of drill holes be bored, not only to locate the meteoritic mass, but to determine the depth of the true bottom of the crater. Barringer, who apparently favored sinking another shaft with more sophisticated methods, only reluctantly agreed.[37]

In all, five holes were put down under a succession of more or less inept drill foremen, two in the late summer and autumn of 1904 and three in the spring of 1905. None struck what was considered to be the main mass, although sizeable objects were encountered, at least one of which, from its extreme hardness, its metallic ring, and magnetic attraction, was thought to be a large fragment of meteoritic iron. In some cases, these objects could be dislodged or bypassed by the drill; but eventually in the first, second, and fourth holes, obstructions were encountered at 480, 305, and 600 feet, respectively, that forced their abandonment. Attempts to drill through these obstructions or shunt them aside, by jetting large quantities of water into the bottom of the hole or by dropping the heavy drill tools on them, were to no avail. In hole number 4, three sticks of dynamite were detonated on top of the obstruction, but the explosion only drove the object two inches deeper.[38]

Hole number 3, abandoned at 485 feet, apparently was a drilling fiasco, being at one point "too small to get the casing down," with consequent cave-ins and flooding by water and quicksand. Barringer castigated Holsinger for the "mistakes" made in drilling this hole and for "throwing away . . . last month's work and $1,500."[39]

Only the fifth hole penetrated to what, from an abrupt change in the hardness of the material and the difficulty in drilling, appeared to be a rock in place—the brown sandstone of the Red Beds or Supai formation underlying the Aubrey (Coconino) sandstone in the region. Early in June the drill bit entered the Supai at 890 feet and continued in it to a depth of 1,003 feet, where drilling was stopped, having reached what Barringer, who was on hand at the time, thought must be the actual bottom of the crater.[40] As Tilghman explained later: "The change from white to reddish brown sand is quite marked and sudden, and if this material had been stirred up by the passage of a projectile through it, it would be so mixed as to be indistinguishable, or at any rate certainly not have a definite boundary between the two materials."[41]

In all of the holes, small bits and pieces of what was believed to be meteoritic material were recovered below the quicksand from levels between 300 and 500 feet. This, analyzed by H. H. Alexander, metallurgist for the U.S. Smelting, Refining and Mining Company's Globe smelter at Denver, and by Mallet at the University of Virginia, consisted of tiny flakes and particles of magnetite, sand grains coated with, or cemented together by magnetite or covered by a film of iron silicate, and small globules of iron enveloped in the magnetite. The magnetite, it was found, resembled the magnetic iron oxide or iron shale that occurred on the rim and on the surrounding plain except that its nickel content was lower, the result, they concluded, of leaching by water percolating through it from the quicksand zone above. That leaching had taken place also was indicated when Alexander found small percentages of nickel in sand and rock fragments collected at levels well below the lowest point where the so-called meteoritic material had been found.[42] This finding, incidentally, caused Barringer to worry briefly that the drill bit or the drilling mud might have "salted" the holes until investigation showed that neither the bit nor the mud contained any nickel.[43]

The small, magnetite-coated globules of iron, which Barringer later likened to tiny shale balls, in particular proved to be a strong argument for the meteoritic, or at least the alien, nature of the material. "While it is conceivable that silicate of iron and magnetite might occur in the wreck of terrestrial strata of the character found in this locality," Tilghman later pointed out, "it is extremely improbable, because there is no trace of any of this material in the unpulverized rock forming the strata in question. But it is absolutely inconceivable that those little metallic spheres with their coating of magnetite could exist in ordinary sedimentary strata."[44]

If, however, the drilling program had been productive from a scientific standpoint, from Barringer's standpoint it had been something of a disaster. As early as April, after the abandonment of the third hole, he had confessed that he was "sick of the whole business of drilling," and declared his intention of sinking a new and larger shaft.[45] The program had failed to find the meteoritic mass and had yielded "no really satisfactory conclusion," as he later informed Alexander. "I should have never given my unwilling consent," he added, vowing to "never repeat the mistake."[46] Moreover, the large objects encountered in the holes had raised fears that the meteorite might be composed of relatively small pieces of iron imbedded in a matrix of rock—"like plums in a pudding," he remarked, unknowingly using the same analogy Gilbert had used to describe Howell's similar suggestion as to its possible character.

"I have always had this fear," he had confided to Holsinger back in April, and to Mallet he had confessed: "I wish I could be as sure that the mass is workable and possesses commercial value as I am that it lies at the bottom of the crater at Coon Mountain or somewhere in it."[47] Finally, the drill holes had cost nearly $30,000, or four times his initial estimate of $1,500 per hole, a sum that had been raised through increasingly frequent $3,000 assessments, with Barringer himself, of course, paying the lion's share.[48] Late in June he ordered a halt to all further drilling operations and turned his attention to the sinking of a new and larger shaft, advising Tilghman of his decision, incidentally, only after the event.[49]

Two other developments, unrelated to the drilling, undoubtedly influenced his decision.

First, he now believed that the meteoritic iron was even more valuable than he had first thought. Initially, he had estimated its value on the basis of its nearly 8 percent nickel content alone, but Alexander and Mallet had since reported finding traces of platinum and iridium in a twenty-five-pound sample of the iron.[50] While the amounts were small—Mallet later set them at 3.63 grams of platinum and 14.96 grams of iridium per ton[51]—they were sufficient to more than double the value of the iron as an ore at then current prices, assuming as he did that the precious metals could be extracted profitably. "If that stuff is worth $200 a ton," he now vowed, "I will make the interior of the crater look like a rabbit warren before I give up my search for the projectile that certainly caused that hole. I cannot believe it is over a thousand feet deep, but I am ready to believe it is scattered far and wide and broken into millions of pieces."[52]

The second development was as bizarre as it was disquieting, and again raised his fears that the "plums in a pudding" theory might be valid and that the meteorite had been composed primarily of worthless rock. During his visit to Coon Mountain in June, he had found a small stone about a mile and a half northwest of the crater that seemed to be unusually heavy, was unlike any other rock he had found in the vicinity, and, moreover, showed a thin coating of iron oxide. He immediately suspected that it was an aerolite, or stony meteorite, and promptly turned it over to Alexander for analysis. Now he fumed to Bennitt that "the confounded stone which I picked up on my way to the Cañon a few days before I left is almost certainly a stone meteorite and is full of meteoric iron. I cannot understand how it belongs to the other fall, and I am not at all sure that it does."[53] Alexander reported that, while it contained 22.3 percent iron and 1.65 percent nickel, it was composed primarily of oxides of silicon and magnesium, with small percentages of aluminum and calcium oxide and elemental sulphur.[54] From this and the fact that the "miserable" aerolite contained no platinum or iridium,[55] Barringer decided that it was "unlikely that it was part of or connected with the big meteorite."[56]

A NEW AND LARGE SHAFT

The new shaft was to be a far more extensive undertaking than the previous shaft, for Barringer was determined to sink "a double compartment shaft 1,000 feet if need be," using heavy timbering, a fast-acting steam hoist in place of the horse whim, and a larger and more experienced crew of workmen who would be under the charge of a seasoned mining superintendent.[57] It would also be a more expensive undertaking. "It is going to cost us heavily," he warned Bennitt in raising the July assessment to $5,000. In August the figure was doubled, and henceforth assessments would be $10,000.[58]

If all went well, the new shaft, by revealing the meteoritic mass, would also provide a crowning argument for the scientific papers Barringer and Tilghman had now decided to prepare marshaling the evidence they had gathered and proclaiming the impact origin of Coon Mountain and its crater. Barringer completed a draft of his paper early in September, and immediately arranged for the eventual publication of the papers with S. G. Dixon, president of The Academy of Natural Sciences in Philadelphia, of which both he and Tilghman were members.[59] On September 25, apparently at Dixon's suggestion, he addressed a formal letter to the academy to establish priority for their claim that Coon Mountain had been formed by meteoritic impact.[60]

Through the summer and autumn of 1905 preparations for sinking the new shaft proceeded at a furious pace, and although the demands of his other mining interests were now especially heavy,[61] Barringer concerned himself with even the most minute details. The new steam hoist was ordered and in due time installed, with Barringer worrying that there would not be enough heavy manila rope on hand to lower it safely down the steep crater walls.[62] Dams of tarpaper, canvas, and clay were thrown up across Canyon Diablo and the reservoirs behind them fenced against the unwanted intrusions of cattle grazing in the area. Indeed, until the fences could be put up, Barringer ordered Holsinger to "blast the cliffs near the cattle trail so that the cattle cannot get into the dams."[63] The reservoirs were sealed, not too effectively, against leakage,[64] an entrenched pipeline was run from them to the crater, and windmills were set up to pump the water, when there was any, to 20,000-gallon tanks on the rim and in the crater itself.[65] The camp on the north rim also was expanded to accommodate additional workmen and equipment.

Holsinger's role in this new effort was to be peripheral. He was to have nothing to do with the actual sinking of the new shaft, which was to be in the charge of a superintendent who was to be "executive head of all mining operations," but was to restrict his activities to keeping accounts, making reports, relaying orders, and performing various other tasks that Barringer assigned him outside the crater proper.[66] Holsinger's earlier pessimism, in fact, had deepened and Barringer apparently now felt, or perhaps even hoped, that he would quit the enterprise altogether. In July, for example, he wrote his friend Gifford Pinchot, head of the Bureau of Forestry in Washington, D.C., to "put in a good word" for Holsinger who, he added, "may wish to soon reenter the service of your Bureau."[67] Six months later, indeed, Holsinger's continued "croaking and dismal prophecies" led Barringer to ask Pinchot directly to take him back, but Pinchot would not have him.[68]

Work on the new shaft began in July 1905, and by August the workmen were down to the 100-foot level.[69] Barringer's plan was to go down slowly and carefully to a depth of 180 feet, or just above the quicksand, where the work would be suspended until the new machinery was installed and all was in readiness for a massive assault on the quicksand itself.[70] This, he realized, would require a hopefully brief but intense effort on the part of all concerned, with the machinery operating continuously and the men working in shifts around the clock for days, or perhaps even weeks at a time. "The whole secret," he explained patiently to Holsinger at one point, "is rapidity of the work."[71]

This assault was initially scheduled for October, but a series of problems large and small forced repeated delays. Not the least of these was finding a superintendent who met Barringer's exacting standards. By late October, after two candidates had been tried and found wanting, he finally hired one James Nevill, a veteran mine supervisor experienced in the sinking of shafts.[72] Nevill, however, did not arrive at the crater until mid-November, and an impatient Barringer now instructed him to begin the assault on the quicksand "before the first of December," adding pointedly that "the slow way of doing things has prevailed for too long."[73]

Two factors added urgency to the shaft through the quicksand at this time. First, the onset of winter weather at Coon Mountain was imminent and could only complicate what was already a difficult mining operation. Second, he and Tilghman were now convinced that they must publish their conclusions regarding the crater as soon as possible for, inevitably, the bustling activity there over the past few months had attracted attention, and speculation, in the press. These conclusions, moreover, would be greatly strengthened if they included a report of the actual discovery of a meteoritic mass beneath the crater's floor.

On November 21 Barringer advised Dixon that there was "no good reason for keeping the matter secret any longer," and urged publication of their papers in the "next Proceedings" of the Academy.[74] To Mallet, he confided that "we may be compelled to publish . . . within the next week. The cat is out of the bag," he added, explaining that there had been "many newspaper accounts of our enterprise and one of these has been copied in the Scientific American."[75] A few days later, he ordered Holsinger not to answer mailed inquiries about the crater and to "chase off visitors lest we lose the credit."[76] And to Bennitt, he reported that he and Tilghman felt "obliged to publish," even though they "had hoped this could be deferred until some important discovery in the shaft."[77]

As both papers were quite lengthy, however, and as their publication would require time, Barringer arranged for Dixon to make a brief preliminary announcement of their conclusions immediately. Consequently, at the Academy's December 5 meeting, Dixon proclaimed Barringer's and Tilghman's "discovery that the crater of Coon Mountain or Coon Butte . . . is an impact crater and not a crater produced by a steam explosion, as has been supposed since the examination made of it by members of the United States Geological Survey. They have proved," he continued, "that the large crater and elevation known as Coon Mountain is the result of a collision with the earth of a very large meteorite or possibly a small asteroid, fragments of which are well known to the scientific world by the name of the Canon Diablo siderites. . . . Mr. Barringer and Mr. Tilghman have presented to The Academy for publication two comprehensive papers in which they set forth in full their reasons for the above statements."[78]

The assault on the quicksand had, in fact, begun a few days earlier; but after two weeks, and by mid-December, it had become obvious that no progress could be made through "simple timbering and rapid continuous hoisting."[79] Various other methods were now explored, and at Barringer's suggestion, a pressurized caisson was rigged up that they hoped would stem the inrush of quicksand into the bottom of the shaft.[80] Other preparations, it was found, were also necessary. A gasoline pump was obtained to supplement the sometimes unreliable windmills at the reservoirs, and a steam pump was installed in the crater to handle the water problem in the shaft.[81] The increasingly cold weather, as well as blowing silica dust on the crater floor, now also necessitated the construction of a large fireproof and dustproof shaft house to protect both the men and the machinery.[82] And finally, a telephone line was strung to facilitate communications between the shaft head and the north rim camp.[83]

Not until mid-January 1906 were these preparations completed, and the attempt to push the shaft below the 200-foot level resumed. But almost immediately the news from the crater was grim. On January 19, as Barringer anxiously awaited word that the first twenty feet of the quicksand had been penetrated, Nevill telegraphed him that it was "impossible to sink shaft any depth."[84] The pulverized silica was holding water "much like a wet towel," Barringer noted, and "falls off from behind the timbers in great masses."[85] Moreover, the quicksand tended to suck the heavy timbers downward and into itself, there being nothing substantial to anchor them in place beyond a thin layer of encrusted calcium carbonate ($CaCO_3$), or caliche, that the workmen had encountered at the very top of the quicksand zone.[86] Barringer now sought the advice of outside experts and made some suggestions of his own. New equipment was rigged up or brought in—a "Tilghman shield" devised by Tilghman and a "Cameron sinker." But these, as the other measures, proved ineffective against the rapidly oozing silica at the bottom and sides of the shaft.[87] Early in February Barringer confided ruefully to Mallet that "the ground has caved dreadfully."[88]

Barringer now had also become convinced that the crater was, in effect, a vast sealed basin that was holding water well above the level of the water table of the surrounding country. "I am afraid," he wrote, "we shall have to go outside the crater and commence at the base of the mountain and sink a long incline of 30 or 40° to get underneath the bottom and make a great sump, and pump perhaps for a year."[89] A few weeks later, he proposed a somewhat less grandiose scheme to "pump that crater dry in six months with four or five wells at a cost not to exceed $2,000 a month."[90] By mid-March, certain now that it was a "case of having to pump first and sink afterwards," he assigned Holsinger the task of boring "four wells at the four corners of the shaft." He even wondered if the railroad would be willing to build a pipeline from its Sunshine flag stop so that water from the pumping operation could be sold to them."[91]

The attempt to sink the shaft through the quicksand, however, was already doomed. Tilghman had visited the crater during the first week in March and, with Bennitt, had assessed the situation there, returning a highly pessimistic report.[92] Nevill and the workmen were allowed to continue their efforts for a few more weeks, but on March 26 Barringer ordered the shaft abandoned; and a few days later, he advised Nevill that "I don't see the necessity of your remaining longer."[93] The attempt to sink the shaft had cost some $55,000.[94]

THE CONTROVERSY COMMENCES

Barringer's and Tilghman's papers appeared in print in the Academy's *Proceedings* March 1, 1906, and their publication marks the beginning of what has here been called the "Coon Mountain controversies."

Henceforth, Coon Mountain's origin would be a subject of a lively and sometimes bitter debate that continued for more than twenty years and, indeed, lingered on in some quarters for more than fifty years. Not until the late 1950s, for example, did U.S. Geological Survey officials concede, and then only tacitly, the impact origin of the crater, although as will be seen, some of its individual members privately had granted the point many years before.[95] Some writers, moreover, have since tended to credit Gilbert, a Survey scientist, with the discovery on the grounds that an impact origin was one of the hypotheses he had considered during his 1891 investigation. His rejection of an impact origin is passed over lightly, when it is mentioned at all, and it has been suggested that had he been able to sink shafts and drill holes into the crater, he too would have opted for its impact origin.[96] But, as will be seen, much of Barringer's and Tilghman's evidence came not from shafts or drill holes but from routine observations that Gilbert, as Barringer contended, could have made. Gilbert, Barringer declared, had adopted his steam explosion hypothesis "on what seems to be very insufficient evidence. Perhaps it is more accurate and just to say that he has adopted this theory because of an inadequate examination of Coon Mountain . . . for had he examined the surface carefully, it does not seem possible to me that any experienced geologist could have arrived at such a conclusion."[97]

Strong words, especially when applied to a scientist of Gilbert's great stature by someone outside the scientific establishment, and perhaps deliberately provocative words, although in fact they provoked no response from Gilbert himself. Whatever Gilbert may have thought of Barringer's and Tilghman's findings, he never commented on them publicly; his reply to their pertinent criticisms was to be the powerful argument of silence.

Of the two papers, Tilghman's is somewhat longer and is at once less argumentative and more scientific in tone.[98] Both, of course, described and dealt with the same basic phenomena, but each emphasized different aspects of the data. Barringer, for example, gave considerable attention to refuting the volcanic and steam explosion hypotheses and to his finding of the aerolite and arguing for its coincidental occurrence, devoting pages to matters that Tilghman discussed in paragraphs. Tilghman, on the other hand, concentrated on setting forth in great detail the evidence for the crater's meteoritic origin, emerging as a pioneer in the study of the mechanics of impact cratering in the process.

Barringer and Tilghman, of course, had seen by and large the same phenomena at Coon Mountain that Gilbert had seen, but beyond this they had seen a significance in these phenomena that Gilbert had plainly missed. A case in point is the pulverized silica that Gilbert had referred to only briefly as a "white sand" and a "soft white earth."[99] While Barringer could not estimate "how many millions of tons" of this powdered silica existed in and around the crater, he pointed out that it composed "a great part of the enormous rim, over three miles in length measured around the base of the mountain. . . . The amount of it within the crater is absolutely unknown; for it has been found by means of drill holes to a depth of 850 feet. . . . If one digs down through the surface soil a foot or more, almost anywhere on the outside of the rim, among the fragments which have been thrown out of the hole, he will come into this silica, and a great number of trenches and shafts have shown it to continue downward certainly to the solid or rather more or less shattered rock upon which all of the fragmentary material forming the rim rests."

Barringer noted particularly the massive, washcut exposure of pulverized silica on the south rim, declaring that it is "difficult to understand how this exposure could escape the eye of any careful geologist making a circuit of the crater." Upon examination of it and appreciation of its great quantity, its presence could be explained only one way. Barringer said, "it seems impossible to me that this silica could be produced by volcanic action, or by a steam explosion, and I assume that it could be produced only by the pulverizing effect of an almost inconceivably great blow."

The pulverized silica, too, Barringer wrote, was "plentifully admixed" with broken fragments of the limestone and sandstone, and trenches and shafts dug into the rim "have shown that the silica carrying with it these broken fragments, especially those of smaller size, had evidently welled out of the crater almost like liquid mud, or perhaps, more accurately, like flour when it is poured out of a barrel."

Tilghman expanded on this point, declaring that it made "any theory of the explosive formation of the hole utterly impossible." When a "great projectile" impacts on a "hard, brittle and practically incompressible material," he explained, a "cone of material with an apex angle of about ninety degrees is compressed downward into the solid mass of the material from the point of impact. This cone . . . crushes into powder under the force of the pressure or blow, and this powder being still further compressed transmits the pressure upon it in all directions, somewhat like a fluid. . . ." In other words, "The pressure . . . seeks relief and forces a yielding of the solid material around it which, of course, occurs along the line of least resistance, and bursts the surface upward and outward into a cone-shaped crater around the point of impact. . . ."

The greater the force of the impact, the greater the size of the crater, he added; thus, "the depth of the crater always bears a definite relation to its width, and in large craters it is found that the crater is always surrounded by a cone of cracked and shattered material, which would have been the next material to have been expelled from the crater by the velocity imparted to them by this inelastic silica powder; thus when both rock fragments and powder have progressed far enough to free themselves from the pressure of the penetrating projectile they fly on together, mixed powder and rocks, at the same velocity." The powder, he explained, was not dust in the ordinary sense of the word, "but is almost unmixed with air in solid masses, particle to particle, like flour in a barrel so to speak, which masses obey the laws of projectiles and falling bodies, irrespective of the minute particles of which they are formed, and are thus deposited in the rim in mixture with, and under and over the solid rock masses which accompanied it in its flight. . . ." Had the velocity of the expelled rock debris been due to compressed elastic gases, he added, the gases "would have instantly assumed a much higher velocity than the heavy rock particles . . . and, sweeping by them, would have carried with them every particle of silica powder."

Barringer and Tilghman also dwelt at length on the "very large quantities" of magnetic iron oxide, or iron shale, which, like the meteoritic irons themselves, had been found all around the rim and on the surrounding plain. "This is so abundantly distributed," Barringer declared, "and is so apparent to the casual observer, that it seems wonderful to me that Mr. Gilbert and his associates did not make any reference to it in their report."[100] The iron shale seemed to be different from any substance in nature, he added, "and had they taken the trouble to have it analyzed they would have found that the large pieces almost invariably contain nickel . . . to the same extent, proportionately speaking, as it is found in the Canyon Diablo meteoric iron." Moreover, "if they had merely broken open some of the larger pieces of this magnetic iron oxide, which it seems to me they could not have failed to see, they would have observed in some of the specimens the characteristic green hydroxide of nickel."

Barringer suggested that the iron shale had been produced "by the heat generated from friction while the great iron meteor passed through the earth's atmosphere." And from the fact that "the surface of the surrounding country for perhaps several miles . . . contains minute particles of this iron shale, either in the shape of fragments or as spherules," he theorized that these tiny particles "once constituted a portion of the great luminous tail of the meteoric body which, in our belief, by its collision with the earth made the crater."

Tilghman again enlarged upon the point. A search of the area with magnets, he noted, had shown that the presence of this minute magnetic matter was "absolutely universal . . . inside the hole and out for as far as observed. . . . It consists of a blackish-gray rather fine-grained powder, strongly attractable by the magnet, crystalline in structure, but not so in shape, being in small torn irregular masses with generally intensely fine grains of silica powder adhering so firmly to its surface as to suggest adhesion while in a state of fusion. Of very rare occurrence among it are absolutely round balls with a fused polished surface like intensely fine shot. These, it is supposed, have had time to solidify in the vacuum behind the flying meteor free from the fierce rush of air . . . and they were enabled thus to assume the usual shape of liquid drops." The material, he added, "is not a usual substance and, so far as known, is not a constituent of any of the rocks in the neighborhood. . . ."

Both Barringer and Tilghman declared that the magnetic iron oxide, or iron shale, was "meteoric" in origin, Tilghman listing four reasons for this conclusion. First, it had been found attached to, and in the cavities of some of the larger irons. Second, some of the larger pieces contained centers of "metallic meteoric iron." Third, the proportion of the metals it contained was in "very close agreement" with the composition of the irons. And finally and more speculatively, its general appearance was "different from any terrestrial magnetite known and closely resembles what would be thought, *a priori* to be the appearance of such a product of iron melted and burned on the surface of a great meteorite in its passage through the air."

Given its meteoric nature then and its similarity to the irons, the iron shale also demonstrated the synchronism of the fall of the irons and the formation of the crater itself. For in the trenches and shafts dug into the crater's rim, Barringer reported, "we have found nearly one hundred pieces of meteoric material, some of them as much as fifty pounds in weight, a number of feet beneath the surface in the silica, overlaid and underlaid in no particular order by the various kinds of rock fragments. . . ." One shaft, forty-eight feet deep, yielded "vertically one above the other no less than seven quite large specimens of meteoric material or iron shale; the first being found twelve feet . . . and the last twenty-seven feet beneath the surface," with "the usual admixture" of silica and rock fragments in between. "On no conceivable theory other than the one we have adopted can the facts described above be explained," he declared.

For Tilghman, this synchronism was also illustrated by the distribution of the meteoritic material all around the crater, with somewhat greater amounts being found to the northeast. If the shower of meteoric iron and magnetite occurred after the formation of the crater, he pointed out, "it is inconceivable that the densest portion of the shower of each material should coincide accurately with the northeasterly rim of the hole, and yet none fall into it," while if the shower occurred before the formation of the crater, "it is equally inconceivable that the fallen material could be found most thickly on the surface of the rim, composed of material ejected from the hole."

The cuts and shafts in the debris of the rim also produced another pertinent fact, for there was indeed an order to the vertical distribution of the rock debris, as Barringer noted:

> In one case . . . we found a large piece of meteoric oxidized material or "iron shale" about six feet beneath the surface in the silica, directly underneath an angular fragment, several feet in diameter, of red [Moenkopi] sandstone. On top of this red sandstone was a piece of limestone, and on top of the limestone was a still larger piece of white sandstone. I merely mention this case as it is interesting to reflect that the white sandstone comes from a depth of at least 400 feet below the surface, and yet it is found

on top of the red sandstone fragment (the surface rock) and the limestone fragment which, when the geological order of the rocks is considered, lie above it.

Tilghman had also noted this reversal in the vertical order of the rocks of the area in the debris beneath the crater floor and now offered an explanation for it. In one of the early development shafts, "This formation suggested the idea that the surface material, having first received the impact of the meteorite, started first in its aerial flight, followed by the lower materials in turn as they were reached, and retained this order when falling back into the hole. . . ."

Both Barringer and Tilghman presented a number of other points in favor of meteoritic impact and against a volcanic or steam explosion origin for Coon Mountain. Like Foote and Gilbert before them, they had found no naturally occurring volcanic rock in the vicinity of the crater, the nearest exposed eruptive rock, as Barringer pointed out, being at Sunset Mountain, some nine miles to the southeast. Unlike Foote or Gilbert, however, they had recognized the apparent bilateral symmetry of the crater, emphasizing the more violent upthrusting and shattering of the rock on the southern rim and the larger amount of powdered silica there. "The inference is unavoidable that the cause which produced the crater acted with somewhat more violence in the southwest, south and southeast than in the opposite direction," Barringer declared.

A major problem posed by the data, for Barringer at least, was the fact that the "so-called shale balls are to be found beneath the surface on the outside of the rim, and admixed with the fragmentary material . . . and the angular pieces of meteoric iron have been found to date *only* on the surface or in the shallow soil which overlies the rock fragments and the silica . . . or on the surrounding limestone plain [Barringer's emphasis]." To resolve this, he offered an elaborate and speculative explanation based on the chance observation of a "very brilliant meteor" that both he and Holsinger had seen late in the afternoon of April 11, 1904, in southern Arizona.[101] "The head of the meteor, he recalled, "was blue-white in color; from this head there seemed to dart from time to time, and almost from the moment the meteor became visible, many jets of bluish-colored light. Behind the meteor was a glorious comet-like tail, the color of which was generally yellow. From behind the meteor and out of this tail there appeared from time to time, and after the meteor had been visible for an appreciable length of time, great flaming drops, not unlike drops of burning tar. These rapidly fell behind the meteor, being distanced by it. . . ."

Applying this observation to the case at hand, he declared that "I am inclined to believe that many of the thousands of pieces of metallic meteoric iron which have been found distributed around Coon Mountain . . . were pieces that were torn loose from the surface of the meteor . . . by the violent strains set up because of the intense coldness of the main body of the meteor . . . and the intense heat immediately generated upon the entrance of the meteor into the earth's atmosphere. This would explain the darts of light which Mr. Holsinger and I saw going out of the front of the meteor referred to above. . . ." These pieces of iron, falling behind the meteor, "probably reached the earth after the collision had taken place and all of the material had been thrown out of the crater by the impact," he added.

The shale balls, so his reasoning went, were analogous to the "flaming drops" that he and Holsinger had seen falling out of the tail of the southern Arizona meteor. "As the front surface became more and more heated," he suggested, "it is possible that fewer of these irons would be thrown off, and almost certain that more of the iron would be melted and would naturally run back to the sides or the rear surface of the meteor, from which from time to time it would be detached. This burning iron would then drop behind . . . and form the shale balls above referred to." This, he noted, explained the laminated structure of much of the iron shale, and the "flaming drops" he and Holsinger had seen, as well as "why they were not seen until the meteor had been visible for an appreciable length of time."

Moreover, these burning, and thus oxidizing, pieces of iron "probably continued to drop off from the Canyon Diablo meteor . . . until the very moment of collision. . . . It is natural, therefore, to conclude that some of them must have been caught before they reached the surface of the earth by the outgoing rock fragments and silica which poured out of the hole at the moment of collision." Most of these burning pieces of iron would have been entirely converted into iron oxide, he added, but "some of them seem to have been smothered out when covered up by the silica and the rock fragments included in it. This would perfectly explain why some of them have iron centers. . . ."

Barringer also dealt at some length with his finding of the aerolite, stressing that chemical analysis of the stone and the meteoritic iron "shows a wide difference between the two." But he also related "a story which is at least very interesting, for as a coincidence, if such it is, it is very remarkable." About January 15, 1904, he reported, two employees at the crater "saw a meteor fall somewhere northwest of the mountain, between them and the railroad." Then,

> if we have been able to fix the dates correctly, on the same evening, at the same moment, a few minutes before nine o'clock, the hour being fixed by the train schedule, Dr. A. Rounsville of Williams and Dr. G. F. Manning of

Flagstaff, Arizona, were traveling to Canon Diablo station, where Dr. Manning had been called to visit a patient. Just before the train stopped, Dr. Rounsville saw from one of the windows, on the south side of the train, a blazing meteor fall in the direction of Coon Mountain. . . . It is very probable that this was the same meteor that was seen by our employees at Coon Mountain. . . . As accurately as I can determine, it was very near a spot near the intersection of the two lines of sight, a spot which of course they could not locate exactly, that I found the above described meteoric stone. . . . I have stated the facts as accurately as possible," he declared, "and I have no opinion as to whether or not these involve anything more than a coincidence.[102]

Barringer and Tilghman were in close agreement on Coon Mountain's age. Both cited a minimum age of 700 years from the age of the junipers growing on its outer rim, while Barringer concluded that "it is not more than 2,000 or 3,000 years old, and perhaps much younger,"[103] and Tilghman, reasoning from the degree of erosion of debris from a nearby butte, set its age at "something well inside 5,000 years," with 10,000 years as "the utmost possible limit."

Tilghman added several points to Barringer's exposition. At the outset, he challenged the validity of the "so-called crucial experiments of Professor Gilbert," explaining that "if these can be regarded as definitely settling the matter in the negative, there is no use in bringing forward facts looking towards its probability, no matter how plausible they may be."

As to the "alleged identity of the contents of the rim and the hole," Tilghman noted that from his own surveys he was "very sure that the content of the rim . . . is short by many, at least several million, cubic yards of the quantity necessary to fill the hole at all." The reason for this, he added, was simply that "in the time since the impact the rim has been reduced to its present dimensions by erosion," but the circumstance nonetheless "entirely destroys the weight of Professor Gilbert's reasoning. . . ."

The absence of a significant magnetic anomaly in the area was "on its face a much more serious objection, as it undoubtedly proves the absence of any *one* large mass of iron near the locality [Tilghman's emphasis]." However, he pointed out, it would have "no bearing whatever as to the presence or absence of a mass of magnetized fragments each having sufficient coercive force of its own to be independent of the earth's inductive action, to the extent at least of retaining its own proper polarity irrespective of the position in which it is placed in regard to the terrestrial magnetic field. Such a mass of polarized fragments would form a series of closed magnetic circuits with practically no external field whatever."

To test this, Tilghman experimented with two cubes of the magnetite, each about a half an inch on a side, and both having essentially the same effect on a compass needle. One of these cubes he then ground up into grains "about the size of coarse sand," and encased in paper.

> It was found that this had to be approached to within an eighth of an inch of the compass needle to produce the same deflection that the original piece did at eight inches. Not only this, but it was found that one single grain of the sand-like fragments of the pulverized magnetite had more effect upon the compass needle when taken alone than the whole mass of them when taken together. If the attraction of the mass of fragments of the supposed iron meteorite could be reduced in this proportion to its normal attraction when in a single piece, it might, on Professor Gilbert's own figures, lie within a very short distance of the surface of the present bottom of the hole.

Moreover, the meteorite itself should be found in this fragmentary condition, and its fragments would possess the requisite inherent magnetism, he argued. "The overwhelming majority of the fragments picked up on the surface, probably ninety-eight percent, do have this much magnetism, and some much more. . . ." And again, the velocity and shock of impact would be "enormous," and

> ordinary soft iron at the temperature of liquid air is of about the brittleness of glass under the shock of a blow. Now, as it is practically certain that the body of such a falling mass would be at the actual absolute zero of space beneath its incandescent exterior, it seems much more than probable that the result of such a collision would be to reduce the projectile to an extremely fine state of subdivision in comparison with its original size.
>
> If these conditions of subdivision and magnetism are present, and it seems much more than probable that they are, the crux of the second crucial experiment is also escaped, and we may proceed to consider the question on its merits, as nothing forbids us from allowing the possibility that the wreck of a great iron meteorite may underlie the bottom of the crater at Coon Butte.

Tilghman also found an argument for impact and against a volcanic or steam explosion origin of the crater in the narrow zone of crushed rock at the top of the white sandstone stratum and just below the Aubrey (Kaibab) limestone on the crater walls that Gilbert had noted in his description of the crater.[104] "It is unquestionably due to excessive pressure," Tilghman declared, but "if this cone and crater are due to any form of volcanic action, it is difficult to see how this crushing occurred." For the accumulation of magmatic or steam pressure would be a gradual process, and "then comes the giving way, and the explosion, and the result to the remaining rock left around the hole is a relief from pressure and not an increase of it. . . ." The crushing of strata and shattering of the rock of the walls, then, was "the direct and obvious result of the blow of a great projectile. There is almost immediately generated an overwhelming pressure deep down in the rocks, tending to lift the surrounding strata 1,000 or more feet per second. The great weight and inertia of these strata oppose an enormous obstacle to this sudden movement, and the crushing strains developed crush up the weakest rock until the necessary yielding and velocity have been imparted to the overlying strata."

Tilghman, indeed, found it "difficult" to discuss the steam explosion hypothesis at all "for the reason that nobody has ever seen one or known with certainty of any such action, except the blowing off of the tops or sides of ordinary volcanoes . . . which is different as possible in its effects from the so-called maars. There are," he added, "a lot of holes, not very uniform nor congruous among themselves, which for want of a better explanation of their formation, have been ascribed to this source, and to which class Coon Butte has been assigned by Prof. Gilbert, as a result of his investigations."

Tilghman also briefly discussed the probable size of the meteorite, a subject which would be hotly debated in the later stages of the Coon Mountain controversies. The problem, however, "contains too many unknowns to make calculation much, if any, better than guesswork," Tilghman pointed out. Yet artillery tables indicated that the penetration of a projectile was "something less than two diameters in solid limestone rock for shot at about 1,800 feet per second," and as the meteorite at Coon Mountain had penetrated about 900 feet of rock, "on the whole considerably softer than limestone," its diameter thus would be "considerably less than 450 feet." Moreover, while the velocity of the Coon Mountain meteorite could "only be guessed at," it must have been much greater than 1,800 feet per second. The velocity of such bodies "was well known from astronomical considerations and it probably struck the atmosphere at between nine and forty-five miles per second," depending on its motion relative to the earth. "We know that this excessive velocity is soon dissipated in the smaller meteorites and that they strike the earth with a very moderate velocity; but could such a thin layer as the atmosphere deal in the same manner with a large body?" he wondered. "The author is of the opinion that it could not, and that this body probably struck with a large part of its planetary velocity, and that it was extremely small in comparison to anything that would be deduced by assuming for it any such striking velocity as we have ever seen produced in a terrestrial projectile."

But for Tilghman, "extremely small" was a relative term, and both he and Barringer made this clear in summarizing their conclusions in their respective papers. "We believe," Barringer declared, "we have proved the following facts:"

First. That a great meteor, the whole or at least the outside of which was metallic in nature, did fall to earth at this locality, and that it was so large that portions of it became fused and were detached.

Second. That this great hole in the upper strata of the Aubrey formation was made at the instant of time when this meteor fell upon this exact spot. Having proved these facts, the conclusion is unavoidable that this hole, which as we have seen cannot have been produced by a volcano or by a steam explosion, was produced by the impact of a meteor, which, even admitting that it retained some large proportion of its planetary speed, must have been of great size.

And Tilghman concluded:

In view of these positively established facts, the author feels that he is justified, under due reserve as to subsequently developed facts, in announcing that the formation at this locality is due to the impact of a meteor of enormous and hitherto unprecedented size.

FIVE

Proving the Claim

> Interest, almost to the point of sensationalism, has been revived . . . through the publication by Messrs. D. M. Barringer and B. C. Tilghman of Philadelphia, who have shown the meteoric hypothesis of the origin of the crater to have had a much more substantial basis than many, including this writer, were at first disposed to admit.
>
> GEORGE PERKINS MERRILL, 1907

Controversy in science is distinguished from mere debate by an emotional component and by the fact that the point at issue is almost always sensational in nature. The disputes over Coon Mountain qualify as controversies on both criteria. The emotional undercurrents, as will be seen, were deep and strong. And certainly, the idea that a large meteorite could strike the earth with sufficient force to excavate a crater three-quarters of a mile wide and more than 1,000 feet deep was sensational at the time. To George Perkins Merrill, head curator of geology at the U.S. National Museum, the idea was "startling";[1] to the editor of the *New York Herald* it was "incredible."[2]

The question of why these controversies persisted so long in time is an interesting and complex one. The answer certainly lies, in part at least, in the unorthodox nature of the idea of impact vis-à-vis the entrenched belief in the volcanic origin of all natural crateriform features, and the tendency on the part of most scientists when faced with unorthodoxy to follow the conventional wisdom on matters on which they have no direct personal knowledge. Acceptance of an impact origin for Coon Mountain by individual scientists, in fact, correlates well with the amount of firsthand data they acquired. Those few who actually visited the crater and studied the phenomena there were quick to agree with Barringer's and Tilghman's general proposition, although they did not always agree on some of their corollary claims, while those who viewed the problem secondhand and from afar tended to favor Gilbert's steam explosion hypothesis or some other explanation based on more or less familiar geological processes.

But the answer surely also rests in large part in the human factors in these controversies. Such motives as pride, prestige, and profit clearly influenced the thinking and actions of many participants, and plainly such interests were not always compatible with those of science.

THE "ENEMIES' FORCES"

For Barringer, the principal antagonists were Gilbert, the U.S. Geological Survey, and its geological constituency—"the enemies' forces," as he would soon call them.[3] He was at first puzzled and then frustrated to the point of fury by the refusal of Gilbert and Survey officials to publicly acknowledge even the evidence, let alone the conclusions, that he and Tilghman presented. In time, and with encouragement from others who were hostile to the Survey, such as geologist John Branner, he came to believe that this strategy of silence was a vast bureaucratic conspiracy to deny him the credit he felt was his rightful due for having demonstrated the impact origin of Coon Mountain. In time, too, he came to believe that the Survey was plotting covertly to shift the credit for the discovery of the crater's meteoritic origin to Gilbert or to others, such as Merrill, who were associated with the Survey, and thus to the Survey itself.

Gilbert, of course, was also not without pride and might well have reacted somewhat differently had Barringer and Tilghman simply presented their data and conclusions unembellished by their provocative criticisms concerning his earlier investigation at Coon Mountain. Certainly, he could not have appreciated the plain implication in their papers that his examination of the crater and its environs had been slipshod and superficial. Given his mild manner and his distaste for controversy of any kind, it is also probable that Gilbert, who had advanced his steam explosion hypothesis with the caveat that new evidence might alter his conclusion, felt that Barringer's and Tilghman's new evidence spoke for itself, as it were, and thus no further comment on his part was required. Branner, who subsequently discussed Barringer's and Tilghman's findings privately with Gilbert, concluded this to be the case.

For officials of the Survey, refusal to comment publicly on the papers seems to have been a matter of bureaucratic prestige, a strategy designed not only to protect the well-deserved reputation of a highly revered colleague who had uncharacteristically erred, but to preserve the fictional infallibility that bureaucracies everywhere and at all times have tacitly claimed. Astronomer Harlow Shapley perhaps understated the point many years later when he gently reminded a furious Barringer apropos the Coon Mountain controversies that one must be "a bit optimistic if he believes that the government will ever admit that it is wrong."[4]

Some individual members of the Survey, of course, agreed with Barringer's conclusions regarding Coon Mountain; and a few of them told him so, although they made it plain that their opinions were not for public consumption and were offered privately and confidentially. Thus Merrill felt constrained to conclude a long supportive letter to Barringer with a scrawled plea to "Please hold this as personal. I do not want to be quoted as slandering my associates."[5] And Merrill's National Museum colleague, metallurgist Wirt Tassin, found it advisable to qualify his own stated agreement by noting that "This is my opinion and of course is binding only on myself."[6] Somewhat later, Smithsonian Institution paleontologist William H. Dall, who had studied the freshwater fossils found in the crater's fill, explained the situation in which Survey members found themselves to Barringer with full candor. "As a member of the U.S. Geological Survey," he wrote, "I have no right to express an opinion for publication without the authority of the Director, but, as between private individuals, I am obliged to confess that I think your explanation of the crater seems to be the only practical one."[7]

The profit motive, of course, had been present from the beginning; and Barringer himself had called attention to it somewhat obliquely in his paper, declaring in a concluding footnote that "this paper treats only of such facts as are of interest to the scientific world, and has no reference whatever to the commercial value of the discovery."[8] Despite this disclaimer, however, or perhaps because of it, the papers proved to be of interest in the commercial world as well. Barringer, indeed, may well have been fishing for new capital to help finance the already costly venture and using the footnote as bait. If so, he quickly caught what he called "a very old trout." Only days after the papers appeared, eighty-year-old Joseph Wharton, a multimillionaire Philadelphia financier with mining interests of his own, called Barringer to say that he had read them "with great interest and care,"[9] and a few days later the two men discussed Coon Mountain over lunch. But while Barringer was now anxious to attract new investors, he soon realized that Wharton's interest in the crater was inimical to his own. "I have registered a vow," he subsequently reported to Bennitt, "that as long as I live the Standard Iron Co. must never lose control of this enterprise and we must never lose control of the Standard Iron Co. Wharton owns control of the International Nickel Co., and if he could get control of the crater he would shut it up like a jack knife."[10]

Tilghman, parenthetically, did not participate in these controversies, although his differences with Barringer over policies for exploring and exploiting the crater would soon develop into a particularly bitter private dispute and would result in 1910 in his angry withdrawal from the enterprise. Unfortunately, documentary evidence for the specific nature of these differences is incomplete, a circumstance that is perhaps explained by a letter Barringer sent Tilghman in February 1910, shortly after Tilghman had relinquished his interest in Standard Iron. "I herewith enclose certain letters to me," Barringer wrote, "so that you can, as you said, burn them up. I will gladly do the same in regard to any letters of mine that you may wish to return to me. We will then drop and forget the whole matter."[11]

Nevertheless, several points of disagreement between them seem clear. For one thing, Tilghman did not share Barringer's concern over retaining control of Standard Iron, and thus of the crater, apparently feeling that an opportunity to acquire control would make investment more attractive to men of capital. "The outsiders," he wrote Barringer early in 1908, "should be rich men interested in science and the consideration should be a mixed one, the certainty of the development of scientific facts of very great interest and the possibility not only of doing this for nothing but of reaping very large rewards in case of complete success (This is the only thing that will catch them). Therefore, we ought to let the question of control go as of no earthly matter."[12]

Another difference seems to have involved the question of where to drill the new exploratory holes that Barringer, in view of his straitened finances and ignoring his earlier vow, now agreed was the only economically feasible way of pursuing the search for the meteorite. Over the next three years, twenty-three holes would be drilled—three in 1906, sixteen in 1907, and four in 1908—all within a few hundred yards east, southeast, or south of the center of the crater and all in an area less than one-twentieth of the total area of the crater floor.[13] Tilghman seems to have taken the more active role in this new drilling program and to have selected most, if not all, of the drill sites on the theory that the meteorite struck at a relatively high angle and would thus be found close to the center of the crater, with a "spatter about 500 feet."[14] Barringer, on the other hand, quite early began to consider "the possibility of the meteoric mass being out nearer the cliffs rather than in the center of the crater."[15]

By July 1908, after hole number 28—the final hole as it turned out—was abandoned, the dispute seems to have come to a head. "As to differences in opinions," Tilghman now wrote, "I can hardly express it more forcefully than to express my willingness to join you in a hole located according to your ideas—which according to my ideas is entirely hopeless, but one word of warning I would now put in and that is that I cannot now afford to begin a new groping in the dark without any clear or definite idea except that it must be somewhere and by elimination of sufficient ground we are sure to find it."[16] Less than a month later, Barringer ordered all further drilling stopped. To Holsinger, he reported that Tilghman was "terribly cut up by our failure to find the wreck of the meteorite,"[17] and to Bennitt he confided that "I am afraid that Ben Tilghman has lost a great deal of interest since his theory did not work out. I feared all along that he was too much wedded to it, but I believe that sooner or later he will recover from his disappointment and will be just as optimistic as ever."[18]

But Tilghman did not recover, and negotiations soon began for his withdrawal from Standard Iron. More than a year later, however, the matter had not been settled, to Barringer's growing exasperation. "I have done everything possible to expedite the settlement with Mr. Tilghman," he wrote Holsinger as the year 1909 ended. "I wonder if he realizes that he has tied up his former associates for over a year by his refusal to do anything in this matter. I think because of his selfishness and self-interest he is blind to the rights of others. I cannot believe that the man is bad at heart, notwithstanding his unjustifiable vilification of me. He is simply, as his father and uncle described [him] to me as being 'cock sure of his own opinion and impervious to reason,' and it is therefore impossible to deal with him on any business basis."[19] Two weeks later Tilghman finally signed the agreements formally relinquishing his interest in Standard Iron.[20] In little more than a year, he would be dead, a victim of Bright's disease.[21]

Nor did Gilbert or the Survey participate publicly in the controversies, although Barringer certainly became convinced that Survey officials were working quietly behind the scenes, as it were, on the Survey's and Gilbert's behalf. With the publication of his and Tilghman's papers, he confidently expected these officials to make some comment, even if critical; and he was somewhat surprised and disappointed when none was forthcoming. "The United States Geological Survey is going to stand out for its theories, right or wrong, until it is compelled to give in," he opined to Bennitt a few weeks after the papers appeared,[22] and to Mallet he complained that the Survey "does not seem to have heard of our papers."[23] Consequently, late in March, he brought them directly to the attention of the Survey's director, Charles Doolittle Walcott. "We naturally have been very reluctant to take issue with the conclusions of the United States Geological Survey based upon the report of Mr. Gilbert," he wrote, "and have done so only because the evidence all seems to be one way. I appreciate that the Survey, in reaching its conclusions, did not have the advantage of the information which has been furnished by our extensive development work. . . . I would be very much gratified to have the Survey take an interest," he added. "If any member of the Survey should wish to visit Coon Butte, kindly let me know in advance."[24] Walcott did not reply.

Gilbert also did not answer Barringer's letters,[25] nor was Barringer able to learn his reaction to the papers indirectly through Gilbert's friends and associates. At least Henry A. Ward of Ward's Establishment in Rochester, and Merrill, both of whom were quite close to Gilbert, were unable, or perhaps unwilling, to enlighten him on the point.[26] Only Branner, who did not know Gilbert well and who was, in fact, waging his own public war with the Survey at the time,[27] managed eventually to give Barringer some indication of what Gilbert was thinking.

Branner's association with Gilbert was, in fact, a chance one and resulted from their mutual service on a state commission formed in California shortly after the great San Francisco earthquake of April 18, 1906, to study the causes and effects of the disaster. Late in April, Barringer urged Branner to persuade Gilbert to return to the crater. "I am almost sure," he wrote, "that Gilbert was strongly inclined to the belief that it was an impact crater, but was hooted at and ridiculed by members of the Survey and had to manufacture some other explanation or lose his job as 'crazy.'"[28] Branner was willing, but pessimistic. "I agree that Gilbert seemed somewhat disappointed at the outcome of his study," he added, "but the Survey people have a way of knowing it all that is quite convincing to themselves and to a large part of the rest of the world. That you dare to call into question the conclusions of a member of the Survey will be looked upon with suspicion and strong disapproval you may be sure."[29]

Branner found no immediate opportunity to talk to Gilbert, however; and by September, Barringer complained that he had not heard "a peep" from either Gilbert or any other member of the Survey.[30] Branner replied that he had not seen Gilbert since the previous spring and suggested "a better plan"—to invite the well-known geologist Thomas C. Chamberlin of the University of Chicago to visit the crater. "The phenomena you have around there will be of the greatest interest and use to Chamberlin in his geologic theories," he explained. But he also warned Barringer not to criticize Chamberlin for his treatment of Coon Mountain as a steam explosion crater in his widely used textbook, *Geology*. "I don't think you should call what Chamberlin says . . . *his* theory of the crater. Gilbert's paper was tacitly accepted by geologists up until the publication of the papers by you and Tilghman. No other views had ever been expressed in the geologic literature, and he makes no claim to personal knowledge. The theory is Gilbert's and was simply and necessarily accepted by Chamberlin when his book was published."[31] Chamberlin, however, apparently was not as interested in Coon Mountain as Branner had expected and subsequently declined Barringer's invitation.[32]

Not until mid-October did Branner have an opportunity to talk with Gilbert about Coon Mountain, and he promptly reported the gist of the conversation to Barringer. "Mr. Gilbert was here yesterday, and we discussed Coon Butte," he wrote. "I ought not, perhaps, to undertake to give his views, but he left me with the impression that he considered you and Tilghman had brought forward evidence that entirely changed the conclusions he had drawn regarding the origin of the crater."[33]

Barringer was jubilant. "The commander-in-chief of the enemies' forces has been won over!" he exulted, again urging Branner to press Gilbert to return to the crater.[34] But Branner remained pessimistic on this point. "I am reasonably sure of what he will say: that it is not necessary, for he accepts the new facts brought forward by you and Mr. Tilghman as facts. I am sure he fully realizes that his conclusions as published are erroneous, and I am confident that if he attends the New York meeting of the G.S.A. [Geological Society of America] he will frankly admit it."[35] This meeting, held late in December, would indeed be an important one for Barringer's and Tilghman's thesis, but not because of any statement made there by Gilbert. Branner's assessment of Gilbert's position, however, seems to have tempered Barringer's antagonism toward Gilbert personally, and henceforth his criticisms and complaints would be directed primarily at the Survey.

During 1906, Branner proved to be Barringer's most effective ally in his attempts to win acceptance for the meteoritic impact origin of the crater. After visiting the crater early in June, Branner summarized Barringer's and Tilghman's papers and his own conclusions later that month at a meeting of the Geology section of the American Association for the Advancement of Science in Ithaca, New York.[36] Barringer was "greatly pleased" that this presentation "excited the interest it deserved,"[37] and indeed it drew favorable comment from some of the geologists attending. Edmund O. Hovey of New York's American Museum of Natural History and secretary of the section, later reported in *Science* that the fact that meteoritic material had been found interbedded with lithic debris at the crater "lends strong support to the idea that the depression was made by a meteor striking the earth."[38] Michigan state geologist Albert C. Lane also agreed with the crater's meteoritic origin, but thought it probable that the meteorite had bounced out, or been blown out of the crater after the impact. Barringer, of course, could not accept this and promptly wrote Lane a long letter arguing that the main mass of the meteorite must still lie somewhere beneath the crater's floor.[39]

Hovey, incidentally, became an ardent convert to the meteoritic hypothesis, subsequently congratulating Barringer "on the fact that continued investigation of your wonderful hole in the ground has served to establish so conclusively the correctness of the theory as to its origin which seemed so improbable. . . ."[40] A few years later, Hovey himself visited Coon Mountain, spending "two delightful and instructive days" there and advising Barringer on his return that he was "quite prepared to accept and advocate the meteoric origin of the crater."[41]

Other comments on Barringer's and Tilghman's papers published about this same time, however, were quite critical and favored Gilbert's steam explosion hypothesis or some other geological explanation. The most important of these, perhaps, was a lengthy article in *Nature* by L. A. Fletcher, the keeper of minerals for the British Museum and a leading authority on meteorites.[42]

Fletcher reviewed the various investigations of Coon Mountain at some length, emphasizing in particular the "minuteness" of Gilbert's work, and disagreeing with Barringer's and Tilghman's conclusions on four specific points. First he declared that "all the oxide, magnetic or not, is the result of weathering. There has been plenty of time for this action, for cedars now 700 years old are growing on the rim of the mountain." Moreover, while the "lumps of the meteoric oxide" that had been found buried beneath the blocks of limestone and sandstone "must have taken up their present position at the same time as the blocks themselves," he argued that "it seems probable that they had been buried, possibly a long time, in the upper layers of sandstone, and were ejected with the rock fragments when the crater was formed. . . ."

Fletcher also questioned Barringer's and Tilghman's contention that the millions of tons of pulverized silica in and around the crater could only have been produced by the impact of a massive projectile. It was "difficult to see why the crushing of the grains could not have been produced by an enormous pressure of steam, such as must have preceded . . . the formation of the crater," he declared, adding that the pressure of the steam that forced lava to the summit of volcanic Mount Etna "must certainly exceed 1000 atmospheres."

Two other points concerned the "meteoric material" that Barringer and Tilghman had found in the depths of the crater—the hard, metallic obstructions encountered by the drill and the tiny particles of iron oxide mixed with the rock deep beneath the crater floor. "But the presence of some small masses of iron beneath the crater is to be expected if all the masses were lying embedded in the sandstone before the crater was formed," Fletcher noted. "Those which were projected nearly vertically upwards must have fallen back into the large hole and be deep down among the debris."

Nevertheless, Fletcher conceded, it was at least physically possible that a giant meteorite or an asteroid could create such a crater. Noting that it "is found as a matter of experience that meteorites on striking the ground have a comparatively small velocity," he attempted to calculate the retarding effect that the atmosphere would have on meteors. His results indicated that an iron meteorite 100 meters (328 feet) in diameter would have an impact velocity of 8.3 kilometers (5.2 miles) per second, and one 1,000 meters in diameter would strike at 25.5 kilometers (15.8 miles) per second. "It would then appear," he declared, citing the penetration data used by both Gilbert and Tilghman for artillery projectiles traveling only 1,800 feet per second, "that a meteorite of large size would not be prevented by the earth's atmosphere from having a penetrative effect sufficient to produce such a crater."

A more bizarre explanation appeared in November in *Science* when William P. Blake, a Yale University geologist, proposed that Coon Mountain's crater was simply a limestone "sink" and compared it with a feature known then and now as "Montezuma's Well," some fifty miles southwest of Coon Mountain in Arizona's Verde Valley.[43]

Montezuma's Well, Blake explained, was "evidently the result of caving-in, a falling down, of the roof of a cavern formed by running water in the nearly horizontal limestone strata," and the phenomena there "suggest that the remarkable crater-like cavity known as 'Coon Butte Crater' may have similarly originated." The suggestion, he added, was "made with some reluctance, inasmuch as I have not studied the locality."

Blake, in fact, had written Barringer regarding the crater before publishing his article; and Barringer, in his reply, had stressed the significance of the presence of the irons and the iron shale.[44] Blake, however, considered this circumstance coincidental and the result of an earlier fall. "The subsidence of the area," he explained, "would carry down with it any meteoric fragments which were on the surface or in the soil, and to considerable depths." After reading the article, Barringer declared that Blake had made "a fool of himself"[45] and sent him a searing letter setting forth some of the phenomena at the crater that his limestone sink hypothesis did not explain. "You should have gone to Coon Butte," he fumed. "Your theory of its being a sink-hole is absurd. . . . The only other man that held this theory is a mining engineer of this city who was sent out to examine the crater [and was] blind or demented."[46]

Blake's explanation was soon disputed by geologist F. N. Guild of the University of Arizona who, unlike Blake, had visited the crater briefly in the summer of 1906 while studying the craters and lava flows of the neighboring San Francisco Peaks volcanic field to the northwest. Guild, also writing in *Science*, noted that the "endeavor to explain the crater by other than volcanic agencies had led some writers to suggest that it may have been produced by solution. This hypothesis cannot be applied to Coon Mountain, however," he declared, "for the reason that it leaves unexplained the most noticeable feature of the phenomenon, namely, the upturned strata which forms the rim. This could have been produced only by means of forces working from below."[47]

But Guild did not think that the meteoritic impact hypothesis was any better. Although citing Barringer's and Tilghman's "admirable description" of the crater, Guild belittled their conclusions as to its origin, declaring they were merely "the result of the adoption and elaboration . . . of the local common talk of the inhabitants of the immediate neighborhood of the mountain. Here it is religiously believed that an immense meteor nearly one half mile in diameter buried itself in the earth, forming a deep cavity with an upturned edge or rim very much as when a bullet is allowed to fall into soft mud."

Actually, Guild had first thought that the crater had been formed by "a gradual upheaval caused by the approach of an igneous mass, followed by a sudden collapse," and had written Barringer to this effect at the time of his visit to the crater.[48] But he had since changed his mind and now opted for Gilbert's steam explosion hypothesis, comparing Coon Mountain with Sunset Crater, an indubitable volcanic cone a few miles east of the San Francisco Peaks. "In the minds of the uninitiated," he wrote, Sunset Crater "must have been produced by quite different agencies than were at work at Coon Mountain, whereas probably the difference lies in the degree and not the kind of action. It would seem quite probable that on the border of a region of such extreme volcanic activity as has given rise to the most lofty mountains in Arizona, there might have been an explosion lacking the energy necessary to bring the igneous mass or even fragments of it to the surface. . . . It would seem, then," he concluded, "that the phenomenon exhibited here can be satisfactorily explained as having been produced by an explosion followed by an entire lack of volcanic activity, as first explained by G. K. Gilbert, of the United States Geological Survey. The meteorites found here," he added, "probably had nothing to do with the formation of the depression. . . . It is to be concluded, then, that these two striking phenomena were simply coincidences and should not be interpreted as cause and effect."

The resumption of drilling had begun in April 1906 while Barringer was at the crater assessing the situation in view of the failure to sink the shaft. Holsinger, who had been banished from the crater early in February for, among other things, "throwing cold water on the enterprise,"[49] was now restored to partial grace and put in charge of the small drill crew. Despite his earlier shortcomings, he soon proved himself competent to the task. "Holsinger," Barringer noted in August after two holes had been completed and the third was drilling ahead, "has finally learned to be a wonderfully good drill man . . . two hundred feet a day seems to be an easy thing for him to do and he has made as high as three hundred and forty feet in twenty-four hours."[50] Like the earlier hole number 5, all three holes drilled in 1906 penetrated the red Supai sandstone to depths below 1,000 feet; and Tilghman, after testing them, found that the Supai at this depth was free of nickel and that "it looks like bottom."[51]

The new drilling program was not without its problems, however; and the water situation at the crater in particular would soon take an ironic turn. For in September, drilling had to be suspended for the year after only three holes had been completed because of a shortage of water. "While it has rained terrifically during the past summer," Barringer explained to Alexander, "the canyon has never flowed once. This is almost unprecedented . . . and we have been obliged to close down until our reservoirs are full, for without water we cannot possibly drill. . . . It is wonderful how ill luck seems to pursue us in connection with this enterprise. . . . I am not over sanguine as to ultimate success."[52]

Barringer's hopes for the new drilling program, in fact, had never been high. "I regret to say that there is nothing new at Coon Butte," he reported to his cousin Paul B. Barringer on his return from the crater early in May.[53] "It is absolutely impossible to sink the shaft through this finely pulverized material with the amount of water which is now contained in it. We are going to put down a few more drill holes in the hope of striking pieces of the meteor which we know is somewhere in the bottom of the hole. This is merely being done to induce capital to help us sink the long incline from the outside which is the only practical method that I can see of getting our hands on the bird. . . . However, we have learned something . . . which we could not have learned in any other way, and frankly, I did not think it possible that the crater was holding water as it is doing. I am not sure of anything now in connection with this miserable crater."

Uncertainties large and small would, in fact, beset Barringer for the next few years. Not the least of these was his deteriorating financial situation. His funds were becoming seriously depleted and, while assessments for the exploratory work were now less frequent, at $10,000 they still represented a considerable drain on his resources. As the new drilling program progressed, he warned Bennitt that Holsinger "must economize,"[54] and to Holsinger he explained that "I have very little money left to spend. I may have to give up my home on account of this business."[55] Indeed, he thought it probable that the successful pursuit of the search for the meteorite might require another $200,000 and hoped that the new drilling program might produce enough new evidence for the presence of the mass in the crater to entice some new investors into the enterprise.⁵⁶ He now also began to consider seriously the possibility of marketing the pulverized silica and exploiting the crater's potential for tourism. Shortly after the shaft was abandoned, he suggested to Tilghman that they could sell silica to "make cannel soap, polishing powder, dentifrice, glass, paint, etc." and "charge tourists $3 a head to get our money back."⁵⁷

Barringer's other enterprises, too, were not faring at all well. By 1906, for example, the once-rich Commonwealth mine had run out of high-grade ore; and Barringer hoped that it could be kept in production by cyaniding the "extremely low grade" ore that remained.⁵⁸ In 1907 Barringer, along with the rest of the nation, suffered through the financial "panic" of that year; and in 1908, his placer mining venture in northern California received a severe setback when fire destroyed its newly modernized dredge.⁵⁹ Not until early in 1909 did the outlook begin to brighten. "I have been greatly worried this winter," he confided then to Holsinger, "but I know you will rejoice with me when I tell you that it looks as if my financial children are going to get well, although they are still pretty sick."⁶⁰ By this time, however, all exploration at the crater had been suspended.

MYSTERIOUS METAMORPHIC ROCKS

Other uncertainties concerned Coon Mountain itself and the interpretation of the evidence that he and Tilghman had found there. Of major concern was the identification of two forms of metamorphosed, siliceous rock that they had recently discovered and which, for want of better names, they had called "Variety A" and "Variety B." The first was a slaty, laminated sandstone, somewhat denser than the ordinary sandstone at the crater; the second and much rarer form, a hard, opalescent, pumiceous rock so light that it was found floating on the surface of the water at the bottom of the shafts and drill holes. Interestingly enough, Gilbert had also found these metamorphosed rocks in 1891 and had even had the pumiceous form analyzed chemically.⁶¹ Apparently, however, he did not consider the results significant, for he did not mention them in his published discussion of Coon Mountain.

During his visit to the crater in April, Barringer suddenly became afraid that these strange siliceous rocks might be extraterrestrial in origin and thus might be evidence for the "plums in a pudding" theory. Seeking reassurance, he immediately sent specimens to Mallet for analysis, explaining that "I have identified every rock found in this locality except A and B, and if you will wire me that they are of terrestrial origin I shall be greatly relieved. I may be frightened at shadows after the trying experiences we have had here and may be making a fanciful mountain out of a fanciful molehill, but I do not like nor do I understand this new discovery. . . . Could it be possible that it was an immense aerolite after all. . . . An astral siliceous pudding with iron plums scattered through it?"[62]

Two weeks later, Mallet telegraphed him that the two rock forms were "undoubtedly terrestrial,"[63] but Barringer had already overcome his fears. "I was sure you would say that A and B were terrestrial," he replied immediately, "and there is absolutely no doubt in my mind that they existed as finished products before the collision. . . . I was temporarily a little scared," he admitted, "for I am strongly disposed to grasp at straws which seem to afford even slight evidence in support of any other theory. . . ."[64]

But Barringer here was grasping at another straw, for when Mallet's more detailed report arrived a few days later in the mail, it contained a disquieting suggestion. While Mallet had not been able to identify specifically the rock forms, he had concluded that "they have (especially B) the general character of siliceous sinter or geyserite from hot springs, and so seem to furnish some evidence of a kind (which I had supposed from your account to be entirely lacking) in support of Dr. Gilbert's steam explosion theory."[65]

Barringer "was not sure that either one is geyserite," but he shipped the specimens to the eminent petrographer, Samuel L. Penfield of Yale University, for a second opinion.[66] Two weeks later, both he and Tilghman journeyed to New Haven to learn Penfield's findings firsthand. Penfield too, as it turned out, could not specify the identity of the two rocks; but he concluded that Variety A, at least, was nothing more than ordinary sandstone "absolutely ruined in situ," as Barringer reported to Branner,[67] and thus "an impactite," as he wrote to Mallet.[68] While the nature of Variety B was "still uncertain," Barringer now became convinced that it had also been formed by impact, explaining to Mallet that A had been caused by a shock wave that had "passed through and shattered the particles without displacing them," while B was "the result of the action of steam on pulverized silica."[69] These conclusions, as will be seen, were later supported by other geologists. Variety B, in particular, proved to be of unusual interest to mineralogists and eventually would be identified as a stable, high-temperature form of silica. Merrill, for example, would soon recognize its unusual nature by comparing it to fulgurite formed by lightning striking quartzose sand.[70] In 1915 this form would be named "lechateliérite" after Henry Le Chatelier, a French chemist well known for his studies of silica; and in 1927, the president of the Mineralogical Society of

America, Austin F. Rogers of Stanford University, would proclaim its occurrence at Coon Mountain "unique."[71]

In 1906 too, Barringer was forced to revise some of his thinking in regard to the origin of the so-called iron "shale" and the "shale balls," largely because of work by Oliver C. Farrington, curator of geology at Chicago's Field Museum of Natural History, and at that time America's leading authority on meteorites. In particular, Barringer had to abandon his elaborate speculation, based on his observation of the meteor over Tucson in 1904, that the shale balls represented "flaming drops" of melted iron which had been sloughed off by the meteor during its flight and which had become oxidized by "burning" during their fall through the atmosphere. In mid-May Farrington requested samples of the shale for analysis; Barringer quickly obliged.[72]

Farrington found that quantitative analyses of the shale confirmed the "qualitative" judgments of Koenig, as quoted by Foote, and of Barringer "that nickel and cobalt are to be found in the shale in the same proportion as in the meteoric iron."[73] Moreover, the explanation of the laminated structure of the shale was "doubtless furnished, as suggested by Barringer, by the shale balls. Oxidation and hydration proceed from the surface inward," he explained, and these changes "cause an increase in bulk of the layers successively reached in the process, so that they separate slightly from the unchanged material beneath, and interstices are afforded through which the oxygen and water enter to attack the metal."

But Farrington also offered another explanation of why some fragments of the meteorite had become oxidized while others had not. "One reason," he reported, "is probably to be found in the observation of Barringer that the iron shale often occurs beneath the surface while the meteoric irons are found only at the surface. Those meteorites which were covered more or less by soil and rock fragments would receive a larger supply of water and hence would suffer more rapid oxidation and hydration than those at the surface." More important, he pointed out, the shale contained a small amount of chlorine in the form of lawrencite, while the irons did not; and it was well known that meteorites containing lawrencite "'sweat' and rapidly decompose, even in a museum case." Using formulas, he also demonstrated that the chemical reactions involved would produce free hydrochloric acid which "would obviously likewise exert a decomposing action."

The oxidation and hydration processes required time, however; and Farrington thus concluded that the shale and the shale balls could not have come to their present condition during their brief fall to earth. "In the present writer's view," he declared, "the oxidation occurred subsequently to the fall of the meteorite, and was so gradual that the production of the shale can be explained only by assuming that the fall took place many years ago."

Tilghman quickly accepted Farrington's explanation,[74] but Barringer was reluctant to abandon his "flaming drops" hypothesis, confidently assuring Branner that he would be able to convince Farrington of his error.[75] As will be seen, however, Barringer would soon concede that, on this point at least, the error was his own.

QUALIFIED SUPPORT

Branner's efforts on behalf of Barringer's and Tilghman's thesis had one other important result in 1906. In August he advised Barringer that "I have just induced Prof. [Herman Leroy] Fairchild to go to Coon Mountain," and indeed Fairchild arrived there within a week, spending several days studying the crater and its environs.[76] Fairchild, then secretary of the Geological Society of America and a geologist on the faculty of Gilbert's alma mater, the University of Rochester, was a longtime friend of Gilbert's and indeed had proposed him for an honorary LL.D. degree that Rochester had conferred upon him in 1898. At Gilbert's death in 1918, he also became one of his principal obituarists noting, after listing Gilbert's many achievements, that his conclusion regarding the origin of Coon Mountain "was one of the very few times in which his theory has been proven wrong."[77]

Fairchild now became the first geologist not closely associated with Barringer to accept the impact origin of the crater and to state his acceptance in print. He apparently had become convinced during his visit to Coon Mountain, for in September he had favored the impact theory in an informal discussion before the Tenth International Geological Congress in Mexico City.[78] In October he advised Barringer that he planned to make a more formal presentation at the Geological Society of America's meeting late in December.[79]

Fairchild's initial concept of the impact process, however, differed from Barringer's; and in fact he favored what Barringer called the "percussion cap" theory—the idea that the impacting meteorite, not necessarily very large, had triggered an explosion of steam deep within the rock strata and that both forces thus had contributed to the excavation of the crater.[80] Gilbert had considered this same idea and had rejected it as involving "a coincidence of what may be called the second order."

Fairchild visited Philadelphia early in November and Barringer invited him to dinner at his suburban home,[81] later complaining bitterly to Branner that Fairchild "still clings to his theory."[82] Branner's reaction was philosophical. "I am sorry that people bother you with fantastic theories regarding the Coon Butte phenomena, but it is inevitable. We seem to be made that way. The main thing is to keep up a lively interest in the subject, and these various theories will all help to do that."[83]

Barringer now sent a letter to Fairchild detailing his arguments against a steam explosion component. It was not necessary, he pointed out, for it could not account for varieties A and B of the metamorphosed sandstone nor the fact that the pulverized silica and the larger, heavier rock debris were found together on the crater's rim.[84] Fairchild's reply was apologetic. "I fear I made my suggestion of explosion more prominent and argumentative than I really felt. My notion seriously considered only a supplementary or reaction effect of the blow." Nonetheless, he added, "there must have been great heat through a large volume of the powder, with considerable water and air, and then what would happen? I don't know, but in a letter to Mr. Tilghman I have explained how I think wet dust might puff out and carry out fragments of the meteor. My idea is to get ready to explain the absence of the meteorite in the pit in case you don't find it."[85]

As the dynamics of impact were more in the area of Tilghman's competence, Barringer now decided, wisely as it turned out, to "stand out of the way and let you and Mr. Tilghman have it out."[86] Tilghman's reply to Fairchild's letter proved persuasive,[87] for within days Barringer could note that Fairchild "has finally positively abandoned his theory of a 'mild steam explosion' having assisted," and accepted "simple impact" as the explanation of the crater's origin.[88]

Nonetheless, Fairchild's presentation on December 29 at the Geological Society meeting probably was not all that Barringer hoped it might be. For while Fairchild unequivocally rejected the volcanic steam explosion hypothesis as "untenable," he hedged his acceptance of the impact theory with qualifications; more particularly, his discussions of the oxidized iron shale and of the probable disposition of the meteorite were quite speculative and inconclusive.[89]

In disposing of the volcanic hypothesis, Fairchild first pointed out that the drill had found the red Supai sandstone underlying the crater undisturbed, and that this "shows the absence of any chimney or volcanic pipe of considerable size and positively rules out any explosion from beneath." Nor was an explosion of pent-up steam in the white Aubrey sandstone above the Supai admissable, he argued, citing Tilghman's "unpublished writings." For the Aubrey sandstone was quite porous and "the pressure of any heated water or vapor sufficient to throw the mass of rock out of the pit must have existed through an immense horizontal extent of the strata, and a long time would be required for entire relief of the pressure through the vent, and with decided hot spring or fumarolic or solfataric phenomena. . . . Further, the pulverized character of the rock debris . . . is not consistent with the idea of an explosion nor with the digestive action of heated waters. The explosive or volcanic theory," he concluded, "has not a single fact to stand on."

On the other hand, there were "several remarkable facts which are well explained by the impact theory of crater genesis," chief among these being the various forms of altered and metamorphosed sandstone found in and around the crater. Varieties A and B, he thought, were of "special significance." The slaty A "seems to be the direct product of intense compression," he opined, while the cellular or pumiceous B "would seem to be the natural product of the intense heat of compression on portions of the sandstone which contained sufficient water, perhaps along the joints of the strata, to effect an aqueo-igneous fusion." All three forms of the altered sandstone, "the powdery, the slaty and the cellular," were what "should be expected from a tremendous crushing blow with resulting heat."

Moreover, the presence of "iron oxides of meteoric origin" favored the impact theory, as their occurrence in the debris on the rim and deep within the crushed sandstone in the crater itself was "very significant of an association in time." But it was also "a remarkable and critical fact" that there were "two varieties of siderites" at Coon Mountain—"the well-known Canyon Diablo and the chlorine-rich or oxidizing irons." He wondered: "Do they represent two distinct falls or are they only variations in the decomposition of one huge meteorite?" There were also two forms of the iron shale, "the limonitic and perishable coatings found on the surface of the oxidizing irons, and the firm, compact, and more permanent form, represented by the fissured nodules [shale balls]. Probably these oxides shade into each other," he suggested, "but nevertheless they must indicate variations in the composition of the meteoric material or different circumstances during the hydration process."

Fairchild was even more uncertain about what might have happened to the meteorite after its impact. And here, to what must have been Barringer's consternation, he brought up E. E. Howell's original "plums in a pudding" theory as outlined by Gilbert in his 1896 paper. "The explorers," Fairchild noted, "have not been unmindful of this suggestion, but thus far their search has not discovered any stony material foreign to the local rock. Nevertheless this suggestion may give clue to the fate of the bolide."

On the other hand, if the oxidizing irons and the Canyon Diablo irons belonged to the same fall, "then the latter probably represent merely a few unoxidizable fragments from the enormous mass of the wrecked meteor, and a part of the wreck may still lie in the crater." Howell, he recalled, had told him in a conversation many years before that the crater had been formed by impact and that the meteorite might be partly buried in its depths. "He seems to have been correct in the view as to the genesis of the crater, and may yet prove so as to the disposition of the meteor."

Perhaps Fairchild's most lasting contribution to the history of Coon Mountain was to give it a new name. At the very end of his paper, he formally proposed that the crater henceforth be known as "Meteor Crater," pointing out that its existing names, Coon Mountain or Coon Butte, were "wholly inappropriate" and the particularizing adjective, "Coon," was meaningless. The term "crater," on the other hand, was both applicable with reference to form and "significant and accurate" under either the volcanic or impact hypotheses of its origin. The adjective "Meteor" was also appropriate, he argued, both because of the presence of the Canyon Diablo irons there and the "reasonably certain conclusion that the crater is the effect of impact." But more important from a nomenclature standpoint was the fact that the nearest U.S. post office, located at the railroad's Sunshine flag stop six miles north of the crater, was named "Meteor, Arizona." This geographical reference, he declared, "gives official standing to the word."

The name seems to have been Fairchild's own idea; at least it did not come directly from Barringer, who was still referring to the crater as "Coon Mountain" or "Coon Butte" in his correspondence. In October Fairchild had advised him that "I do not like the name 'Coon Butte,' which is meaningless. At the Congress I used the name METEOR CRATER. Can you think of a better one?"[90]

Whether Fairchild was aware of Barringer's role in the establishment of the "Meteor" post office early in 1906 is problematical. But Barringer's motives at the time were practical rather than promotional. First, a post office at Sunshine would be far more convenient to the crater than the existing Canyon Diablo post office, some ten miles away. Second, as he was forced to cut expenses at the crater and had reduced Holsinger's pay, he planned to install Holsinger as postmaster. This, he explained to Holsinger, would "supplement your salary" and insure "that your compensation would be at least equal to what it is at present."[91]

Barringer used his considerable political influence to assure that the new facility would indeed be established. In February 1906 he sent the necessary petitions to Theodore Roosevelt, requesting the president to "Please hand the enclosed petitions to the Postmaster General."[92] This Roosevelt did, but two months later no further action had been taken on them; and Barringer, fearing that Volz, the postmaster at Canyon Diablo, might be blocking the application, now enlisted the aid of another powerful friend, U.S. Senator Boies Penrose of Pennsylvania.[93] Within days, on April 27, 1906, the "Meteor, Arizona" post office was open for business, with Holsinger as its postmaster.[94] Barringer, incidentally, does not seem to have been particularly enthusiastic about its name. In thanking Penrose for his assistance, he remarked that "I had forgotten that this name was requested, but I suppose one name will do as well as another."[95]

Fairchild's acceptance of an impact origin for Coon Mountain was at best conditional; "if the phenomena are not meteoric in origin," he had declared, "then they constitute the most interesting geological puzzle of the present time." But nonetheless, given Fairchild's standing as a geologist, this very speculative tone served to stress the fact that there were indeed a number of important questions regarding the crater that remained unanswered. These questions would now be addressed by another geologist of high standing, George Perkins Merrill.

THE TURNING POINT

Merrill had first written Barringer concerning the crater in May 1906, requesting a sample of the stony meteorite that Barringer had found there for study, and adding that he was "greatly interested in the accounts of your work at Canyon Diablo, doubly so from the fact that I was on the ground for a portion of a day at the time Mr. Gilbert was making his studies."[96] Barringer immediately sent him several thin sections of the aerolite, along with an invitation to revisit the crater with Branner and himself early in June.[97] Merrill politely declined, and the two men subsequently corresponded only sporadically and on superficial matters.

Early in January 1907, however, and only weeks after Fairchild's presentation, Merrill expressed renewed interest in the problems at Coon Mountain, requesting a representative collection of the various materials found there for a new petrographic and metallurgical study.[98] Barringer was now wary, for he was aware that Merrill was both a friend and long-time associate of Gilbert's and that, along with his post at the National Museum, he was a member of the U.S. Geological Survey. He also knew that the museum was a unit within the Smithsonian Institution and that the Survey's uncommunicative director, Walcott, had recently been appointed to the influential position of Smithsonian secretary and was thus above Merrill in the Smithsonian's bureaucratic hierarchy. Indeed, as Merrill later noted, his plans for a reinvestigation of the phenomena at Coon Mountain had been submitted to and approved by Walcott.[99]

But beyond this, the varied reactions of recognized scientists to his and

Tilghman's thesis had by now disillusioned Barringer about scientists in general, while the lack of any reaction at all from Survey officials, reinforced by Branner's vitriolic criticisms of the Survey itself as a "national bureau grown into a scientific trust," had left him with an "unmitigated contempt for a great many of its members."[100] Thus, before complying with Merrill's request, Barringer sought Mallet's opinion of Merrill's qualifications. "Before entrusting such a serious matter to Dr. Merrill," he explained to his cousin Paul Barringer, "I would like to know whether Dr. Mallet considers him a first class all around man who is entitled to be called a scientist. My experience is about 9 out of 10 of the people who are referred to by this appellation are not entitled to it, and I think 99 out of 100 would be nearer the mark."[101] Mallet apparently reassured him on the point for the requested specimens were soon shipped to Washington.

Barringer, indeed, had other reasons for encouraging Merrill's proposed investigation of the Coon Mountain phenomena, as he revealed a few months later in a long letter to Bennitt's wife, Emma. "I am anxious to convert through him the U.S. Geological Survey," he wrote, "so that we may at least not only get the credit for having made this wonderful discovery which is our due, but what is really important to me just at present to make it possible to persuade the United States Government to buy the property from us and to set it aside as a great curiosity, greater by far than the wonders of the Yellow Stone Park or the Yosemite, or even the Grand Canyon or Niagara Falls. In this way we may get out of it the money which we have put in it in the event of our failure to find the old bird which we have hunted so long and so faithfully, and which undoubtedly made that hole, whatever may have become of it since then."[102]

Merrill's investigation, conducted in association with his colleague, Tassin, was carried out primarily at the National Museum's laboratories, although Merrill, again at Barringer's invitation, did visit Coon Mountain in May 1907, spending several days there studying the crater and its environs and watching as Barringer and others excavated some of the shale balls from the debris on the rim that Barringer had been "reserving . . . *in situ* for Mr. G. K. Gilbert and other 'doubting Thomases.'"[103]

The first significant findings, however, emerged from Tassin's metallurgical studies of the irons and iron shale and, incidentally, provided the *coup de grâce* to Barringer's "flaming drops" theory of the origin of the shale ball irons. "Mr. Tassin," Barringer wrote Merrill early in February, "has shaken my confidence in the line of reasoning which I have been following but he has not utterly destroyed it. I still think that it is extremely possible that many of these irons 'flaked off' in the upper atmosphere as soon as the meteor entered it, and I also think that at least the small 'magnetite' pear-shaped shale balls

... were once the blazing, burning drops of iron which had been detached from the main mass. . . ."[104]

Only a few weeks later, however, he had capitulated. "We are finding really uncanny things in our shale balls," he now reported to Alexander:

> They are widely different from the ordinary Canyon Diablo irons. The Washington chemists have not found platinum in them, but they have found as high as 25 percent nickel and good-sized diamonds, metallic silicon among other things (which has never been seen before), pure graphite, graphitic iron, etc. etc. Also they have come pretty near convincing me, Dr. Mallet to the contrary notwithstanding, that all of our iron shale has been produced after the fall by reactions at or near ordinary atmospheric temperatures!! In other words, that these shale balls are now slowly dissolving into shale. . . . But one thing appears certain—they were never melted, or melting and burning drops, as we have all along supposed. . . . A very interesting discovery.[105]

Tassin announced his discovery of elemental silicon in the irons and its "first occurrence in nature" at the spring meeting of the National Academy of Sciences[106] and reported the presence of graphitic iron in the meteorites in a brief note in the National Museum's *Proceedings*.[107] But three papers published subsequently by Merrill regarding their joint investigation held greater significance for the Coon Mountain controversies.

In the first of these,[108] Merrill discussed the various siliceous rocks found in and around the crater, declaring at the outset that his paper "has little to do" with the question of Coon Mountain's origin. "It is sufficient to say that Mr. Gilbert, after discussing various hypotheses, was led to regard that of an origin through explosive volcanic action as most plausible," he noted, adding that Barringer's and Tilghman's investigation was "based on the theory that the depression is due to the impact of a gigantic meteorite." He then proceeded to describe the forms of sandstone found at the crater in detail, concluding that all of them, from the pulverized silica sand or "rock flour," as he preferred to call it, and the "ghost sandstone," through what Barringer had labeled Varieties A and B, were progressive phases of a metamorphism apparently resulting from brief but intense pressure.

The first phase, he explained, involved the more or less pulverized sandstone in which the quartz granules were simply crushed and distorted "as though they had been struck a sharp blow with a hammer." The second phase, represented by what he termed the "crystalline variety" (Variety A), "on casual inspection might readily pass for an old siliceous or calcareous sinter," but to Merrill bore a closer resemblance to a "decomposed chert."

Under the microscope, he discerned features new to him "which are at variance with our ideas of the stable character of quartz sand. . . . The appearance indeed is such as to suggest that the granules have been subjected to pressure while in a putty like or plastic condition."

From this phase, the metamorphism progressed to the more or less vesicular, pumiceous forms (Variety B) "showing to the unaided eye all the features of an obsidian pumice, but of a white color." Under the microscope again, this form was resolved into "a colorless vesicular glass, more or less muddied through dust-like material and showing here and there residual particles of unaltered quartz. This glass does not, however, resemble the glass of a pumice," he added, "nor is it like that obtained by the artificial fusion of quartz in the geophysical laboratories of the Carnegie Institution. So far as the writer's observations go, it more closely resembles fulgurite glass, formed by the lightning striking in siliceous sand." As to the composition of the three forms—the crushed, the crystalline variety and the pumice—there was "no essential chemical difference," he pointed out, citing Tassin's analysis.

As to "the exciting cause of this metamorphism," however, Merrill remained equivocal. "So far as the writer has information," he declared, "no more satisfactory theory has been advanced than that of Messrs. Barringer and Tilghman, who ascribe it to the impact and incidental heat of an enormous mass of meteoric iron which constituted a part of the well-known Canyon Diablo fall. Startling as it may seem," he acknowledged "it must at least receive consideration, for the simple reason that nothing else seemingly worthy of consideration presents itself." That there must have been intense heat and only for a brief period was "certainly manifest." On the other hand, the majority of ordinary meteorites entered the atmosphere at relatively slow velocities and thus their force of impact was "not great," and their depth of penetration "but slight." If, however, "we conceive a mass—as from the Leonids—meeting the earth head on, as it were, it would enter our atmosphere with an initial speed of 45 miles a second. If such a mass were of sufficient size to escape anything like complete destruction through burning, its force of impact would be enormous." Nonetheless, he added: "Whether it could or did produce the effects described is, perhaps, yet an open question."

Merrill was more successful in avoiding the question of Coon Mountain's origin in his second paper.[109] In this, he confined himself to an extensive discussion of the shale ball irons and the iron shale found at the crater, relying on Tassin's detailed chemical analyses which were published as Part II of the paper. Tassin's work confirmed that there were indeed two kinds of meteoritic iron, one the ordinary Canyon Diablo iron and the other the iron in the centers of the shale balls which contained not only chlorine but iron phosphides, primarily in the form of the mineral schreibersite and sulphides such as troilite and thus were "peculiarly susceptible" to the oxidation process. In addition, the etched surface of the shale ball iron "shows a structure quite unlike that of the typical iron . . . a difference so striking that one unacquainted with the conditions under which they were found would certainly be justified in pronouncing them independent falls."

These differences led Merrill to several conclusions. First, the shale ball irons were not, as Farrington had suggested, simply ordinary Canyon Diablo irons that had become oxidized more rapidly by burial in the ground or in the debris of the rim. On the contrary, that incompletely oxidized shale balls with unoxidized iron centers were found only beneath the surface was "due to the protective action of the dry soil in a region of great aridity." As the buried shale balls became exposed over time through erosive action, they quickly fell apart, accelerating the oxidation of the iron centers. This circumstance, in turn, explained the occurrence of the thousands of fragments of the laminated iron shale on the rim and the surrounding plain.

Second, Merrill rejected Barringer's hypothesis that the shale balls might be solidified drops of fused iron that had been stripped off the main mass of the meteorite by friction during its fall through the atmosphere. "Recognizable traces" of the original crystalline structure of the irons [i.e., Widmanstätten figures], he reported, were found in the "more solid parts of wholly oxidized shale balls and, as recent work by Tassin and others had shown, "the heating of meteoric iron, even at temperatures far below the point of fusion, completely changes its structure." Thus the shale balls could not have been the "melting and burning drops" that Barringer had postulated. "We are forced," Merrill declared, "to conclude that these forms are also products of terrestrial oxidation of small sulphur-chlorine-rich individuals."

As to the relationship between the two types of meteoritic iron, Merrill proposed what, in effect, was a modification of the "plums in a pudding" hypothesis—iron "plums" in an oxidizable iron "pudding." "The possibility of the irons being but residuals out of a large and coarse-grained stony meteorite or pallasite has often been considered and is discussed by Mr. Barringer," Merrill noted. "He fails to find any evidence in favor of such a supposition. With these conclusions the writer agrees." However, Merrill pointed out, from a comparative study of a large number of irons, "we have been able to trace a tendency toward gradation of one form into another. . . . we found here and there apparent intermediate structures and portions rich in iron phosphide, with thin particles of shale adhering. We have therefore come to the conception of a large heterogeneous mass of nickel-iron with segregation masses rich in chlorides, phosphides and sulphides." It was possible, he added, "that these susceptible portions would oxidize and wholly disappear, leaving only the refractory to be found later. This would account for the almost constant association of shale and irons . . . at various points out on the plain."

MERRILL'S LANDMARK PAPER

Merrill's final paper, published early in 1908,[110] is something of a landmark in the history of the Coon Mountain controversies, for it set the direction these controversies would follow for many years to come. Henceforth, the major point at issue would shift progressively from the origin of the crater to the fate of the meteorite that had created it, and to the dynamics of meteoritic impact.

The bulk of Merrill's lengthy exposition was devoted to a detailed summary of all the investigations that had been made of the crater and the various phenomena there since Foote's first visit in 1891 and including, of course, the results of his own and Tassin's mineralogical and petrographic studies. In this, he offered little that was new; his description of the geology and physiography of the crater and its environs borrowed heavily from published sources, and especially from Barringer's and Tilghman's papers; and his discussions of the altered and metamorphosed sandstone, and of the meteoric irons, shale balls and iron shale, not only drew on Barringer's and Tilghman's published and unpublished work, but included extended passages taken directly from his own earlier papers.

On the subject of the origin of the crater, which he also addressed at some length, Merrill was curiously ambivalent. While the assumption of a meteoritic impact origin was implicit in his discussion, and indeed in the title of his paper, he did not explicitly endorse the hypothesis in so many words. His position, perhaps, was best described some months later by a fellow geologist who heard him expound on "Coon Butte or Meteor Crater"* before a meeting of the Geological Society of Washington. "While not committing himself definitely to the theory of origin through the impact of a gigantic meteorite, and while pointing out the seeming objections to such an hypothesis, the speaker showed nevertheless that no other conclusion seemed possible."[111]

Merrill's references to Gilbert's investigation of Coon Mountain were respectful and entirely devoid of any hint of criticism. The volcanic steam explosion hypothesis he credited to Gilbert's Survey colleague, Willard D.

---

*Merrill, in his published paper, declared that the name "Coon Mountain" was a "singularly inappropriate use of terms" and suggested that the crater "in the existing condition of our knowledge might be renamed Meteor Crater" without, however, mentioning Fairchild's formal proposal of this name a year earlier.

Johnson, and he explained Gilbert's acceptance of it on the ground that he had simply had no other choice. Gilbert, he pointed out, had considered an impact origin, "but the facts then available failed, in his opinion, to substantiate so startling a conclusion." Later he elaborated somewhat on this point.

> "Mr. Gilbert, in the systematic and conservative manner for which he is noted, considered the problem from various standpoints, and particularly with reference to the theories of its having been formed by the (1) "collision and penetration of a stellar body"—i.e., by a meteorite, and (2) by steam explosion. In light of the evidence then available, Mr. Gilbert did not feel justified in adopting the meteoric hypothesis, and, as the only alternative, was forced to adopt the second mentioned above, though those who knew Mr. Gilbert thought to read in his report a strong leaning toward the first mentioned, abandoned only because not borne out, so far as he could see, by the facts.

From the new evidence made available by the drill, he stated at another point, "we are enabled to gain a record entirely inaccessible at the time Mr. Gilbert made his studies, and which throws such light on the subject as to justify us in reverting once more to the original hypothesis as set out in Mr. Gilbert's paper and advocated by Messrs. Barringer and Tilghman."

Merrill, similarly, was "forced" by the facts to entertain the possibility of an impact origin for Coon Mountain. From its shape, he declared in summarizing his conclusions, the crater could have been "formed equally well by blowout or impact," but the character of the ejected material pointed "strongly" to an origin by impact. "It is difficult, if not impossible," he conceded, "to conceive of the smashing and metamorphism of the sandstone on any other ground." The presence of fused quartz—Barringer's Variety B—also indicated great heat which "might be due either to impact or vulcanism." Yet the "slightly disturbed and unchanged condition of the deeper-lying sandstone [the Supai] seems to prove the superficial character of the phenomena. . . . This apparently prohibits consideration of some deep-seated cause" and as there was nothing in the beds themselves that would produce such a result, "one is forced to the consideration of an extraneous source; and, if extraneous, I can conceive of but two—the electric and meteoric. Of these, only the latter seems worthy of serious consideration."

But to Merrill's mind there was a major objection to the impact hypothesis that grew out of his conception of the dynamics of meteoritic impact. For while evidence from recorded falls was "extremely contradictory," he noted, it nonetheless suggested that the violence of impact and the depth of penetration of meteorites in general was relatively slight. The greatest penetration of which Merrill was aware involved a fall in the 1860s near Knyahinya, Hungary, where a 660-pound stone, striking with unknown velocity and at an angle of about 27°, had excavated a small crater eleven feet deep. On the other hand, much heavier masses, such as the 37.5-ton Cape York, the 20-ton Bacubirito, and the 15.6-ton Willamette irons "have been found under conditions as to lead one to infer they scarcely buried themselves."

At Coon Mountain, however, there was "unquestionable proof of a force of impact sufficient to crush a mass of limestone 300 feet in thickness . . . and of sandstone 500 feet in thickness . . . and this, too, with a production of heat equivalent to 3,900°, or the fusing point of quartz." But beyond this, such a force, applied in only an instant of time, should have produced "an amount of heat so vastly greater than 3,900° that its expression in figures would be utterly meaningless and incomprehensible, and in the writer's opinion, the greatest difficulty in accepting the meteoric hypothesis lies in the absence of sufficient evidences of such extreme temperatures."

Merrill himself, however, now offered a way around this difficulty, suggesting that there had, indeed, been a steam explosion. Postulating a meteorite 500 feet in diameter which, by Tilghman's estimate, would produce a crater about the size of Coon Mountain's crater, he suggested that the heat generated by the penetration of such a body through rock "would produce fusion and probably a partial volatilization, and where sufficient moisture was present, other conditions being favorable, would give rise to the pumiceous structure found in the altered sandstone." The pumice itself was evidence for the presence of moisture, and "the effect of impact would be to convert almost instantly this moisture into steam with an enormous explosive power."

As a result, he argued, quantities of debris—the amount depending on the amount of water—and including even portions of the meteorite itself, would be ejected "and thrown back above the crater rim and scattered widely over the plain. It would seemingly be safe to assume a temporary pseudovolcanic condition." This would also account for the intermixing of the rock flour, fragments and blocks of limestone and sandstone, and even the shale balls in "moraine-like mounds" around the crater. "Their association may be best explained on the assumption that all were poured out together over the crater rim, perhaps in the condition of mud, during this pseudo-volcanic stage."

What proportion of the meteoritic irons scattered over the surrounding plain "are the result of this secondary effect one can only surmise," Merrill added. But the fact that there was seemingly no regularity to their distribution, no sorting out by size along the supposed line of flight, and no evidence of atmospheric friction on their surfaces, "favors the idea that *all* were thrown out [Merrill's emphasis], and it is not impossible that practically the entire mass not dissipated by volatilization was thus ejected."

Moreover: "The failure thus far to find a large intact mass within the crater might further be explained on the ground that a considerable portion of it was volatilized by the intense heat generated at the moment of striking the surface, and the comparatively small residual remaining has largely succumbed to oxidation."

Barringer was not ready to give up his search, however, because the stakes were still too high and the uncertainties too great.

SIX

UNDER THE SOUTHERN CLIFFS

> Why should anyone try to find this great meteor? As a commercial matter, of course, for the iron, nickel, platinum and diamond it contains.
>
> ELIHU THOMSON, 1912

As a commercial matter, Barringer realized, Merrill's suggestion as to the ultimate fate of the meteorite must not be allowed to stand unchallenged. Volatilization on impact and the scouring of the crater by a subsequent explosion implied that no large mass of nickeliferous iron lay buried at Coon Mountain and, from a practical if not from a scientific standpoint, that the search for such a mass was futile. Men of capital, he knew, were not interested in chasing chimeras.

As a scientific matter, however, Barringer also realized that the uncertainties of the problem left Merrill's suggestion open to challenge. Merrill himself, indeed, had stressed these uncertainties by citing a particularly pertinent disclaimer by an English scientist, H. E. Wimperis, who had attempted to calculate the minimum initial mass that a meteor must possess to survive destruction from atmospheric friction during its earthward plunge. "I am aware," Wimperis wrote of his conclusion, "that the whole structure of the investigation rests on the evil principle of extrapolation, but until man is capable of experimenting with velocities of 10 or 20 miles a second, and surviving thereafter to record his results, no other manner of investigation seems possible."[1]

THE PROBLEM OF VOLATILIZATION

The idea that impacting meteorites might be volatilized had been suggested before, although not in regard to terrestrial phenomena. Proctor, it will be recalled, had declared in 1878 that "almost every mass which thus strikes the Moon must be vaporized by the intense heat excited as it impinges upon the Moon's surface," and he apparently considered this statement sufficiently self-evident not to require elucidation.[2] In the 1880s English astronomer J. Norman Lockyer made the collision and resulting volatilization of meteors moving in primordial swarms fundamental to his short-lived "Meteoritic Hypothesis" of the origin and evolution of celestial bodies.[3] At the time his basic premise received some support from speculative researches into the dynamics of meteor swarms by George H. Darwin, who concluded that it was "*a priori,* probable that, when two meteorites clash, a portion of the solid matter in each is volatilized."[4] Violent impact, too, was implicit in the more durable and popular "Planetesimal Hypothesis" advanced in 1900 and thereafter by geologist Thomas C. Chamberlin and astronomer Forest R. Moulton. It was used to explain the genesis of the solar system and its planets and satellites through the accretion, the falling together, of bits and pieces of matter, or "planetesimals," torn out of the sun in swirling, spiraling streams by gravitational effects resulting from the near collision of the sun and another star.[5]

Merrill undoubtedly was familiar with Lockyer's and Chamberlin's and Moulton's work,[6] but almost certainly his application of the idea of volatilization to the phenomena at Coon Mountain derived from geologist Nathaniel S. Shaler's 1903 study of the surface features on the moon, a study which Merrill had formally reviewed for the Smithsonian Institution and had approved for publication.[7] Shaler, although rejecting an impact origin for the lunar craters or "vulcanoids," had clearly recognized the implications inherent in the relation, $e = \frac{1}{2}mv^2$*, for meteoric masses impacting at planetary velocities, as well as the many imponderables in the problem. "The effects due to the great speed at which meteorites would usually fall upon the Moon cannot be accurately determined; certain of them can, however, within limits, fairly be conjectured," he had pointed out.

> Assuming the rocks of the lunar surface to have the average resistance of pumice, it seems evident that any meteoric body such as we know to fall on the earth would not only penetrate to great depth, but would probably be volatilized by the very high temperature it would attain. . . . [W]e may

*Where e=kinetic energy, m=mass, and v=velocity.

fairly assume that the mass would, in effect, explode, the gaseous products being cast forth from the opening made. The temperature produced by the arrest of the movement at a rate of twenty miles a second would vaporize the mass.[8]

Shaler, here, was speculating in a generalized way about meteoritic impacts at planetary velocities on an airless moon; in the case of the earth, of course, retardation caused by atmospheric friction must also be taken into account. But in both cases, the actual effects of impact of a particular meteorite cannot even "fairly be conjectured" unless a number of other factors, such as initial velocity, mass, configuration, and direction and angle of fall, are known with some degree of accuracy. At Coon Mountain these data were lacking, and the velocity of impact of the meteorite that fell there was, for practical purposes, anybody's guess. Barringer had agonized over the problem even before Merrill began his investigation and had concluded that this velocity must have been, relatively speaking, low.

Barringer had been concerned about volatilization of the meteorite since the summer of 1906, when his cousin Paul had raised the possibility in discussing the continuing failure of the drill to locate any sizeable meteoritic mass under the crater floor. "My cousin," Barringer had written to Branner then, "does not think we are entirely justified in applying rules governing high speed projectiles, and that it is quite conceivable that the celestial projectile was not appreciably cushioned on the earth's atmosphere and was dissipated into metallic mist."[9] And to Holsinger, he confided that "I am beginning to fear that we have been giving the meteorite too small an amount of speed. If so, it was a smaller body than we have been figuring on and was vaporized or dissipated."[10]

Tilghman, an authority on ballistics, and Mallet both rejected the possibility out of hand,[11] but Barringer continued to have reservations. "Now unless the object bounded out of the hole, or was dissipated into mist, or was gassified, or has been dissolved away since it fell, it is still in the hole," he wrote Bennitt in April 1907.[12] And a few weeks later, he spelled out his doubts in detail in a letter to Bennitt's wife, Emma. "Don't think I am unduly discouraged," he explained, "but I have no opinion one way or the other as to whether we will ever make a commercial success of it."[13]

> In fact, I have felt all along that we are densely ignorant as to the one great factor in the matter, namely, the speed at which the projectile was coming. If it came from the Leonids, as Dr. Merrill suggests, it met the earth head on and the speed when it entered the earth's atmosphere was 45 miles

a second. It therefore may have struck at something like 20 miles a second, and if so we may never know much more than we know at present. Dr. Mallet and Dr. Tilghman strongly urge that if this were the case there would be many more evidences of fusion than we find. . . . But to this I simply say *quien sabe*. If it struck at 10,000 feet a second or less than 2 miles a second, then it would certainly seem that we ought to find it somewhere, and that the vast majority must still be in the hole.

Despite his private reservations, however, Barringer objected vigorously to Merrill's published suggestion of volatilization, citing Tilghman's and Mallet's point on the lack of evidence of extensive fusion at the crater. Merrill, deferentially but deftly, threw the argument back at him. "There is nothing in my paper that conflicts with your idea that there still lies in the bottom of the crater a big mass of meteoric iron," he replied to Barringer's initial protest.[14]

> In fact I state definitely . . . that the work thus far done does not disprove it. The 'gassification' is speculation. My 'theory of steam explosion' is one of the means to account for conditions as they exist. There may be others just as good. . . . Strangely enough, you fail completely to see my point in the suggestion of volatilization. This idea was introduced to explain the absence of fusion phenomena. . . . an absence which on the meteoric theory is otherwise incomprehensible. . . . and which you yourself recognize.

Merrill subsequently expanded on this point, stressing that he had suggested only partial volatilization of the meteorite. "It seems perfectly fair . . . to assume that a portion of the iron may have been volatilized and through the sudden expansion of these gases the hole made so much larger than the bolide itself that all traces of contact phenomena other than the smashed and fused quartz have disappeared. . . . Personally, I wouldn't give a rap for the opinion of any man who has not actually been on the ground, as to the presence of a meteorite at the bottom of the crater. It would, at least, seem as though a part of it must be there."[15]

What was needed, Merrill added pointedly, was proof.

> It seems to me that the only point with which you are seriously dissatisfied in my conclusion is that I do not come out flat-footed as the saying is, and declare that the meteorite is still in the hole. But my saying so would not put it there, so what's the use? . . . its presence must be proven. So apparently it is up to you and your associates. You do not need to be reminded that the problem at which you are at work is one utterly unlike anything that has been undertaken before. You are in the very front rank.

... and after you are through it will be easy work for any of us, however far we may be in the rear, to come forward and say I told you so.

Merrill revisited the crater in May 1908, and Barringer confided to Branner some weeks later that Merrill "is coming along all right along the road that Tilghman and I have blazed out for him. . . . He struggles and remonstrates, but nevertheless he is coming steadily along and we should not complain. . . ."[16] Barringer, indeed, found Merrill's endorsement of an impact origin for Coon Mountain, however qualified, to be useful in his efforts to interest financial backers in further exploration of the crater. "You may or may not know," he now wrote mining magnate James Douglas, for example, "that the government people, through Dr. Merrill, have thrown up the sponge and admitted that we have been entirely right all along in our conclusions."[17] He was dissembling here, of course, for he was well aware that while a few "government people" had privately conceded the case for impact, only Merrill had done so in print. It would be many more years before another government scientist would opt publicly for the impact origin of Coon Mountain.

Merrill's 1908 visit to the crater, parenthetically, was reported briefly in the press; and Barringer saw in these reports a subtle attempt to "steal" the credit for demonstrating the crater's impact origin for the government through Merrill. His sensitivity on the point undoubtedly had been heightened by Branner's warning a year earlier, apropos Merrill's status, that "you'll be lucky if they [government scientists] don't manage to show that they had said so all along and that you are the one to oppose the truth!"[18] But the reports did indeed tend to give the impression that Merrill had done all the work and drawn all the conclusions. *The New York Times,* for example, declared that while at the crater, Merrill "witnessed the boring of wells" and that "These and other studies have tended to confirm the conclusions reached by him [Merrill] last year that the crater must have been caused by a meteor." The report, which emanated from Washington, did not mention Barringer or Tilghman.[19]

"Has Dr. Branner returned?" an irate Barringer now inquired of one of Branner's Stanford colleagues. "Tell him that his prophecy is coming true, namely that there would be an attempt to steal from us the credit for making the remarkable discovery in Arizona. . . ."[20] And to Mallet, he complained that "Several determined efforts have been made recently to shift the credit, in the public's mind at least, from Dr. Tilghman and myself, to Dr. Merrill or to the United States Government through Dr. Merrill, for having properly explained the origin of Coon Butte and its crater. . . . I think, however, that I have successfully headed off this attempt."[21]

Barringer, indeed, had contacted the Philadelphia *Public Ledger,* which had then published a long article, illustrated with five photographs and a sketch, under a full-page headline which proclaimed: "Meteor Crater Discovered by Philadelphians."[22] Citing a "dispatch from Washington a few days ago," the anonymous writer declared at the outset that "the impression given by the dispatch was that geologists connected with the Government had made the discovery. . . . The truth of the matter, which is of more popular interest than scientific subjects generally, is that the origin of the crater was worked out by two Philadelphians five years ago. . . ." Then, after reviewing Barringer's and Tilghman's 1906 papers at length, the writer declared that the "chain of evidence" they had collected for the meteoritic impact origin of the crater "is regarded as conclusive." This theory "may now be said to be universally accepted and the only wonder is that it was not originally proposed by the United States Survey when they examined the field years ago. One suggested reason for the neglect is said to be the unpopularity of the view, and 'lunar crater men' in the geological survey are looked upon with suspicion. There is something entirely too popular in this line of thought and, consequently, an extraterrestrial hypothesis for anything concerning the earth's surface is usually discarded as unscientific."

Barringer followed up this article with a letter to the editor;[23] and some months later, when the issue arose again, he wrote letters to the editors of both the *Public Ledger* and *The Iron Age* protesting "an effort to claim undue scientific credit for the Smithsonian Institution or for the government."[24] He did not, however, challenge Smithsonian Secretary Walcott's subsequent report in the Institution's *Annual Report* for 1908 in which the former director of the U.S. Geological Survey also described Merrill's researches and conclusions without reference to Barringer or Tilghman.[25] Perhaps he did not see the report, or perhaps he had now adopted Branner's bitter attitude toward such matters. "What did you expect?" Branner had replied to his complaints about the newspaper reports. "These people will belittle everything and anything done by anyone not of their own organizations; and if they can appropriate a problem that an outsider is at work on, they never hesitate to do it. They always suggest to my mind a lot of chickens in a barnyard: If one finds a morsel he should swallow it quietly; otherwise the whole pack are down on him." [26]

BARRINGER'S NEW THEORY

The summer of 1908 marks a major turning point in the history of Coon Mountain. Barringer's straitened finances, combined with Tilghman's disenchantment with the venture, forced suspension of the drilling program in August; it would be another dozen years before the physical exploration of the crater would be resumed. During these years Barringer would work tirelessly to convince eminent men of science of the impact origin of the crater and eminent men of capital that potential profits lay buried beneath the crater floor. From time to time, he would also pursue the possible industrial uses of silica sand with various manufacturers of silica-based products.[27]

More immediately, in an effort to make Holsinger's new caretaker status self-supporting and thus reduce expenses at the crater, Barringer tried briefly to induce the Santa Fe Railway to promote the crater as a tourist attraction.[28] After all, he pointed out to Branner, the crater "has become one of the most interesting localities on earth."[29] But when the railroad did not agree, he quickly dropped the idea and would never seriously entertain it again. In fact, he did not favor having tourists at the crater, if only because of their habit of collecting souvenirs. Many years later he candidly stated his position to a Flagstaff correspondent who had asked to include the crater in a tourist guide to points of interest in northern Arizona. "I hope you will understand when I tell you that we are strongly opposed to encouraging visits by tourists," he had written then. "The amount of meteoric material, valuable to us for scientific as well as commercial reasons, that tourists have picked up . . . is unbelievable. When I first went to the crater the whole country for several miles around was strewn with oxidized meteoric iron (the so-called iron shale). Moreover, a friend and I went out one afternoon and picked up 177 *iron* meteorites. Today it is next to impossible to find an iron meteorite and only occasionally is a piece of iron shale found."[30]

By the summer of 1908, too, Barringer had done some hard thinking about the problems at Coon Mountain in light of the failure of the drill to find any meteoritic mass and of Merrill's challenge that the presence of such a mass must be proven. He now began to evolve "a rather new theory . . . to account for all the facts we have observed" which "fully explains our failure to have struck the meteorite up to date."[31] He would expound his new theory in a lengthy paper before a meeting of the National Academy of Sciences held at Princeton University in November 1909.[32] In this, he would argue that the projectile had not been a single, solid body but rather a compact cluster or swarm of meteors, perhaps the head of a small comet, that had plowed into the earth at an oblique angle from the north. Thus its mass, in the form of thousands of individual oxidized and unoxidized iron meteorites, must lie wedged under the southern cliffs, an area of the crater so far wholly unexplored.

Barringer's new theory, incidentally, may well have contributed to the final break with Tilghman at this same time. Originally, they had assumed that the mass would be found beneath the center of the crater on the basis of the crater's circular shape which, as Barringer once remarked to Merrill, "is almost conclusive to my mind that the crater was made by a body as we now think falling almost vertically."[33] But he now decided that the physical evidences for a north-south bilateral symmetry at the crater should be given more weight, and that these contradicted their original assumption.

The most important evidence of this symmetry was the extreme shattering and massive uplifting of the rocks on the south rim. There, a half-mile-long segment of the strata had been vertically uplifted 105 feet, on a level and as a mass, from its original position. "The cause of the uplift of this tremendous weight (probably some 20 to 30 million tons) may be found in the fact of the meteoric mass having wedged itself underneath, or in the development underneath it of a vast amount of shattered rock, or . . . of great quantities of what we know as Variety A and Variety B of the metamorphosed sandstone."[34] Moreover, directly across the crater on the north rim was "a very noticeable low place or depression" which "corresponds quite closely in length with the great vertical uplift of the strata in the southern wall."

There were other evidences of this rough north-south symmetry as well. Most obvious, perhaps, were the "two remarkable fields of big boulders . . . weighing hundreds and even thousands of tons, opposite to each other, one on the eastern side and one on the western side of the crater."

Bilateral symmetry was also shown by the fact that the quaquaversal, or outward dip of the strata around the crater's rim increased progressively from 5° on the north-northwest to 90° on the south-southeast and south-southwest where two great faults bounded the great uplifted section of the southern rim. All of these points were graphically illustrated on maps of the crater and its environs that he presented with his paper.

The amount of rock ejected from the crater Barringer now estimated at more than three hundred million tons, and he added that a "vast amount of the surrounding rock in addition has been violently disturbed." But he conceded that "it is practically impossible to estimate accurately the amount of work done or energy expended; almost as impossible as it is to estimate the speed at which the projectile was moving when it struck." It was thus "extremely difficult" to estimate the weight of the projectile, although it "must have been very great" and probably "in excess of a million tons."

Barringer also reviewed briefly developments at the crater since the publication of his earlier paper. A total of twenty-eight holes had been drilled in the central portion of the crater, and in fourteen of these "undoubted meteoric material" had been encountered between 450 and 680 feet, although this material was "very sparsely distributed" in most of the holes and never comprised "more than a very small percentage of the fragmentary material brought up by the drill." He also discussed the discovery of Variety A and B of the metamorphosed sandstone, citing Mallet's, Branner's, and Penfield's studies of these forms, rather than Fairchild's or Merrill's. He stressed that in Variety A in particular, the "bedding planes are still well shown, and an interesting fact is that they are not always at right angles, but are frequently at different oblique angles, to the slaty structure. . . . This fact alone practically necessitates this metamorphosed rock having been made by a projectile advancing through the rock strata of which these are fragments."

Variety B, he felt, had not been heated much higher than A "but, owing to the presence of water, semi-fusion or, in the case of the opalescent quartz, complete fusion could take place." There was also reason to believe that Variety A and B had been formed close to the advancing projectile "since fragments have been found which are stained with the oxides of iron and nickel." There was no other evidence of the action of steam at the crater, however; and thus it was "not necessary to presuppose an explosion on a large scale of any sort. As I read them, the facts are all against any explosion except those represented by small and relatively unimportant puffs of steam where there was water in the sandstone."

Now he theorized that most, if not all, of the meteoritic material found in and around Coon Mountain had originally been in the form of the shale ball iron that had fallen, moreover, as individual masses. Here he stressed the "remarkable fact" that the shale ball irons "are never angular . . . but always seem to be rounded, and globular or oval in shape." Moreover, shale ball iron had been found on occasion adhering to or in close association with ordinary Canyon Diablo iron and thus it was "certain that these two classes of iron have frequently formed part of the same mass." From this, he envisioned the crater rim and surrounding plain as being "strewn with perhaps thousands of these unoxidized shale ball iron meteorites a few minutes after impact," and suggested that all of the shale ball irons, and most of the Canyon Diablo irons as well, "represent the residual unoxidizable portions of such shale ball iron meteorites."

There were, however, difficulties here, he admitted. While shale ball irons were found buried in the rock debris around the crater, "so far no piece of unoxidizable Canyon Diablo iron has yet been found deeply imbedded in the silica on the slopes of the mountain, but always on top of it, that is, on or very near the surface. . . . I have no completely satisfactory explanation to offer for this fact." Again, he conceded that the finding of ordinary Canyon Diablo irons weighing more than 300 pounds, and in one case 1,000 pounds, which were not associated with any iron shale, "does not seem to be in agreement with this theory." Nor had any ordinary Canyon Diablo irons been found in the remains of shale balls that had disintegrated into iron shale since being found. "Notwithstanding these facts," he argued somewhat speciously, "the evidence is quite strongly confirmatory of the theory that many of the smaller, and especially the very small Canyon Diablo siderites have been derived from the disintegration of shale ball meteorites great and small. There is a characteristic shape among them that points to a common origin. . . ."

From these considerations Barringer now drew a major conclusion—"that the crater was not made by a single giant meteorite, but rather by a compact cluster or swarm of many thousands of shale ball meteorites and also possibly of other iron meteorites . . . traveling together as the head, or part of the head, of a small comet."

This theory, he noted, explained the distribution of the irons and iron shale around the crater, as well as the burial of some of the shale balls in the ejecta of the rim. The outriders and stragglers would be subject to greater retardation by atmospheric friction, and perhaps even slightly diverted from their course, and would thus reach the earth after the main cluster had struck. The meteoritic material on the rim and the plain could not have been blown out of the crater after the impact, he argued, because it showed no evidence of having been highly heated. The shale balls, when sectioned and etched, "show the Widmanstätten figures beautifully, as do the ordinary Canyon Diablo meteorites wherever found, and this, I am told by those who have made the experiment, they would not do if they had been heated beyond 700 degrees or 800 degrees Centigrade."

The absence of any magnetic anomaly associated with the crater was also consistent with the hypothesis that the projectile had been a cluster; thus it was not necessary to postulate, as Tilghman had done, the shattering on impact of a single metallic mass. Here, Barringer cited magnetic experiments made by Princeton physicist William Francis Magie, a former Princeton classmate who had spent ten days at the crater with Barringer in August 1909.[35] One such experiment, Barringer reported, involved placing a box half full of shale balls and Canyon Diablo irons weighing two hundred pounds "directly under the dip needle of an excellent instrument . . . and within three feet of the instrument itself, and yet no movement of the needle was noticed."

Extrapolating from this, he declared that it "would seem justifiable, therefore, to conclude that a vast number of similar masses of meteoric iron may be hidden in the depths of the crater, five to seven hundred feet below the surface, without the needle of the instrument showing any variation or dip because of their presence." And here, as Tilghman had done, he cited the "intrinsic magnetism" of the individual fragments of iron shale and shale ball iron "from which it is argued that the irregularly distributed masses in the bottom of the crater would neutralize each other's fields."

Nor was it possible that any appreciable portion of the mass had been volatilized on impact, he contended, because of insufficient evidence of heat inside or outside the crater. The absence of staining by iron and nickel oxides on a grand scale is absolutely opposed to such a theory." And again: "It must always be borne in mind that if the projectile struck so hard as to be totally vaporized, not only, in my opinion, would there have been a very shallow crater and greater fusion of the target (the arenaceous limestone was perfectly adapted to the making of slag), but this vapor would have condensed instantly and most of the rock fragments and silica, instead of being snow white, as in the case of the silica, and wonderfully free from iron, would necessarily be abundantly stained by oxides of iron and nickel. . . . There is no escaping this conclusion."

To further discuss the origin of the crater was "idle," Barringer concluded, and the only real question remaining at Coon Mountain was what had become of the meteoritic mass that had created it. For Barringer the answer was clear. The mass that had formed the crater, he declared, "lies in the bottom of it somewhere." Moreover, he predicted that when found, it would be partially oxidized "after the manner of the buried shale balls, in which event it would still possess great commercial value." He repeated his belief that most of it would be found in the southern or southeastern portions of the crater beneath the cliffs which form the southern wall.

Barringer's paper proved to be significant on two counts. First it established the position that he would consistently hold in the Coon Mountain controversies for the rest of his life; indeed, as he now confided to Holsinger, he was "disgusted with the errors in my old paper."[36] Second, it brought a new group of scientific allies to the cause of the impact origin of the crater—allies who did not always agree with his corollary thinking but who supported his main thesis to the bitter end. In addition to Magie, these new recruits included Princeton's Henry Norris Russell, one of the most widely known and respected astronomers of the twentieth century; and Elihu Thomson, a founder and chief scientist of the General Electric Company and an inventive genius who amassed more than 700 patents during his long career.

Russell, after reading Barringer's paper and agreeing with its main conclusions,[37] made a suggestion that Barringer included, with Russell's approval, as an appendix in its subsequent publication.[38] This was that the meteoric mass "would cut out a 'wad' of air going through the atmosphere like a charge of shot going through a board in front of the muzzle of a shotgun. The total mass of this entrained air might be as great as that of the original meteorite," and thus "the blow at the earth's surface would be to a certain degree cushioned, which would help account for the absence of notable evidences of volatilization in the iron. The mass of hot air would be fairly driven into the sandstone and limestone strata by the first impact, and would then escape upward around the advancing projectile and be very efficient in removing the pulverized and shattered rock, and so in digging the crater." Russell's "wad" of compressed air preceding the projectile would in time become Barringer's key argument for the presence of a mass of nickeliferous iron in the crater's depths.

Barringer's unelaborated suggestion that the mass that made the crater might have been the head or part of the head of a small comet, parenthetically, is of more than incidental interest, for he would subsequently develop it into a bizarre theory to explain the brightness and visibility of comets in general, reasoning from the phenomena at Meteor Crater. The idea that a comet might lie buried there, indeed, had occurred to him as early as April 1906, when he had remarked to Mallet that "I wonder if any mortal man ever wished to own a comet! It sounds like diviling idocy [sic], so I shall forebear."[39]

But such an idea was no longer "diviling idocy" in 1909. That year, the subject of comets was of high interest not only to scientists but to the public generally because of the imminent apparition, after an absence of seventy-five years, of Halley's famous comet, which was due to make its perihelion passage around the sun in May 1910. Astronomers were predicting that the earth would pass through a part of the great comet's tail, and dire speculations as to the possible consequences of such an encounter had already begun to appear in the popular press.

Almost certainly, however, Barringer's revival of the idea that the Coon Mountain meteorite had been part of a small comet derived from some speculations at this time by Harvard astronomer William Henry Pickering, a prolific and wide-ranging writer on astronomical topics in both professional and popular journals. Pickering had written him in January 1909 requesting copies of his and Tilghman's 1906 papers and sending him a copy of his own "What Is a Comet?" that had appeared that same year in *Harper's* magazine.[40] In June, Pickering published an article in which he proposed that not only was the Coon Butte meteorite part of the head of a comet, but that seven other large iron meteorites found within a 900-mile radius of Coon Butte might have been parts of the same comet's head as well.[41] Pickering noted that the "extraordinarily slow speed" at which the Coon Butte meteorite had apparently struck "seems to be characteristic of the large iron meteorites," adding that "this slow speed indicates that the comet and the earth were moving in the same direction, and since the comet was near perihelion, it must have been moving the faster of the two and, therefore, overtook the earth."

Barringer's concept of the structure of a comet's head, however, differed from Pickering's, as the latter pointed out just three weeks before Barringer presented his paper at Princeton. "Regarding your theory that the crater is due to the impact of a compact mass of materials," Pickering advised him, "I would say that the idea does not conform to the present astronomical conceptions. The different meteorites in a comet's head are supposed to be separated from their nearest neighbors by a distance of several miles. This conclusion is based on the small mass of comets and the fact that stars are seen through their heads.... Of course," he added, "you may be right, but your theory is not supported by any present astronomical knowledge."[42]

Barringer, however, clung to his idea that comet heads were made up of tightly clustered meteoric bodies, and for a reason. For while he had not developed the idea in his paper, he had decided that the rounded shape of the shale ball meteorites at the crater could be explained by the abrasive action of the bodies in such a cluster grinding against each other as they traveled together through space. In May 1910, as Halley's Comet staged its most spectacular display,[43] he broached his idea to Russell, noting that the shale ball meteorites were rounded "like pebbles," arguing that the more angular Canyon Diablo irons might have been "stragglers," and wondering whether the particles of matter resulting from such abrasive action might account for the formation of a comet's "tail."[44] Russell rejected the idea "vehemently," Barringer later noted,[45] and Barringer apparently was never able to interest other astronomers in it. Nonetheless, and to anticipate events here somewhat, he would propose the theory in a highly speculative paper read before the Academy of Natural Sciences of Philadelphia in 1916 and subsequently published in the Academy's *Proceedings,* along with a strongly supportive addendum in the form of a letter from Elihu Thomson.[46]

In this, he suggested that the "inconceivably gentle abrasive action" between the bodies composing the head of a comet "may be sufficient to wear off infinitesimal particles of matter, considering the enormous periods of time available.... May not these fine particles of matter, possibly electrified, also form the so-called tail as they are swept away from the main body of the comet...?" There was "no doubt," he added, that the shale ball meteorites found at Meteor Crater "show evidence of what seems to have been abrasive action," and the inference "was unavoidable that they may have been subjected, during perhaps billions of years, to such action, inconceivably slight, it is true, but nevertheless sufficient to bring them to the shape in which we found them. May they not be in fact 'celestial cobblestones,' and may not the milling action when they rub against each other, even very gently, account in some way, not perfectly understood, for not only what we call the brilliant head of a comet but for its tail as well?"

In his addendum, Thomson declared that there was "little doubt" that Barringer's explanation of the rounded shale balls "is the true one" and objected only to his suggestion that the abraded particles might have been electrified, pointing out that "the vacuous space in which the tail moves is so high a vacuum that no conduction would be possible."[47]

Nonetheless, astronomers were no more receptive to Barringer's concept of comets as compact clusters now than Pickering and Russell had been. Barringer subsequently sent a copy of the paper to John A. Miller, the director of Swarthmore College's Sproul Observatory, only to find that Miller also disagreed with his ideas. "If the stones were near enough to collide," Miller pointed out as Pickering had done before him, "the comet head would need to have a greater mass than is found." Moreover, he added, the theory did not account for the bright emission lines in the spectra of comets.[48]

PUBLICITY, GOOD AND BAD

In the years following his National Academy paper, however, such brief and sporadic ventures into speculative astronomy were the exception for Barringer rather than the rule, for a host of more mundane matters relating to the crater occupied his time and his energies. The crater and the controversy over its origin were now becoming more widely known to scientists and the public as a result of a spate of articles concerning it that began to appear following the publication of Merrill's final paper. Barringer certainly was not aware of everything that was printed about the crater, but those articles that he did read evoked reactions of frustration, fury, and, in one case, fear.

In April 1910, as Barringer hurried to publish his National Academy paper "before Halley's Comet appears, as the crater was apparently caused by a small comet,"[49] Magie reported on his investigation of the crater to the American Philosophical Society.[50] He not only described his magnetic tests and experiments involving firing rifle bullets into silica sand, but he attempted what Barringer, in his paper, had declared to be "practically impossible"—the determination of the work done, or energy expended, in the formation of the crater, and thus the mass of the meteorite.

Magie estimated that penetration, lifting, ejection, and crushing had involved at least 500 million tons of rock; and from the relatively sparse evidence of fusion and the presence of Widmanstätten figures in the irons, he concluded that this material also had been raised by shock of impact to a "general temperature" of 625° C. On this basis, he calculated that the work done in forming the crater amounted to more than 60 trillion foot-tons. Then, taking 3 and 48 miles per second as the "outside limits" for the velocity of the meteorite, he computed its mass for these limits at 15 million and

60,000 tons, respectively. "Manifestly the extreme velocities estimated are not probable, so that the mass is probably neither so large nor so small.... A mass of 400,000 tons moving with a velocity of 18 or 20 miles per second would bring in the estimated amount of energy. In the absence of other evidence," Magie added, "this seems a reasonable mass to assign to the buried meteorite."

His estimate was not reasonable to Barringer, who would, indeed, soon raise his own estimate of the mass to "not less than two million tons."[51] The 400,000-ton figure, he immediately informed Magie, was "too small,"[52] and to Russell he complained that Magie "underestimates the energy required, and therefore, the mass."[53] It would take him more than a year, however, to persuade Magie that the mass might be larger through persistent arguments that the velocity had been low, that a mass of 400,000 tons was inconsistent with a cluster of iron meteorites 200 or more feet in diameter, and that such a high-velocity impact would have produced more fusion and perhaps volatilization but could not have penetrated more than 1,000 feet of rock.[54] On this last point, incidentally, he drew on his big-game hunting experience to contend that there was an optimum velocity at which greatest penetration occurred. Maximum penetration occurred at an intermediate speed, he advised Magie, explaining that "I could never shoot through an elk at 40–50 yards, practically always at 200–250 yards, and not at all at 400."[55] How well Magie was eventually persuaded is evident perhaps in a letter he wrote some years later in response to Barringer's request to state his views on the crater for potential investors. Magie declared then that "in view of the immense amount of rock moved and broken, the mass of the large meteorite, or of the swarm of small ones which made the crater, cannot be less than ten million tons."[56]

Another paper in 1910, this one in *Science* by a John M. Davidson, proved to be less a problem. Davidson argued that the pulverized silica at the crater "does not look like the result of a crushing blow from above or a steam explosion from beneath.... It rather suggests long-continued deposition of this powder, with occasional pieces of rock, by geyser action, and a final explosion or series of explosions that closed the drama."[57] For an analogy he cited the geysers and hot mineral springs at Yellowstone Park. Barringer's reaction was terse and to the point. "Your article on Coon Butte in the present 'Science' is all wet," he advised. "Visit the crater!"[58]

Of greater concern, although from a wholly different standpoint, was a highly romanticized version of the discovery of the Canyon Diablo irons published in *The Pacific Monthly* by a minor Forest Service official in Washington, D.C., named Will C. Barnes, who had once herded sheep in the vicinity of the crater.[59] Barnes wrote that in February 1891, while working at a sheep camp near Canyon Diablo, the camp cook, one "Frederick Krapf," had discovered the irons; that he, Krapf, and others had then formed the "Canon Diablo Iron Mining Company"; that numerous claims had been located in and around Coon Butte, and that later, after learning that the iron was meteoric and contained diamonds, a shaft some 75 feet deep had been dug into the crater floor.

In Barnes's colorful, rambling story, "Krapf" parallels camp cook "Craft" in Gilbert's 1896 account to which Barnes, however, made no reference. Like Craft, Krapf had also received a grubstake in Albuquerque—$1,500 according to Barnes—which he proceeded to spend on that town's "fascinating inducements" in what Barnes indicated was a monumental, three-day carouse. Meanwhile, a cowboy known only as "Pink," who had accompanied Krapf, had hurried back to Coon Butte, "quietly located the entire country surrounding the mountain," hired some Mexican herders and wagons, and had "searched with a fine tooth comb, every piece of the meteoric iron 'float' being thrown in the wagons as fast as discovered until it is estimated that several tons of the stuff were picked up." The "stuff" was then sold to museums and other collectors and, Barnes added, the "original locators of the claims, having cleaned up several thousand dollars by gathering and selling the pieces of float iron, had no further use for the locations . . . and practically abandoned them."

But soon thereafter, Barnes related, he and others at the camp learned of "an assay made by someone in Washington of a piece of the iron which showed it to contain diamonds—pure South African diamonds. . . . So, after considerable discussion, an open shaft was started in the bottom of the crater, the plan being to dig down to the great body of the meteor and mine it out, like any other mineral deposit. . . . After digging some seventy-five or eighty feet and finding no indications of nearness to the meteor, our funds and patience gave out at the same time. . . . The doings of the Canon Diablo Iron Mining Company then became an obscure matter of history."

Holsinger called Barringer's attention to Barnes's article in July 1910 and, typically, offered some comments on it. He had met Barnes several years before in Phoenix, he wrote, and Barnes had "surprised me not a little by asserting that he had been one of the first discoverers of Meteor Crater or Coon Butte as they called it then, and that he and several associates had at one time dug a shaft some seventy feet in depth in the crater. . . . Of course you know that there is no possibility of his statement being true." Moreover, Holsinger reported that one Cattelano Ybarra, who had come to Canyon Diablo in 1882 and had collected irons for Volz, the trader there, "assures me that no one ever dug a pit or shaft in the crater as much as two feet deep. . . . These faked write-ups of the crater are annoying," he concluded, "but after all quite harmless."[60] Holsinger had written Barnes asking for specific details, and when Barnes replied by merely repeating the statements in his article, he advised Barringer that Barnes's story was simply "a plain case of reckless lying."[61]

Barringer, however, did not see it quite that way; rather he saw in Barnes's published claims of prior location and the sinking of a shaft a possible threat to the Standard Iron Company's title to the crater and the land surrounding it. He was already aware, in fact, that Barnes had been telling some such story in Washington scientific circles. Early in February 1908, Merrill had casually remarked in a letter that "at a little company at [Alexander] Graham Bell's last night," one of the guests "had brought with him that Mr. Barnes who was one of the original party interested in locating an iron mine at the crater."[62] The story, Barringer now feared, "is believed by Alexander Graham Bell, possibly by Merrill, and others. . . . For the sake of my boys more than anyone else I wish to make it impossible for anyone to throw any doubt . . . that it was I who first of all sufficiently believed in the impact theory to acquire title . . . and to expend large sums of money."[63]

Over the ensuing months, Barringer accumulated evidence to refute Barnes's story, directing Holsinger to search the records at the county courthouse in Flagstaff for early claim notices, and gathering affidavits from Holsinger, Ybarra, and others as to the pristine condition of the crater's floor at the start of Standard Iron's operations there in 1903. Not until August 1911 and, indeed, only a few weeks after Holsinger's death from tuberculosis,[64] did he challenge Barnes directly. He then pointed out that of the eight claims that had been recorded at the crater prior to Standard Iron's claims, all had been filed by Volz or one of his employees in June and July 1891 shortly after Foote's first visit there, and all had since been allowed to lapse. As to the shaft, he declared that if Barnes could prove its existence and point out its location, "I will pay your expenses to the crater and give you $1,000."[65] Barnes never accepted his offer.

Still another paper on the crater that appeared late in 1910 caused Barringer great consternation. This was a U.S. Geological Survey *Bulletin* written by geologist Nelson Horatio Darton, one of Gilbert's junior colleagues, which to Barringer's mind at least put the Survey on record squarely behind Gilbert's volcanic steam explosion hypothesis of the crater's origin. Darton, in 1905, had published a paper on Zuni Salt Lake, a crateriform feature in northwestern New Mexico which, although there were two volcanic cinder cones in its center, he had considered to be "comparable in some respects to 'Coon Butte,' a great depression in eastern Arizona which, Dr. Gilbert has shown, is due to explosion."[66]

In his 1910 *Bulletin*, Darton discussed Coon Butte in greater detail.[67] "Altogether," he now declared, "the features suggest a great bubble with the top blown off and the ejected material piled up on the margin." The meteoric hypothesis, he conceded, was not inconsistent with some of its features, "but does not accord with the all-important fact that no meteor is present, as has been demonstrated by many borings." Darton also found Merrill's suggestion that a meteorite might have been partially volatilized "difficult to accept." On the other hand, Gilbert's volcanic steam explosion hypothesis "appeals to me most strongly, notwithstanding its gratuitous character. The competency of such a cause is well illustrated by explosion craters in various parts of the world." Here he cited Gilbert's point that Coon Butte was "in the midst of a volcanic region" and added his own conclusion that Coon Butte's crater had been formed by the explosion of "an accumulation of steam."

> This steam, permeating the pores of the sandstone, would on explosion rend the particles into rock flour such as that which exists in large volume in and about the crater, and it is difficult to understand how any other force could have produced so much fine material. The slight traces of metamorphism which some of the sandstone particles exhibit doubtless were caused by increased heat due to friction along some of the zones of movement. It appears probable that about 500 feet of the Supai were involved in the explosion, and as little of this material appears in the detritus there is a strong suggestion that steam and hot water may have bleached it so that it is now represented by part of the white rock flour.

Darton did not include the Canyon Diablo irons or the iron shale in his analysis, noting only that the "occurrence of a few tons of meteoric iron in the vicinity and mingled with some of the debris on the rim is an enigma."

Barringer's reaction to Darton's *Bulletin* was one of quick anger followed by slow frustration. Hospitalized at the time while his physicians considered surgery for some unidentified ailment, Barringer demanded bitterly of Branner: "Why is Gilbert not man enough to state that he was mistaken?"[68] And a few weeks later, he inquired: "Can't you get Gilbert to retract? Darton agrees with him."[69] To Holsinger, he wrote asking when, if at all, Darton had visited the crater in the past six years and directing him to "make an affadavit that every statement of fact in my last paper is correct."[70] Two months later he wrote Darton, sending him a copy of his National Academy paper, criticizing Darton's conclusions in general and singling out his reference to the Supai sandstone in particular. "You haven't studied it [the crater] on the ground and so are in error," he declared. "About 500 feet of the red sandstone was *not* involved in an explosion or was bleached [Barringer's emphasis]."[71] Darton answered tersely that "I have read *all* the previous literature and never could see any substantiation for the meteor theory. Moreover, I do *not* share the belief that the meteor was volatilized [Darton's emphases]."⁷²

Darton's Gilbertian position was reinforced in papers presented by geologist Charles R. Keyes before the Geological Society of America and the Iowa Academy of Science.⁷³ "Contrary to recently expressed views regarding this remarkable crater," Keyes declared, "the most critical evidences seem to indicate that this feature of the local landscape is only one of many manifestations of the explosive type of volcanism so prevalent through the region." But Keyes, unlike Darton, did not consider the presence of meteoritic irons at Coon Mountain an enigma. Rather he suggested that they had fallen coincidentally and could be accounted for simply by the fact that meteorites in general were better preserved and more readily discovered in the dry climate and barren terrain of desert areas.

To Barringer's exasperation, Darton ignored his subsequent letters, and more than a year later Barringer complained to Thomson that "I have written to Mr. Darton asking him to kindly refer to his notes and let me know on what dates he visited the crater. He has not seen fit to reply."⁷⁴ Early in 1914, however, Barringer encountered Darton by chance on the Princeton campus and challenged him directly on the point. "Mr. Darton told me," he reported later, "that the matter had been put up to him (from his manner you might have thought he was the supreme arbiter to whom all such questions are referred) and he stated that in his opinion there was no iron in the bottom of the crater and that Gilbert was right. . . . I tried to pin him down as to whether he had visited the crater since our exploration work had been done. He said he had, but refused to give me the date and said, when I pressed him for an answer, that he had not let anybody know that he was there. It was not possible for him to examine the crater in detail without our men seeing him."⁷⁵

Henceforth, however, Darton was somewhat more equivocal in stating his conclusions, continuing to opt for Gilbert's hypothesis, but stressing that the question of the crater's origin was not resolved. In 1915, in another Survey *Bulletin,* he declared that the crater was "perhaps the most mysterious geologic feature in the West," Although he again looked with favor on the steam explosion theory. But, he was now careful to note that "Such an explosion might not greatly disturb the underlying Supai sandstone if the zone of explosion were in the overlying Coconino sandstone, which is the more porous material."⁷⁶ The following year, Darton expounded on "Explosion Craters" in general in an article in *The Scientific Monthly* "published by permission of the Director" of the survey.⁷⁷ The origin of the Arizona crater, he again declared, "has not been ascertained and it remains one of the greatest enigmas in nature," adding that the meteoric theory "has not yet been substantiated by borings, shaft and the survey of magnetic declination." But, after citing volcanic maars in various parts of the world, the Krakatoa and Kobandai volcanic eruptions of 1883 and 1888, and even the craters formed by exploding land mines in the war then raging in Europe, he still found it "interesting to compare the crater with the illustrations of explosion craters given above, and the close similarity of many features is very suggestive in connection with Gilbert's original suggestion that Crater Mound was caused by a steam explosion."

The name "Crater Mound," parenthetically, is Darton's particular contribution to the Coon Mountain controversies, and indeed, largely through his efforts, the name was formally adopted in 1932.[78] Darton, in his 1910 paper, had agreed that the name "Coon Butte" was meaningless, but had rejected "Meteor Crater," despite its increasing popularity. As late as 1945, Darton still objected strenuously to the name, arguing that this "wonderful rock outburst" was not meteoritic in origin and thus that "the words 'Meteor Crater' are not admissible."[79]

During these years, too, Barringer actively sought to persuade prominent figures in and out of science, individually and collectively, to visit the crater and judge the phenomena there for themselves. These persistent efforts, however, met with little success. In 1910 he invited astronomers attending the fourth conference of the Inter-National Union for Cooperation in Solar Research to stop at the crater on their way by train to or from the conference in Pasadena, California.[80] But while the astronomers were not averse to some sightseeing—they did visit Lowell Observatory in nearby Flagstaff and the Grand Canyon—they passed the crater by.[81] Barringer then sent copies of his National Academy paper to the conference's host, astronomer George Ellery Hale, for distribution to the conferees, but they did not arrive until after the end of the conference. Hale, Barringer later complained, never acknowledged their receipt.[82]

Barringer was no more successful some months later in persuading Theodore Roosevelt to visit the crater after learning that the former president was planning a trip west that included a stop at the Grand Canyon. In a long letter to Roosevelt, he described the crater in glowing terms, touched on his hopes and frustrations, and pointed out that Roosevelt's visit would "undoubtedly attract more attention to the crater than has hitherto been attracted to it."[83] He also noted that the proper exploration of the crater would cost another $200,000 and asked Roosevelt's help in interesting the Carnegie Institution "or some such scientific body of men to assist me in this exploration work." He was, he added, willing to surrender forty odd per cent interest in the property which "may contain the wealth of Monte Cristo." The "body or group of bodies" that had formed the crater, he explained, was "composed almost entirely of nickeliferous iron carrying platinum and abundantly studded with minute diamonds" and "Professor Russell, the astronomer of Princeton, and others have estimated that 'the order of magnitude' called for by the effects produced is probably not less than one million and possibly several million tons weight."

Perhaps it was impolitic of him, but Barringer also complained to the ex-president about the U.S. Geological Survey in general and Darton in particular. "I have proved the origin of the crater beyond the possibility of doubt," he declared. "Now the United States Geological Survey with strange persistence refuses to eat the dish of crow which we have prepared for them. They of course cannot deny the statement of facts contained in my paper. It is a surprising thing, however, that they will not send anyone out there to see whether I am a liar or not . . . . [I]n a late Bulletin a prominent member of the Survey states that he believes the crater to be due to a steam explosion (following Gilbert's old theory proposed in 1892 [sic]), although he has no personal knowledge of the crater other than that obtained from a visit several years ago and before any of our proofs were published. . . ."

Roosevelt politely but firmly declined Barringer's invitation. "I wish I could do as you suggest for I do not know of a more interesting place to visit, except the Grand Canyon. But . . . I simply cannot stop; and I am really very sorry. . . . Your letter is absorbingly interesting."[84] Barringer expressed his regret, but reiterated his hope that "you may after all speak to Mr. Carnegie," and now offered a specific 45 percent interest in the crater property.[85] But Roosevelt was not encouraging even on this point. "When I see you," he replied, "I will explain to you more in detail not merely the difficulty but the impossibility of my taking effective interest in these things."[86]

DOUBTING THOMASES

In 1912, however, Barringer did succeed in persuading Harvard geologist William Morris Davis to include the crater on the itinerary of a transcontinental excursion by some ninety geographers and geologists, many of them distinguished European scientists, being sponsored by the American Geographical Society. Unfortunately, the occasion proved to be a promotional fiasco from Barringer's point of view. Moreover, it marked the return to the crater of Survey member Willard D. Johnson, who had first suggested the volcanic steam explosion hypothesis to Gilbert in 1891 and who, as it turned out, had not since changed his mind.

Barringer began negotiations with Davis early in August, and it was soon agreed that the party would stop at the crater on October 1 on its way by train from the Petrified Forest to the Grand Canyon. From the outset, Barringer stressed to Davis the importance of having the party arrive early enough in the day to allow the scientists sufficient time to thoroughly inspect the crater.[87] But when the final timetable for the trip was set, it was clear that they would have less than two hours of daylight in which to conduct their investigations. "I am disgusted," he fumed to Thomson, "to know that they will not reach the crater until about 4:30 in the afternoon and will leave the same evening for the Grand Canyon. Naturally they will not be able to see very much and a great many may carry away the impression that we have been trying to strain the facts to fit our view as to the theory of its origin."[88] It was too late to change the schedule, he realized, but he nonetheless conveyed his disappointment to Davis in typically irreverent terms. "All I ask," he wrote, "is that the doubting Thomases read my paper in advance, pick out something which they wish to verify and verify it promptly within the time at their disposal."[89] Later he would apologize for the tone of this correspondence, pleading his deep concern over the short time allotted to the visit. "Had the explanation of the crater been obvious," he explained to Davis, "Gilbert and the others would have easily read the riddle many years ago."[90]

Barringer himself served as host at the crater; he had hoped that Thomson could join him, but Thomson found it impossible to make the trip.[91] The occasion also was apparently convivial, for in addition to the intellectual fare of the crater itself, Barringer provided his guests with a sandwich supper, catered by the Harvey House in nearby Winslow, along with forty-eight quarts of beer, forty-eight quarts of mineral water, and twelve bottles of California wine to supplement the "whiskey on the train."[92]

The crater proved to be the high point of the entire western excursion, according to an article in *The New York Times* a week after the event. The scientists, the paper reported, "managed to make at least one discovery of probably great importance in regard to the famous Meteor Crater in Northern Central Arizona, the origin of which has been a matter of much dispute and mystery to scientists. Instead of being the result of a meteoric explosion, several of the scientists concluded, on careful investigation, that the crater was probably the result of a steam explosion or a gigantic geyser."[93] The anonymous *Times* correspondent cited three excursionists, including Johnson, in support of this statement:

> Professor Eugene De Cholnoky of . . . Hungary, who has individual ideas on most subjects, started off on an independent investigation as soon as his arrival on the ground and found a bed of travertine on the western

side of the rim and other material which is associated with deposits made by hot springs. Judging by this and also by the position of the beds of rock, both those which had been uplifted and those which composed the top of the plain outside, he came to the conclusion that the pit had been made by a steam explosion or a geyser on a gigantic scale. . . . After returning to the train, he was greatly pleased to discover that W. D. Johnson of the United States Geological Survey had also found travertine on the eastern side of the ridge and had very much the same opinion as to the cause of the formation of the pit. Mr. Johnson believes that water in the interstices of the underlying rock had been heated to the boiling point by volcanic action still further down, until the accumulated steam burst out in the grand explosion which made the crater.

Professor Eduard Brückner of the University of Vienna also shared this view and compared the crater to "an enormous pit or caldera at Ries, in Wurtemburg, in the Swabian Jura" that was some twelve and a half miles in diameter, and 1,000 to 1,200 feet deep.* In Brückner's opinion, the *Times* reported, the large, dislodged blocks and boulders of granite and gneiss that surround this feature "indicate pretty clearly that the force causing their displacement must have come from below and was probably due to a geyser-like steam explosion."

In the *Times* account, only an otherwise unidentified "Prof. Chalx" questioned this interpretation, pointing out that the rocks at the crater should show more signs of heat, and that traces of any volcanic action should be visible. He was thus, the paper noted, "inclined to the meteoric theory."

In summing up, however, the *Times* correspondent declared that "No definite conclusion regarding this most interesting phenomenon could be arrived at, and it seems that the only way to settle the question would be to have a large number of deep borings carefully made . . . an investigation which the scientists hope will soon be made in the interests of science." Barringer could agree, at least, on this last point.

Barringer was "furious" over the *Times* article and promptly registered his bitter disappointment over the visit itself in letters to Merrill, Davis, and Thomson.[94] To Thomson, he noted the skepticism of some of the scientists, adding that when one had claimed to be "from Missouri," he had told him "to

---

*The Rieskessel, in southern Germany. Interestingly, a meteoritic impact origin for this large crateriform feature was suggested as early as 1904 and in recent years has become widely accepted. See E. M. Shoemaker and E. C. T. Chao, "New Evidence for the Impact Origin of the Ries Basin, Bavaria, Germany," *Journal of Geophysical Research,* 66(1961): 3371–78.

go back there! . . . The behavior of the U.S. Geological Survey men who accompanied the party was anything but polite," he fumed, "and they even went so far as to express doubts, in my hearing and while they were my guests, as to the reliability of the statements my paper contains. One foreigner said openly that he believed that the shale ball irons . . . were planted and the offer I made to him to pay a thousand dollars apiece for any similar specimens collected from any other part of the world did not seem to have any effect on his biased brain."[95]

To Davis and to Johnson, Barringer was somewhat more restrained, pointing out that the "travertine" that Cholnoky and Johnson claimed to have found was actually Variety B. To Johnson, he explained that "this material is simply metamorphosed sandstone" and cited the opinions of Mallet, Merrill, Penfield, and Branner on the point. "I hope you can visit the crater again," he added. "I am sure that you will realize that I have been guilty of no serious misstatements or overstatements in my National Academy paper."[96] He also sent a copy of this letter to Cholnoky in Budapest, advising the Hungarian scientist that "I very cheerfully challenge any geologist or mineralogist to prove that I am wrong in my description of any of the rocks."[97]

All his remonstrances, however, were futile, although Johnson subsequently claimed that Cholnoky had misquoted him about the travertine to the *Times* correspondent.[98] Certainly, his experiences with the junketing geographers and geologists did nothing to improve Barringer's opinion of scientists in general.

THOMSON'S SUPPORT

But despite these frustrations, Barringer could also count some gains. In the early summer of 1911, he traveled to England where, he found, scientists at the British Museum and the British Geological Survey "made quite a lion out of me."[99] The English scientists— "broader gauged, bigger men"—agreed that "the mass lies in the bottom of the crater and is, in all probability, a cluster," he reported.[100]

While Barringer was in London, incidentally, Holsinger discovered the largest Canyon Diablo iron ever found in the vicinity of the crater. Weighing between 1,400 and 1,600 pounds, the siderite had been partly buried some one-and-a-half miles northeast of the crater.[101] It was soon put on display in a small stone museum that Holsinger was still building when he died early in August. Barringer had also directed Holsinger just before his death to "build a monument on the exact spot" where the siderite was found and later proposed to Holsinger's widow that "the big iron always be known by Holsinger's name."[102]

Early in 1912, moreover, the widely known and highly respected Thomson not only publicly declared his support for Barringer's ideas regarding the crater, but did so in terms that surely exceeded even Barringer's most sanguine hopes. Their correspondence begins early in March 1911, when the scientist-industrialist wrote Barringer of his intention to stop at the crater in the course of a trip to the west coast and suggested that Barringer might want to be there at the time. Barringer had other commitments,[103] but he directed Holsinger at the crater to "Lay yourself out" as Thomson was not only "an astronomer of no mean repute" but "very wealthy."[104] Thomson visited the crater in April and returned as an enthusiastic proponent of Barringer's ideas. "Only Thomson," Barringer soon confided to Bennitt, "recognizes that there may be enormous wealth in the crater."[105] The two men quickly formed a friendship that endured until Barringer's death, and henceforth Thomson, with Magie, became Barringer's close confidant in all matters relating to the crater. Eventually, like Magie, he would back up his conviction of the crater's impact origin and the presence of the meteorite in its depths by investing in the final, unsuccessful search for the buried mass.

Over the ensuing years, Barringer frequently relied on Thomson to reinforce his own views in his dealings with "doubting Thomases" in both science and commerce, and Thomson readily obliged by writing Barringer quotable, confirmatory letters which Barringer often used to his purpose. But Thomson's primary public contribution to the Coon Mountain controversies came early with the publication of two lengthy papers, one in March 1912 in the *Proceedings of the American Academy of Arts and Sciences*,[106] and the other, in a more popular vein, in May in the *Scientific American Supplement*.[107] In these, citing some of Barringer's now familiar points along with some of his own, Thomson not only argued strongly for the impact origin of the crater and the existence of a meteoric mass beneath it, but in the latter paper suggested that the commercial value of the metals that lay buried there might amount to hundreds of millions of dollars.

He had visited the crater the previous spring, he noted in his American Academy paper, because "In view of the marked divergences of opinion concerning origin, it seemed desirable to determine, at least to one's own satisfaction, which of the explanations was in accordance with the facts." Attempts to explain the crater either by volcanic action or by a steam explosion he quickly rejected as "rather far fetched in view of the presence and mode of occurrence of the meteoric irons, and the many evidences presented which seem to me to lead inevitably to the conclusion that here indeed we have a huge impact crater and that only." Moreover, from "many considerations which it would take too long to enumerate it seems probable that if a large mass or cluster has buried itself at this spot, it now exists under the southern and southwestern wall." To Darton's argument that borings had not revealed such a mass, he answered that Darton "forgets that only 28 test holes were put down and all near the center of the crater. If a single large mass of 500 feet in diameter fell, to be sure of finding it would demand the sinking of not less than 500 or 600 holes if the whole crater is to be explored." Finally, he declared, "in the case of a large iron meteor there is no such thing possible as an instantaneous vaporization of the whole mass."

To support these conclusions Thomson, after briefly describing the crater, outlined his view of what had happened at Coon Mountain:

> A large mass, or more probably a cluster resembling a small comet, passed so near the earth as to be deflected sufficiently to cause it to strike.
>
> Its velocity, reduced by retardation in the air, may not have been more than two or three miles per second. . . . As it penetrated the dry plain some of the smaller and still more retarded and slower moving pieces following in the rear of the main cluster were diverged therefrom and so fell at various distances around the crater. As soon as the main body reached the wet rock layers superheated steam was generated at high pressure, pulverizing and metamorphosing the siliceous rock and producing an enormous lateral pressure which disturbed and uplifted strata altogether. . . . As the mass advanced the pulverized material in its path and around it would be swept backward almost as a fluid and would expand outwardly at the same time, fracturing the adjacent rock layers, upcasting and folding back the strata, and discharging the enormous mass of pulverized and mixed rock material which now forms the circular rim.

Thomson also described "what must happen during the flight of a meteoric mass through the air," pointing out that smaller bodies, whether stony or metallic, would be consumed by the heat and resulting oxidation or burning caused by atmospheric friction. "If, however, the iron mass or fragment is large, and its velocity insufficient for crushing or further breaking, it may not be entirely burned during its flight and . . . may bury itself in the earth to a depth more or less great." Such a meteorite, he added, would not become heated to a high temperature throughout its mass, and thus would not volatilize on impact, citing the fact that meteorites seen to fall, when found immediately thereafter, proved to be "stone cold or only slightly warmed. . . . It can possess only a thin skin of hot material, a mere film in which the temperature gradient is very steep or abrupt, and this film is constantly blown off or removed as soon as the temperature of fusion is reached." Here he drew an analogy to a block of ice subjected to "a vigorous blast of heated air. . . . The ice melts rapidly and the water formed is blown off as fast as it appears, while what remains is none the less ice to the end of the process."

The indications, then, were "that a body which has so survived will neither be moving at an excessive velocity as compared with that of a projectile from a high-powered gun, nor be a hot body on striking. . . . As such bodies rarely descend vertically, they must pass through many miles of comparatively dense air when their course is more or less horizontal or inclined to the vertical. This and their irregular form renders possible a very great retardation." Thomson added that he was stressing these points "as they entirely negative an idea which has been advanced to account for the failure to find large masses of embedded iron at the Arizona crater. It has been claimed that the mass was vaporized on striking, a possibility which is *impossible,* as I am compelled to regard the matter [Thomson's emphasis]."

In his first paper Thomson only touched on the commercial possibilities at the crater. In his *Scientific American Supplement* article, however, after covering much the same ground again, he dwelt at greater length on these aspects. The proper exploration of the crater would indeed be expensive, he pointed out, for at $2,000 per hole, the cost of the 600 holes needed to find the mass would total $1,200,000.

> But if we assume that one ton of material in the meteor was capable of displacing, say, some thirty tons of rock when it struck, then the mass of the meteor should have been approximately at a low estimate, say, five million tons, mostly iron. Eight per cent, however, would be nickel, and in each ton, by analysis, the average amount of about 0.6 of an ounce of platinum and iridium exists. This would give about three million ounces of platino-iridium, say; which at a valuation, say, of $30 to $35 an ounce, would equal about $100,000,000. If we assume, and of course it is a mere assumption, that one-hundredth of one per cent of diamond exists in the mass, then the 5,000,000 tons would contain about 500 tons of diamond.

There was thus, he concluded, "indeed enough prospect of great value, even in case these figures should be largely exaggerated; and it is no wonder, therefore, that explorations were undertaken for the purpose of locating this great body."

SEVEN

COON MOUNTAIN AND THE MOON

> In fact, the agreement in the appearance of the Meteor Crater of Canyon Diablo and that of a lunar crater is so perfect that through the proof of the meteoric origin of Meteor Crater, the origin of the lunar craters through meteoric impact gains extraordinarily in probability.
>
> FRANZ MEINEKE, 1909

That Coon Mountain's crater and the craters on the moon might have a common origin, of course, had occurred to Grove Karl Gilbert in August 1891 as he sat listening to Foote describe the Arizona crater and report his discovery of diamond-bearing meteoritic irons in its vicinity.[1] But Gilbert quickly found the analogy invalid, for although he boldly posited an impact origin for the lunar craters, he also concluded that Coon Mountain's crater was volcanic. Those who held the prevailing volcanic analogy for lunar craters, however, now found the comparison to their liking. Soon a generation of geologists would learn from Chamberlin and Salisbury's widely used textbook, *Geology*, that the lunar craters were analogous to terrestrial volcanic maars and, in a specific reference to Gilbert's conclusion, to Coon Mountain itself.[2]

The volcanic analogy also continued to prevail among astronomers, particularly among those few, mostly in England, who still gave serious attention to observations of the moon. In 1910, for example, members of the British Astronomical Association, including Walter Goodacre, then and for many years president of its Lunar Section, reaffirmed it in discussions precipitated by astronomer Thomas Jefferson Jackson See's postulate of an impact origin for the lunar craters.[3] See's prescient ideas will be discussed in more detail later; here it is only necessary to note that British selenographers agreed with Goodacre that their effect "has been not to shake my faith in the volcanic theory, but rather to confirm it as being the most probable cause of the origin of the lunar surface formations."[4]

The German geologist Franz Meineke in 1909 seems to have been the first to argue in print for an impact origin for the lunar craters from the evidence adduced for Meteor Crater's origin by Barringer, Tilghman, and Merrill.[5] Most astronomers and geologists, Meineke pointed out, "assume volcanic processes for the origin of the lunar craters wherein they use to some extent the great volcanoes of Hawaii, to some extent the caldera mountain, and to some extent volcanoes of the Somma type, such as Vesuvius." Yet there were "fundamental differences . . . in both the shape and size" between these terrestrial features and the craters on the moon, and thus "the geologist must decisively reject the assumption of a volcanic origin for the lunar craters. . . ." For Meineke, Barringer, Tilghman, and Merrill had demonstrated the impact origin of Meteor Crater; and, because the physical resemblance of its crater and the craters on the moon appeared "perfect" (*vollkommene*) in Meineke's view, the extrapolation from the one terrestrial crater to the thousands of lunar craters seemed eminently logical. That there were "not more Meteor Crater-like features on the earth," he added, was because "the earth has an atmosphere and a hydrosphere which provide protection which the moon lacks."

The year 1909 also marks the first time that the meteoritic impact origin of Coon Mountain was used to argue the possible impact origin of another terrestrial structure. A South African geologist, E. H. L. Schwarz, cited Barringer's, Tilghman's, and Merrill's papers to suggest that the occurrence of amygdaloidal lavas* in the Prieska district of South Africa's Cape Colony might better be explained by the crushing and melting of rock by the impact of a large meteorite than by upwelling and overflowing of molten material from volcanic pipes or chimneys.[6] There was "no reasonable doubt" in Schwarz's mind that Meteor Crater "is actually the result of the impact of a huge bolide and the absence of the large meteor itself is explained by supposing the heat of impact was sufficient to melt and perhaps vaporise its substance." Moreover, the Prieska lavas "cannot be explained on any of the theories of igneous extrusion" for the "field relationships are not compatible with the supposition that the origin . . . was volcanic. . . ." But "The sudden development of a large amount of liquid lava, the permeation throughout of steam holes from water contained in the rocks melted, the ridging up of portions of the periphery of

---

*Porous basalts containing mineralized nodules and geodes that are often almond-shaped, hence "amygdaloidal."

the mass, the absence of true volcanic ash, but the great development of crushed up material entangled in the molten rock—all these phenomena receive an adequate explanation on the meteor hypothesis."

Schwarz's conclusion, however, was overly sanguine, for the "meteor hypothesis" has not been proven adequate in the light of later work, and geologists today still consider the Prieska lavas to be the result of subterranean rather than extraterrestrial forces.[7] Interestingly, Schwarz could not carry his extrapolation beyond the earth; and, although he noted both Gilbert's and Shaler's studies of the moon's surface, he rejected an impact origin for the lunar craters, opting for the volcanic analogy and citing Chamberlin to the effect that the lunar craters "resemble rather outbursts of entangled molten matter during the final consolidation of the moon."

Elihu Thomson, who counted astronomy among his many scientific interests, also was one of the first to draw a direct analogy between Meteor Crater and the craters on the moon; and he, unlike Meineke, reasoned from firsthand observations. "I have often examined the moon through a large telescope," he declared in 1912, "and thought that the lunar craters have the appearance as if something struck there; but when I came to see this meteor crater in Arizona, I felt sure that here we had a lunar crater on the earth."[8]

Thomson, indeed, spelled out the analogy in considerable detail in both his 1912 papers on Meteor Crater, stressing in addition its pertinence to accretion theories of the origin and evolution of the solar system in general, and to the Chamberlin-Moulton planetesimal hypothesis in particular. "Unlike the earth's great volcanoes," he pointed out before the American Academy of Arts and Sciences, "the floor of the lunar craters is in nearly all cases considerably below the general level of the surrounding surface. The Meteor Crater . . . has this same characteristic and in fact if looked down upon from above would appear like a lunar crater transplanted to earth. . . . The general agreement is, to say the least, highly suggestive of a similarity of origin."[9]

THE BIG SPLASH THEORY

The lunar craters, moreover, "do not show great rivers of lava coursing down the slopes and covering the lower levels," and the whitish streaks or rays emanating from such craters as Tycho and Copernicus "bear far more the appearance of instantaneous scattering of material in a vacuum" and "suggest a sudden splash." The central peaks or hills in some of the lunar craters also seemed "more compatible with the idea of a reaction, or 'kick back,' an inrush similar to that which occurs on dropping a stone into a pool of water."

As to the source of the impacting bodies, Thomson pointed out that if the moon had been formed by violent separation from a rapidly rotating earth, as Sir George H. Darwin had conjectured,[10] then the lunar craters might be the scars of smaller bits of matter torn from the earth in the course of the larger cataclysm that later fell back on the receding moon.[11] Or: "If we go farther back in time and think of the possible building up of the sun and the planets of our system by gradual accretions such as are now taking place when a meteor enters our air, but which at sometime, as many millions of years ago, occurred on a grander scale, we can understand the craters on the moon as the last record of such a process, and the Meteor Crater in Arizona as the largest record on earth, all others belonging to earlier geological periods having been obliterated." The accreting fragments of matter in this case might have been hurled into space by forces resulting from "the more or less close passage of two large orbs," a possibility which, he added, "involves a modification of the time-honored nebular hypothesis of La Place and is in line with the ideas of Chamberlin and Moulton in their planetesimal hypothesis."

It was Thomson, in fact, who brought the analogy forcefully to Barringer's mind. Barringer, preoccupied with mundane matters and not especially interested in astronomy or the moon in any case, had not seriously considered the similarities between Meteor Crater and the lunar craters until Thomson detailed them to him following his visit to the Arizona crater in the spring of 1911. Indeed, Barringer was not even aware that Gilbert had postulated an impact origin for the lunar craters until Thomson called his attention to Gilbert's 1893 "moonlet" paper.[12] But Barringer now enthusiastically endorsed the analogy, praising Thomson as "the first to apply the lessons" of Meteor Crater "to the making of worlds and of satellites,"[13] and quickly agreeing that the lunar features were "doubtless impact craters."[14]

Almost immediately, too, he extended the basis for the analogy, seizing on Thomson's reference to the central peaks of the lunar craters and his remark that they were suggestive of a "sudden splash." For Meteor Crater, Barringer suddenly realized, also had a central peak of sorts in the form of a low mound of silica sand rising only a few feet above the otherwise level floor in the northern part of the crater which neither he nor Tilghman had been able to explain and which they had called "silica hill." Both the lunar central peaks and "silica hill," he now concluded, could be accounted for by "the differential settling of dislodged material and fallen back ejecta" after a violent impact.[15]

Not until October 1913 did Barringer have an opportunity to observe the moon and check the analogy with his own eyes. Then he spent an evening with astronomer John A. Miller at Swarthmore's Sproul Observatory "critically examining the lunar craters through the twenty-six-inch [sic] telescope," and concluding that "there is not the slightest doubt in my mind that they were formed in much the same way as our meteor crater,"[16] and that "the central hill or mountain in some cases after all is said and done has its counterpart in our silica hill, the origin of which puzzled Tilghman and me for so many years."[17]

Two months later, his conclusions were reinforced by a book, *A Study of Splashes,* by an English engineer, A. M. Worthington, in which Worthington provided a detailed analysis, illustrated with high-speed photographs and diagrams, of the dynamics of the splashing process for liquid, plastic, and solid materials.[18] Barringer immediately sent Worthington copies of his Meteor Crater papers, declaring that "I am convinced that the most splendid illustration of your admirably thought-out conclusion, that solids under pressure will behave in much the same way as liquids, is to be found in the lunar craters. These are surely impact craters."[19]

Worthington, however, did not agree, arguing for the volcanic analogy and citing the eminent British geologist Sir Archibald Geikie among others as his authority. Barringer now enlisted the aid of both Thomson and Magie in an effort to change Worthington's view, explaining to Magie that Worthington's rejection of an impact origin for the lunar craters was "the result of an overcautious mind," and adding that "Gilbert, in rejecting the impact theory for the origin of our crater, suffered from the same excess of caution. . . ."[20]

Barringer, however, soon found that Worthington's ideas regarding the moon were quite bizarre. "When you urge that you 'utterly fail to find extending from the lunar craters the lava flows which are so characteristic in nearly all terrestrial volcanoes,'" the Englishman had written him, "I find myself asking 'how do you expect to recognize by means of the telescope lava after it has perhaps been weathered, broken up by the roots of vegetation and converted superficially to soil?'"[21] This question seems to have convinced Barringer that further argument would be futile, and he soon terminated the correspondence, commenting dryly to Thomson that "I had not known before that vegetation was possible on the moon."[22]

Nevertheless, Barringer had found Worthington's analysis of splashes useful and had cited it to his purpose in his third paper on Meteor Crater presented to the Academy of Natural Sciences of Philadelphia in July 1914.[23] The lunar analogy and his direct comparison of "silica hill" to the central peaks of lunar craters, however, were not the only arguments he advanced in this paper which, in effect, was a potpourri of arguments old and new. His most telling point now concerned Variety B of the metamorphosed sandstone and the fact that the temperature required to fuse silica was higher than the temperatures produced by volcanic processes. He had discussed the point earlier in his cor-

COON MOUNTAIN AND THE MOON   159

respondence, pointing out to Merrill, for example, that "if my recollection is correct, none of us used this as a knock-down argument—for such it is—in favor of the impact theory rather than the volcanic theory of the origin of this crater."[24]

In his paper, he developed the point at some length. "I am assured by Dr. Merrill and others," he declared, "that there is no record of a sudden outburst of volcanic action wherein the heat generated was sufficient to fuse crystalline quartz. The only case of quartz being fused by a sudden rise in temperature . . . is that of the more or less familiar action of lightning striking sandstone or sand and altering it to what is known as fulgurite glass. No volcanic action, however violent or however long continued, has been known to produce such an effect."[25]

In presenting this argument, Barringer could not resist issuing another challenge to Gilbert, Darton, and the U.S. Geological Survey, declaring that Variety B of the metamorphosed sandstone furnished "incontrovertible proof" of the impact theory.

> It will be in the interest of science if scientific men, and especially those of the United States Geological Survey who deny this theory of origin, will present their reasons for maintaining the hypothesis that the crater was due to some manifestation of volcanic activity. I believe that it will be easy to refute any argument they may advance. No examination of the crater since the exploratory work was done has been made by members of the Survey, to the best of my knowledge and belief. Therefore, unless they can satisfactorily account for the facts which I have stated in this and in my previous papers on the subject on some other theory than that of impact by a great mass of meteoric iron, it would seem that I can fairly claim to have proved the theory that the crater was formed by this agency.

Barringer also contended that the paucity of evidence of extensive volatilization at the crater, and the fact that the projectile had penetrated nearly 1,200 feet of rock, indicated that the "meteoric mass . . . was not moving at very high speed." This mass, which he now thought might have been "as much as 10,000,000 tons," was probably buried "under the southern wall, some *2,000 feet* from where the drilling was done [Barringer's emphasis]," and here he reviewed the evidence in his 1909 paper for a roughly north-south bilateral symmetry at the crater.

As to the analogy between Coon Mountain and the moon, Barringer declared that when "one who is familiar with the Arizona crater examines the lunar craters through a good telescope they are at once seen to show the main features of the former. . . . Even the peculiar conical central hill or mountain which is observed in most of them and which I confidently assert cannot be explained on any theory of volcanic action, has its counterpart in our own Silica Hill at Meteor Crater."

In expanding on the point, Barringer relied on both Thomson and Worthington. Central hills in craters, he suggested, "would seem to be due to the same physical law which we see in operation when we drop a stone into water or soft mud, with which solid rock can be compared if the projectile strikes it with sufficient speed." Here he declared that Worthington's conclusions regarding the behavior and flow of solid substances under "great pressure suddenly applied" seemed to be "fully warranted and also seem to go far toward explaining the presence of conical-shaped hills in nearly all of the lunar craters. Anyone," he added, "who will make a careful study of our Arizona crater and will then read Worthington's book, studying the diagrams he has made, and will then turn his attention to the lunar craters, cannot escape the conviction that the lunar craters are impact craters."

Barringer considered this paper "largely what in sporting parlance might be termed my 'defi' to the U.S.G.S.," as he later confided to George F. Kunz, president of the New York Academy of Sciences. "Either I am an unconscionable and most skillful liar or the case is as definitely proved now as it ever will be. . . ."[26] But the paper evoked no reaction from the Survey itself, nor did Gilbert or Darton respond to his new challenge. A few months after the paper was published, indeed, Barringer reported ruefully to Bennitt that "Mr. Darton of the U.S. Geological Survey still maintains, so Dr. Merrill writes me, that there are other localities which he considers to have been volcanic and which are comparable with Meteor Crater. Without having seen them, I would be willing to bet him $1,000 to his $10 . . . because I feel the need of the $10 very much these days. . . ."[27]

Later this same year, in his published speculations on the nature of comets and the derivation of the rounded shale balls at Coon Mountain,[28] Barringer again followed Thomson's lead by pointing out the pertinence of impact to Chamberlin and Moulton's planetesimal hypothesis. "If the lunar craters and the Arizona crater have had a common origin, as now seems probable, there can be no doubt that our knowledge of cosmogony has been greatly advanced by the discovery of the origin of the Arizona crater, and that there is much stronger reason now than ever before to believe in the general correctness of Chamberlin's and Moulton's theory of the building up of planetary systems," he noted. "Our moon . . . shows evidence of some of the more recent accretions to its mass. . . . Its numerous craters probably merely represent the gathering in of cometary bodies or clusters of meteorites, for they are apparently exactly similar to our Arizona crater, except that most of them are vastly larger."

By 1916, however, ten years into the Coon Mountain controversies, most scientists still held reservations about the impact origin of Meteor Crater, let alone of the craters on the moon.[29] The basis of this skepticism, certainly, was the plausible enough assumption, made by Gilbert even before he had seen the crater, that if a large iron meteorite had plunged into the earth at Coon Mountain, its mass must still lie somewhere beneath the crater's floor. The fact that no such large mass had been located either by the dip needle or the drill had been used by Darton and others as an argument for rejecting the impact theory. And, of course, even those who, like Barringer, accepted the theory took for granted the presence of such a mass; only Merrill had seriously questioned the assumption and had suggested an alternative.

The point was raised again early in 1916 in a belated and not too well-informed review of Barringer's 1909 National Academy paper that appeared in the British journal *Nature*.[30] "It must be confessed, however, that many unsolved difficulties remain to prevent our unhesitating acceptance of the meteoric theory," the anonymous reviewer wrote. "Chief among these is the question of what has become of the vast mass of matter capable of producing the shattering impact. Only scattered fragments of nickel-iron have been detected at the depths reached by the borings, and the existence of a vast mass of meteoric iron at greater depths finds no confirmation from the magnetic observations carried on in and around the 'butte.'"

Barringer certainly was aware of the *Nature* article, for he had supplied illustrations for it.[31] But apparently he did not record his reaction to it in his correspondence. Surely he could not have appreciated it and in fact, he was disputing these same points at this time with Farrington, the Field Museum's authority on meteorites.

Barringer had renewed his correspondence with the geologist in 1915 after Farrington's purely descriptive "Catalogue of the Meteorites of North America" had appeared as a National Academy of Sciences *Memoirs*.[32] Farrington had devoted eight pages in the catalogue to the Canyon Diablo irons and associated materials, citing Barringer's and Tilghman's 1906 papers without comment. Barringer quickly congratulated Farrington but chided him for not accepting the impact theory presented in these papers, noting that additional evidence for the theory had since been found and hinting not too subtly that he, Barringer, at least, was willing to change his mind in the face of new facts. To this latter point, he recalled a letter "a good many years ago indignantly affirming my disbelief in your theory that the so-called iron shale was due to oxidation of what we now know as 'shale ball' iron, and my belief, amounting in those days to almost conviction, that it was due to heat generated by the passage of the meteoric iron through the atmosphere. I now know differently and humbly apologize."[33] Farrington's reply was polite but brief. "I have never cherished any enmity because of the difference of opinion which you held at one time . . . but I am glad to know that you now agree with the opinion which I formed."[34]

Early in 1916, however, Farrington sent Barringer a copy of his newly published book, *Meteorites*, in which he not only described the crater and the meteoritic materials there but discussed the various theories of its origin.[35] And while he credited Barringer with the impact theory, noting that Gilbert had "ascribed the formation of the crater to a steam explosion of volcanic origin," he remained noncommittal, declaring, on the grounds that the drill had failed to find any large meteoritic mass and that magnetic tests had proven negative, that it was "impossible to give final decision as to the origin of the crater."

Barringer now objected somewhat more strenuously. Farrington, he complained to Merrill, "states that although a number of borings have been made in search of such a mass or masses, none has yet been found. Surely he has scientific mind enough to know that it does not make any difference whether the pieces which were found were large or small so long as they were masses of and about meteoric material. . . . the older I grow the more convinced I become that a man . . . has no right to write about things with which he has no personal knowledge."[36]

To Farrington himself, Barringer pointed out that meteoritic material had been found in many of the drill holes, and while Magie had indeed confirmed the negative results of the magnetic tests, "we incidentally found out *why* [Barringer's emphasis] there is no appreciable difference between the behavior of the dip needle in and about the crater." Since Gilbert's "superficial examination" of the crater, he added, "a great many new facts have been ascertained which can have but one explanation."[37]

A few days later, Barringer followed this up with a long letter setting forth his arguments for the crater's impact origin in great detail,[38] but Farrington responded only with some terse advice regarding the proper attitude that a scientist should maintain in such matters. "I have never failed to appreciate the important work you have done in investigating this remarkable occurrence at Meteor Crater," he replied. "Nevertheless I do not feel that I am entitled to censure for not absolutely accepting your theory of its origin. . . . To urge a particular theory is the work of the advocate, not the scientist."[39]

NEVER-ENDING SEARCH FOR FUNDS

Over the next few years, except for occasional references in his correspondence, Barringer would eschew speculations about comets and the origin of the lunar craters in favor of the more pressing problem of raising funds to finance a further search for the elusive meteoritic mass, a project which, he was confident, not only would produce fabulous profits, but would provide the irrefutable evidence of its impact origin that he needed to confound the skeptics in science. He had, in fact, somewhat neglected the crater itself for several years, concentrating most of his attention on his various other business enterprises, and had not even visited Coon Mountain since he had hosted the transcontinental tour of geographers and geologists there in October 1912.[40] In these years, routine matters affecting the crater were handled largely through correspondence with Bennitt in Phoenix. Only two developments in this period are perhaps of some interest here.

First, late in 1914, Barringer asked Bennitt to investigate the possibility of obtaining an option to purchase Holsinger's seventy-five non-assessable shares in Standard Iron from Holsinger's estate for $5,000.[41] Two months later, after negotiations between Bennitt and Holsinger's widow, a five-year option was agreed upon.[42] Just before this option was due to expire, in 1919, and just after Mrs. Holsinger threatened to sell the shares to local ranching interests, Barringer exercised it on Magie's behalf.[43]

Again, late in 1914, to Barringer's consternation, the assessed value on the crater property was abruptly increased nearly sevenfold, from $3,600 to $23,746, and the tax itself more than doubled, from $182 to $373. He immediately directed Bennitt to protest, advising him to argue that the crater had no value "except as a scientific curiosity."[44] Bennitt managed to get the tax reduced to $196, but Barringer himself now protested directly to the county treasurer at Flagstaff that the tax was "still too high,"[45] explaining to Bennitt that "I think we ought to keep constantly before the minds of these people that the crater has *no commercial value whatever* [Barringer's emphasis].... We should never let them lose sight of this fact!"[46]

Barringer's assessment here, of course, was solely for the benefit of the local tax authorities; to men of influence and affluence he told a different story. Shortly before the tax issue arose, he had complained to Thomson that "It certainly seems almost sinful to allow such a vast amount of wealth to lie undeveloped in the bottom of the crater."[47] And shortly thereafter, he explained to Daniel Baugh, a wealthy Philadelphia industrialist, "why this crater is probably one of the most valuable mines in the world," pointing to the nickel, platinum, and diamond in the iron there and suggesting that the gross value of the mass might be as high as $700,000,000.[48] This figure, too, had its purpose, for Barringer was trying to persuade Baugh to put $250,000 into the crater, $25,000 for "added drilling where Thomson, Magie, and I are agreed we are most likely to find the meteoric mass," and another $225,000 for mining and processing the mass once it was found. Baugh, however, was not persuaded. Nor was Barringer's friend, Charles A. Stone, of the prominent Boston engineering and management firm of Stone & Webster, whom he sought to interest in the crater a few months later. Stone, Barringer advised Thomson, "is too busy and we must look elsewhere."[49]

About this same time, Barringer also turned briefly to schemes for exploiting the millions of tons of pulverized silica in and around the crater in the hope of building up a fund to finance further drilling. Over the next few years, he carried on sporadic negotiations with still another wealthy Philadelphian, P. M. Sharples, who manufactured cream separators and milking devices and operated the White Heat Products Company, in which Barringer held a few shares of stock.[50] These discussions led, in September 1916, to an agreement to form a new company to manufacture pressed bricks in Los Angeles, California, from Meteor Crater silica sand.[51] Under this scheme, half of the new company's stock would go to whomever raised $100,000 to build the proposed brick-manufacturing plant, and the remaining shares would be divided equally between the White Heat Products Company for patents and "know-how" and the Standard Iron Company for the raw material. As a candidate for half-owner of the new company, Barringer approached one of Bennitt's affluent west coast friends, a wealthy Los Angeles businessman named Harry Gray.[52] Gray pleaded ill health and a desire to get out of business altogether, but he suggested that Barringer and Sharples should have little difficulty raising the necessary funds from others in the Los Angeles area. He pointed out, however, that a successful brick-manufacturing plant was already operating in Los Angeles and that the cost of making bricks from silica shipped from Meteor Crater, some five hundred miles to the east, might preclude marketing them at a competitive price.[53]

This is about as far as this particular scheme went. Ironically, perhaps, the November 1916 reelection of Barringer's friend and former Princeton classmate, Woodrow Wilson, as President of the United States seems to have provided the *coup de grâce*. Barringer had never been happy about Wilson's economic policies, which he felt discouraged enterprise and investment,[54] and now he could cite his silica brick-manufacturing scheme as a case in point. "The election of Wilson, which I think is nothing short of a national disgrace, has interfered seriously with our plans in this connection," he fumed to Bennitt, "for I have just had a letter from Mr. Gray to this effect, and Mr. Sharples is so disgusted that he intends to build cream separating plants in France, England and Russia rather than endeavor to supply these countries from his present plants in West Chester."[55] A few weeks later, he would dictate a five-page memorandum outlining "Some of the reasons why I am greatly disappointed in Wilson and believe that he is a misfit as President."[56]

Barringer, however, maintained an interest in his silica brick scheme for several more years, continuing an occasional correspondence with both Sharples and Gray.[57] By early June 1917, shortly after America's entry into the war in Europe, he was also exploring the possibilities of using crater silica in the manufacture of war-essential optical glass with his friend Sam S. Woods, president of the Pennsylvania Glass Sand Company.[58] The iron content of the silica was the key here, for an analysis in 1910 showed it to be 0.17 percent, too high for quality glass. But he now asked chemist Howell Furman to make a new analysis, and this showed an iron content of only 0.013 percent, a figure still somewhat high but which Furman assured him could be reduced to 0.004 by treating the silica with hydrochloric acid. "It goes to show," he wrote Thomson in September, "that we need look no further for silica of exceptional purity for the purpose of making optical glass," urging him to "put me in touch with the NRC [National Research Council] member who is interested in finding silica best adapted for making optical glass."[59]

Over the next few months, Barringer also sought to interest William C. Sproul, president of General Refractories, Inc., in which Barringer held stock, and Alfred C. Elkington, president of its subsidiary, the Philadelphia Quartz Company, in using crater silica for optical glass.[60] Supplies of silica sand, however, apparently were adequate and, in any case, were more conveniently available; and these efforts proved futile. Early in January 1918, presumably as a result of Thomson's contacts with members of the N.R.C., Arthur L. Day, director of the Carnegie Institution's Geophysical Laboratory in Washington, wrote to inquire about the crater silica. Barringer replied that the crater was "one of the best sources" of high-quality silica in the country, adding that while he could not afford to ship it to the East Coast in quantity himself, he was willing to let the government do so.[61] The government, however, did not follow up on the offer.

Parenthetically, while Barringer certainly did not intend to lose any money by supplying silica for optical glass, his motives seem to have been strongly patriotic. "As you well know," he had written Thomson when he had first broached the idea, "I am anxious to serve the government in any way that I can in this crisis, and I think I fully appreciate the need for optical glass of the right quality."[62] Barringer, in fact, was quite active in the war effort and at one point, convinced that famine might occur if the war continued, he journeyed to Washington to work on "a sweet potato flour scheme for increasing food production."[63]

Despite these and other activities and concerns, Barringer did not lose sight of his major goal of raising funds for the further exploration of the crater. In July 1917, for example, he persuaded Henry Norris Russell to ask astronomer George Ellery Hale, a wealthy man in his own right, to invest personally in the crater, and specifically to put up $25,000 for a new drilling program.⁶⁴ But Hale, preoccupied with organizing the nation's scientific war effort, apparently misunderstood Russell's message and simply turned his letter over to the National Research Council which, in time, advised Barringer that it was not interested in the proposal.⁶⁵

In September, Barringer stalked bigger game, writing a memorandum of a conversation with his friend J. F. Newsom, a geologist and long-time Branner colleague, who thought he might be able to interest the fabulously rich Guggenheim mining empire in the crater.⁶⁶ In this, Barringer detailed the conditions which, he was careful to note, he had "suggested" rather than "offered" to Newsom as a basis for Newsom's negotiations. These were that the Guggenheims would put up $200,000 for exploration; that if the mass was found and proved to be valuable, the Guggenheims would invest another $300,000 for development in exchange for a 33-percent interest in the property and 50 percent of the net profits up to $33,000,000; that thereafter profits would be divided in proportion to ownership; and, finally, that the Guggenheims would manage the property until they had received $43,000,000. Newsom, under this plan, was to receive 5 percent "of whatever interest I personally should have in the property" as a commission for arranging the deal.

Subsequently, Barringer liberalized these terms somewhat, increasing the ownership to 35 percent, offering 75 percent of the first $10,000,000 in profits, and adding that the Guggenheims could withdraw from the venture after sinking five drill holes to the Red Beds sandstone beneath the crater at locations selected by Barringer and Newsom. In a little more than a month, however, Newsom's contact, Albert C. Burrage, an Arizona copper mining executive who was acting for the Guggenheims, had turned down the proposition, "simply because I would not give title," as Barringer complained to Thomson.⁶⁷

The matter did not end here, however. Pope Yeatman, a widely known and highly respected mining engineer and a Guggenheim associate, seemed interested in the crater, Newsom reported, and Barringer urged Newsom and Yeatman to "go with me to the crater," sending Newsom copies of his and Tilghman's papers, reviewing the evidence for the meteoritic origin of the crater and declaring that "Never in my life have I tried less to make the facts fit any theory which I have previously formed."⁶⁸ Yeatman, as it turned out, could not go; but Newsom was anxious to see the crater himself. Thus late in December, after agreeing to pay Newsom 5 percent of "my income or dividends" should he succeed in "inducing Mr. Pope Yeatman to become interested in the exploration of Meteor Crater,"⁶⁹ Barringer telegraphed O. A.

Hart, a local rancher who had taken over as caretaker from Mrs. Holsinger in 1913, that "my intimate friend Mr. J. F. Newsom of California" would stop at the crater early in January "notwithstanding the unpropitious season."[70]

Newsom reported back that he had spent a "pleasant day" at the crater, but his observations there had raised doubts in his mind. For he had found what he took to be shale balls imbedded in limestone fragments on the crater's rim, and this suggested to him that they might have been present in the rock before the crater was formed. Barringer quickly assured him that "you are in error.... They are *without exception* [Barringer's emphasis] merely oxidized nodules of iron pyrites.... In short, I am absolutely certain that there never has been found or ever will be found a shale ball in place in the limestone."[71]

In March 1918 Barringer also tried once more to interest his friend Stone in the crater. "I believe I can get Charles Stone, of whom I see a great deal these days, and his partner [Edwin S.] Webster, to unite to put up $100,000 for exploration and $300,000 to $400,000 to sink a shaft and to build a smelter," he wrote confidently to Thomson.[72] But in a matter of weeks Stone had turned him down again, this time "because of the prejudiced opinion of his cousin, Mr. Caleb Stone."[73]

A LEASE WITH U.S. SMELTING

Now, however, the pattern would change dramatically, and a chain of events would begin to unfold that, after many frustrations and disappointments, would lead to the exploration of the crater's southern rim by the drill and to what Barringer, at least, would claim was the long-sought meteoritic mass itself. At the same time he was writing so confidently of his talks with Stone, he was also encouraging Thomson and Magie "to unite with me in persuading" the United States Smelting, Refining and Mining Company of Boston and its vice president, Sidney Johnston Jennings, "whom I know ... very well," to undertake the further exploration of the crater.[74] Jennings, he explained, was "a splendid mining engineer and is at present President of the American Institute of Mining Engineers."[75] Soon, Barringer arranged for Jennings to meet with Magie at Princeton to discuss the crater; and on that same day, he presented Jennings with a proposal that was quite similar to the one he had drawn up for Newsom's Guggenheim negotiations, except that U.S. Smelting was to sink ten drill holes, at an estimated cost of $50,000, before they could withdraw from the venture.[76] By the end of March, Jennings had agreed to visit the crater with Barringer late in April.[77]

The prospect seemed promising, but Barringer now realized that his earlier arrangement with Newsom might jeopardize these new negotiations; consequently, he wrote Newsom a long letter spelling out the limits of his [Newsom's] authority. "We have merely discussed some terms in a general way," he reminded him. "When we get far enough along to cause me to think it is time for you to make a definite proposition I will do so. . . . And finally you must realize that you should submit the names and addresses of all the parties with whom you wish to take up this matter before in any way committing us or yourself. We are the only ones to decide as to who shall be our associates in this matter."[78]

Jennings and C. F. Moore, an engineer for U.S. Smelting, visited the crater with Barringer during the last week in April, spending three days there going over the ground and collecting samples of meteoritic material.[79] Barringer confided to Thomson on his return that he was "able to absolutely convince Jennings and Moore of the correctness of the impact theory and to dig up for them a number of shale balls with iron centers on the northern rim."[80] But he also cautioned Thomson not to mention his earlier suggestion about gravity tests and added that there were some other questions which he "feared" might be "confusing" to the U.S. Smelting officials. "I am still as mystified as ever," he pointed out, "as to the real relation between the unoxidized ordinary Canyon Diablo meteorites and the oxidized shale ball meteorites. The evidence is very conflicting as to whether they were part and parcel of the same mass or were separate masses before they reached the earth, and I incline to the latter theory." Moreover, he reported to Bennitt, while Moore agreed that the meteoritic mass had "held together" at impact, Jennings concluded that it had "been scattered far and wide" in the depths of the crater, a circumstance which would make recovery of the mass difficult as a practical mining matter.[81]

A few days later, he wrote Thomson expressing the "hope that you will prevent this idea from growing in Jenning's mind," and urging him to "appeal to their patriotism" in view of the demand for platinum and iridium in the war effort.[82] In a letter to Magie, he discussed "Jennings's impression" at greater length, pointing out that the evidence of a bilateral symmetry at the crater indicated that the "only place" that the mass could be was under the southern cliffs. "In short, I cannot rid myself of the belief that the projectile, even though it were a cluster, would hold together like a load of bird shot fired at close range from a shot gun into a tree, or into stiff mud"—a reference to experiments he had made with shotguns and rifles during his various hunting trips. "I may be obsessed with the idea that it is under the southern wall," he added, "but I cannot rid myself of the obsession, if such it is."[83]

Following their trip to the crater, Jennings and Moore set U.S. Smelting's chemists to work assaying the meteoritic iron and other materials they had collected there, as well as a 25-pound piece of Canyon Diablo iron that Barringer had subsequently shipped to Moore. They also submitted samples to the eminent chemist Robert Edward Lyons of Indiana University for an independent analysis. By mid-May, Lyons found that in addition to platinum and iridium, the meteoritic iron also contained "two or three grams of palladium." Barringer, in passing this information along to Thomson, noted that the price fixed by the government for these metals was now $135 per troy ounce for palladium, $105 per ounce for platinum, and $175 per ounce for iridium.[84] Lyons, however, encountered problems in his study of the metamorphosed sandstone samples, suggesting that the pumiceous form was geyserite. "Apparently Mr. Lyons has gagged on the theory that Variety 'B' is metamorphosed white sandstone," Barringer complained to Magie late in May, "yet nothing on earth that I know of is more susceptible to proof. Even weak-kneed Merrill was bound to admit it. . . . What more can he want? Sudden and intense heat explains everything else in the presence of a certain amount of water."[85]

The chemists at U.S. Smelting, however, were working more slowly, and Barringer was becoming impatient. He continued to argue to Jennings that the mass could not have been scattered by the impact, citing his example of a load of birdshot fired into a tree or into stiff mud.[86] This he followed up with a long letter detailing eight reasons why the mass must be wedged under the southern rim, adding that its weight "cannot . . . be less than ten million tons" and that there would be "no serious difficulty" in mining it.[87]

Not until early in July, and then to his consternation, did Barringer learn that U.S. Smelting's chemists had been unable to find any platinoid metals in the iron and iron shale from the crater. He immediately wrote to Alexander urging him to "prove their chemists wrong."[88] To Magie he pointed out that there was "a drawer full of shale at Princeton, some of which could not be put to better use than to prove that the chemists of the United States Smelting, Refining and Mining Co. are not on to their jobs."[89] Alexander, he reminded Magie, had "found platinum and iridium in *every* piece [Barringer's emphasis] of shale which was submitted to him and certainly no care was taken in the selection of these. . . ."

Barringer now questioned the methods U.S. Smelting's chemists had used in their assays, sending Moore copies of Alexander's correspondence on the subject, his 1906 assays, and a description of Mallet's assay method.[90] He also ordered Hart at the crater to ship 250 pounds of iron and iron shale to Magie at Princeton, where he had arranged for chemist Leroy W. McCay and mineralogist Alexander H. Phillips to make further assays to check U.S. Smelting's results.[91] The problem, it quickly appeared, was that U.S. Smelting's chemists had used hydrochloric acid in their assays; and thus, as Barringer pointed out to Alexander, the platinoid metals had been dissolved in a solution of ferric chloride and discarded in the course of the process. Moore, he added, was "impressed" and "will use your methods."[92] A month later, Barringer could chide Moore that "your chemists are catching on,"[93] and advise Alexander that U.S. Smelting was "at last beginning to find" platinoid metals in the iron.[94]

This small triumph, however, did not resolve Jennings's doubts about the disposition of the mass in the crater depths, or its mineability; and consequently, early in October, he turned down Barringer's proposition. Barringer immediately voiced his disappointment, asking if Jennings would reconsider if he, Barringer, would raise one-third to one-half of the $50,000 needed for the proposed drilling program.[95] But Jennings did not reply to this new offer.[96] "My own opinion," Barringer later confided to Magie, "is that our Boston friends were really afraid of being laughed at by outsiders and criticised by their own stockholders for tackling an absolutely unprecedented proposition. I believe that Mr. Moore strongly recommended the company to go ahead and that Mr. Jennings was more than half persuaded but could not get rid of his fool idea that the mass was scattered around the bottom of the crater and, taking the quicksand condition into consideration, was therefore in such a shape that it could not be profitably mined."[97]

If Barringer was disappointed, however, he was not discouraged, and he immediately renewed his efforts to find other financing. As soon as he heard of Jennings's decision, he had asked Thomson if the General Electric Company might be induced to fund the new venture, now offering a 35-percent ownership interest in the crater for $500,000 expended, 75 percent of the profits until the company had recovered twenty-five times its investment plus 5 percent interest, and then a straight 50-50 profit split.[98] At $70 a ton, he reminded Thomson a few days later, these profits might amount to $500,000,000,[99] but Thomson was not encouraging. In January 1919, Barringer conceived the idea of persuading the trustees of Princeton University to invest in the crater "for the good of Princeton," revising his offer downward somewhat to 75 percent of the profits until only twenty times the investment had been returned.[100] He soon found, however, that Knox Taylor, president of the Taylor-Wharton Steel Company and a key trustee, considered the proposition "visionary."[101] In March, he urged Thomson to broach his proposal to the trustees of the Massachusetts Institute of Technology, where Thomson was a member of the corporation, but Thomson again was not encouraging.[102] Finally in June, he presented his proposition to Henry Fairfield Osborn, distinguished paleontologist and director of the American Museum of Natural History in New York City, and to the museum's board which rejected it a few weeks later.[103]

During these frustrating months, Barringer continued his campaign to convince scientists of the crater's impact origin, sending reprints of his papers to, among others, Dr. Leo H. Baekeland, inventor of "Bakelite," the first of the modern synthetic plastics, and to Sweden's Nobel Prize winning chemist Svante Arrhenius.[104] Arrhenius, incidentally, acknowledging the papers by postcard, called the crater "perhaps the most interesting feature on the surface of our planet," a remark that Barringer and others would often quote in the future in referring to Meteor Crater.[105]

In June 1919, events took another dramatic turn, in part because of his proselytizing among the Princeton trustees. Dr. Louis Davidson Ricketts, a trustee of the university, a widely respected mining engineer associated with major copper mining operations in Arizona, and a former president of the American Institute of Mining Engineers, expressed an interest in the crater. Barringer quickly sent him his papers and invited him to visit Coon Mountain.[106] A few days later, he dispatched instructions for reaching the crater, along with a summary of his arguments for believing that "the cluster of iron meteorites which made the crater must have held together,"[107] and followed this up with an eleven-page letter detailing why the mass must be wedged under the southern rim.[108] To Hart, he sent a letter of introduction for "my old friend" Ricketts, and three days later he directed him to hire "six husky Mexicans" to expose a number of shale balls in one of the old cuts on the crater's north rim in preparation for Ricketts's visit.[109]

In fact, Ricketts was preoccupied with other matters and did not make the trip. But his interest, communicated by Barringer to Jennings early in August,[110] nonetheless served to revive Jennings's own interest in the crater, and soon he and Barringer were again corresponding and discussing various propositions.[111] A few weeks later, Jennings agreed that U.S. Smelting would put up $60,000 for new drilling contingent upon the conclusion of a mutually satisfactory lease agreement for the crater property and Ricketts's favorable opinion of the enterprise.[112] As time passed, however, no further word was received from Ricketts, and Barringer's concern mounted. In September he urged Magie to persuade Ricketts that "if he does turn down the crater, not to do it in such a way as to discourage U.S. Smelting."[113] In October, while Jennings and Moore were inspecting U.S. Smelting properties in the West, he tried frantically to arrange for Ricketts to meet with them at one of the stops on their itinerary,[114] but Ricketts himself was traveling and Barringer could not contact him.[115] In mid-November Ricketts finally responded to one of his telegrams with disconcerting news. He had not gone to the crater himself, he now advised Barringer, but he had arranged for a Minneapolis mining engineer named Hugh M. Roberts to go there and report to him on the feasibility of further exploration.[116] Two weeks later Ricketts forwarded a copy of Roberts's eight-page report to Barringer.[117]

This, to Barringer's great distress, not only recommended that no further drilling be undertaken, but also raised anew the question of the crater's origin. Roberts had decided, after a two-day visit, that the crater was an explosive volcanic feature atop a low anticlinal ridge which, moreover, was situated in an area where there was abundant evidence of past volcanic activity. "Explosions in volcanic areas without accompanying extrusions of lavas are well known," he noted, citing Gilbert to his point. He even declared, though without citation, that "not all the authorities have agreed that the irons are meteoric. . . . The work which has already been done," he concluded, "is a true indication of the conditions which further exploratory work may be expected to reveal, i.e., no mineable quantity of nickeliferous iron may be expected."[118]

Barringer wrote Moore immediately, voicing his hope that the report "will not affect your and Mr. Jenning's judgement,"[119] and then took Ricketts to task. "We had arranged between us that you were to go personally," he remonstrated, ridiculing Roberts's belief in the crater's volcanic origin. There was no evidence of any igneous magma at the crater; the underlying Red Beds sandstone was undisturbed; and the nearest manifestation of volcanic activity was nine miles away at Sunset Mountain. Even the members of the U.S. Geological Survey, he added pointedly, agreed that the irons were meteoritic.[120]

Barringer now sent Thomson and Magie copies of "this fool's report" and urged them to send point-by-point rebuttals.[121] "The misery of it," he fumed to Magie, "is that a fool statement of this sort, bringing into the problem something which has no connection whatever with it, may establish doubt in Mr. Jennings's mind and get him off the right track of reasoning. In short, it would be tragic if this asinine report should cause the United States Smelting, Refining and Mining Company to reverse its decision to go ahead with the drilling. *We simply must not allow it to do so* [Barringer's emphasis]." He also asked for letters of rebuttal on specific points from Merrill; W. B. Scott, chairman of Princeton's Geology Department; E. O. Hovey, geologist at the American Museum of Natural History; and G. F. Kunz of the New York Academy of Sciences.[122] Almost immediately, the returns began coming in, all of them critical of the Roberts report, and Barringer forwarded copies of them to Jennings and Moore as fast as they were received.[123]

How much weight Jennings actually gave to Roberts's report is problematical, but certainly he was impressed by Barringer's massive campaign of rebuttal. Early in January 1920, he informed Barringer that, Roberts and Ricketts notwithstanding, U.S. Smelting was prepared to proceed with drilling at the crater and asked Barringer to meet with him and other U.S. Smelting executives to work out provisions of a lease.[124] The meeting, Barringer reported to Magie a few days later, proved to be "very satisfactory," and to Thomson he noted that "prospects are very bright for drilling this spring or summer."[125] By the end of January, he had received the first draft of a proposed lease, "an agreeable surprise," he confided to Bennitt, and "a good lease . . . better than I ever hoped to get."[126]

Legal details, including clearance of Standard Iron's title to the crater property, consumed another three months, but late in April all was in order and the final agreement was in Barringer's hands. This provided that U.S. Smelting, through a subsidiary called Crater Mining Company to be formed for the venture, would drill up to ten holes or spend up to $75,000 to locate the meteoritic mass. Once found, Crater Mining would then expend $600,000 for development under a ninety-nine year lease, paying Standard Iron 25 percent of the profits as rental until it had recovered its investment, and then 50 percent.[127] Barringer signed the lease and returned it immediately.[128]

There was, however, one more matter to be dealt with. Shortly before the lease agreement was finalized, the influential astronomer William Wallace Campbell, director of the University of California's Lick Observatory, asked to visit the crater late in May with three friends—astronomer George Ellery Hale, the paleontologist John C. Merriam, and Branner, Barringer's old friend and mentor who was now president emeritus of Stanford University. Campbell had, in fact, just published a discussion of the craters on the moon, opting for the volcanic analogy. His interest in Meteor Crater itself dated back to 1915, when he had requested and received copies of Barringer's papers.[129] Late in 1916, Branner had asked if he and Campbell could make a winter visit to the crater, and Barringer had approved enthusiastically, urging Branner to "persuade him [Campbell] that I am not a mighty and unconscionable liar, gifted with a perfervid imagination." The projected visit, however, did not materialize.[130]

But now, with the further exploration of the crater assured and the possibility of finding the meteoritic mass seemingly bright, Barringer's enthusiasm for such visits had cooled, and he tersely advised Campbell to postpone his trip until later in the year "owing to the lease."[131] Campbell protested immediately that May was "the only available time," adding that the "Contents of your papers [are] familiar to me" and that "Your theories [are] probably correct."[132] He followed this up with a letter explaining his reasons for wanting to visit the crater. "I am glad to say that your hypothesis of the meteoric origin of the crater, in my opinion, surpasses all other suggested hypotheses in merit," he wrote. "As you know, the geologists are fairly familiar with the subject. The astronomers of the world know it scarcely at all. . . . I am hoping to make the subject very familiar to astronomers."[133]

But Barringer was adamant. Exploration work was to begin in May, he telegraphed Campbell, and it would be "an inconvenience to have your party visit until summer or autumn," adding that "You will find that my theories have long since been proved to be correct."[134] A few days later, in a long letter, he pointed out that he had spent over $130,000 on the exploration of the crater and reviewed the main arguments for its impact origin, offering to serve as Campbell's personal guide on some later visit.[135] Campbell could only agree to postpone his trip, but he politely rejected Barringer's offer. "I should prefer that my first visit to the crater be under circumstances which would permit all points of view, evidences, and hypotheses to be considered in a judicial manner and not under the constant pressure of an advocate—this notwithstanding my provisional opinion that the origin of the crater is clearly meteoric," he replied. "Later I should be more than pleased to discuss the subject with you in your capacity of advocate. . . . I might as well express myself with entire frankness at once. Having spoken thus, I hope and believe we should get on very well with each other."[136] In June, incidentally, Campbell presented a brief paper to the annual meeting of the Astronomical Society of the Pacific in which he provisionally favored the impact origin of the crater.[137]

Barringer, clearly, had other reasons for wanting Campbell and his party to postpone their visit; but these he confided only to Thomson. For one, he feared that "all four distinguished men" might write papers about the crater "and get the credit." Conversely, he also feared that one or more of them might "throw cold water" on the crater's impact origin and thus dampen U.S. Smelting's enthusiasm for the new drilling venture.[138]

In October, Campbell informed Barringer that he would be traveling to the East Coast late in the month with his wife and asked if he could stop at the crater for a day on his return. Barringer now approved his visit, but only on condition that Campbell would not publish anything about the crater without Barringer's permission and adding that "I require and even insist that in any papers that you or your friends may write, I shall be given full credit for having discovered the origin of the Crater."[139] Campbell again protested, and Barringer now turned to another U.S. Smelting vice president, Neil W. Rice, for reinforcement. "Campbell feels at liberty to write what he pleases," he complained, declaring that the "wishes of the owners of private property should be respected," and pointing out that the crater property might be "taxed enormously" as the result of publicizing the exploration program and its goal.[140] He also urged Thomson to "warn Campbell that he might do us serious injury" and that he must not estimate the size, weight, or value of the meteoritic mass. Campbell, he added, "may have little respect for those of us who must be commercially minded to support our families and educate our children."[141] And to Campbell himself, already enroute to the crater, he telegraphed that the "Condition in allowing you to visit it is that nothing shall be written by you without Jennings's permission."[142]

Whatever Campbell may have thought privately about Barringer's restriction, he nonetheless accepted it in good humor, in part surely through the good offices of Thomson, whom he knew quite well. Early in November he reported to Barringer on his visit and on a discussion of the crater and Barringer's theories that he had held with Thomson a few days earlier in Boston. "In view of your known policy," he wrote, "and in accordance with the dictates of my friendship of long standing with Dr. Thomson, all that I learned from the latter is with me considered confidential. I am not advertising my visit to the crater in any manner, and I shall publish no statement at any time without your permission that has in any manner been influenced directly or indirectly by my visit. . . . The Meteor Crater, I have for many years felt, is the eighth wonder of the world. . . . In whatever I may have to say on the subject of the crater at any time, I shall hope to do you no injustice concerning your share in the establishing of its meteoric origin."[143]

Despite the solemnity of this declaration, however, it is very probable that Campbell was enjoying a private joke at Barringer's expense. For he apparently did not plan to publish anything about the crater in the future and, in fact, he never did. Moreover, he was already aware of what Barringer would soon find out[144]—that Merrill, in October, had published a paper in the *Publications of the Astronomical Society of the Pacific* reiterating in the same qualified terms his 1908 conclusions regarding the crater's origin and the disposition of the meteoritic mass.[145] In these conclusions, as it turned out, Campbell fully concurred.[146] Merrill, indeed, had noted at the outset that his paper had been inspired by "recent conversations" with the Lick Observatory director, although Barringer, in subsequently criticizing the paper to Thomson and Magie, seems to have missed this point.[147]

Some months later, in July 1921, Campbell could not resist chiding Barringer once again about his restrictive policy on publication. He had learned, he wrote Barringer pointedly, that his friend and fellow astronomer, Henry Norris Russell, had recently visited the crater, "and it developed that he [Russell] did not know of any expressed desire on the part of the authorities for the avoidance of publicity."[148] Barringer hastily explained that Russell had been visiting the Lowell Observatory in Flagstaff and that the astronomers there had "induced" him to accompany them to the crater.[149] Campbell was bemused, replying mildly that "I hope you will not think I have a contentious type of mind."[150] But Russell was not, and lectured Barringer at length about his methods of dealing with scientists. Conceding that Barringer's reasons for restricting publication might be valid, he wrote that "I hope, however, that you will take it in good part if I remark that, in expressing them, you do not use the language common among men of science. Professor Campbell is one of the most distinguished scientific men in this country. The friends whom he proposed to take with him are equally well known. It is not customary to use such terms as 'promise' and 'permission' in dealing with men of this standing. . . . The matter is one of form, not substance, but the good opinion of men like Campbell is of importance. . . . Incidentally, let me say that Dr. Campbell is an enthusiastic believer in the meteoric origin of the crater, and hoped to convince his friends on the spot."[151] Barringer seems to have taken Russell's advice to heart, for henceforth he would keep the Lick astronomer abreast of the progress of the drill at the crater.

Campbell, in chiding Barringer on Russell's visit, had also suggested that aerial photographs be made of the crater. "I have reason to believe that the Aviation Service of the United States Army, now operating within a practical distance of the crater, would undertake this," he wrote, adding mischievously: "To this end, I am going to ask you herewith to obtain from the powers that be a statement that those in authority have no objection to this procedure. As soon as I hear, I shall definitely try, with considerable confidence in the result, to have instructions issued to the proper aviation official."[152] Barringer, however, replied politely that he had been in "touch with the inventor and owner" of an aerial camera himself, and here the matter ended.[153] Five years later, it may be added, Barringer only with difficulty obtained copies of the first aerial photographs of the crater, made by Captains A. W. Stevens and L. MacCready of the U.S. Army.[154]

DRILLING DISASTERS

With the signing of the lease, Meteor Crater once again became the scene of frenzied activity. In mid-May, Jennings, now president of the newly formed Crater Mining Company, and other U.S. Smelting officials visited the crater to look over the ground.[155] For more than four years they would be in charge of all operations there and Barringer would be, in effect, only an interested bystander, occasionally giving advice but receiving his information about developments secondhand, usually through Jennings. Early in July, Barringer and Moore spent a week at the crater discussing where to drill the first hole, deciding on a location atop the southern rim. Barringer had preferred a site at the top of the talus on the south wall of the crater; but, as he later conceded to Moore, "owing to the expense, I am content with where we located it, but as near to the edge of the cliffs as possible."[156] Through the summer and into the autumn, crews were busy installing a new pipeline to the reservoirs, strengthening the dams, putting up new buildings, moving in the churn drill rig, making still another inconclusive magnetic survey, and digging exploratory shafts in the debris on the rim, one of which yielded several shale balls at a depth of twelve feet.[157] Late in October Barringer noted ruefully to Magie that "$60,000 has already been spent! And drilling has just commenced!"[158] By mid-December, with the hole barely 200 feet deep and drilling temporarily suspended, Barringer reported that U.S. Smelting had already "more than fulfilled their contract to spend $75,000."[159]

By almost any criteria, Crater Mining's Hole Number 1 at Meteor Crater was a disaster. Numerous breaks in the drill stem, a result of what Barringer considered "very inferior wartime products,"[160] and problems with the casing, and the sludge used to clean the hole and lubricate the drill bit, repeatedly delayed progress. A major crisis developed in mid-February 1921 when the drill stem broke and the drill tools were lost at 311 feet.[161] Despite frantic efforts over the next month to recover them, the tools remained at the bottom of the hole. Barringer and Magie now agreed that U.S. Smelting was using "poor drillers . . . and poor steel."[162] By mid-March, concerned about the cost of moving the drill rig and starting a new hole at another location, U.S. Smelting decided to retrieve the tools by tunneling horizontally nearly 400 feet through the talus and into the south rim, an operation Barringer considered "dangerous" because of the shattered condition of the rock strata of the crater wall.[163] The drift, nonetheless, was undertaken in the late spring and early summer and, in a remarkable feat of mining, succeeded in reaching the tools and clearing the hole. Barringer was briefly encouraged to learn that in the process two shale balls had been found plastered against the rock of the southern wall as the tunnel had passed out of the talus.[164]

Drilling resumed that summer, now with a large rotary drill, and progress, though slow, was steady for the next six months. By December, the drill had reached 605 feet, and by the end of February 1922, it was below 1,100 feet and approaching the level where Barringer expected it to encounter the meteoritic mass.[165] Late in March, indeed, tests of material brought up from the bottom of the hole began showing a positive reaction for nickel, and early in April Barringer thought the bit might be entering the actual mass, as a new sample not only showed "a very strong nickel test," but was "very black and very heavy with pieces of greenish metal showing."[166] The drill, he concluded a few days later, was in the lower portion of the mass and in an area which had been subject to severe leaching.[167]

But while prospects looked bright, the drilling now became more difficult and problems began to multiply. In May the drill stem broke again, and no sooner was the hole cleared of the broken pipe than casing problems caused more delays. Drilling resumed early in June, but within days a "hard object" fell against the side of the drill tools and wedged them securely in the bottom of the hole.[168] After another month's work, the obstructing tools were finally bypassed and by August crews were drilling ahead again, but in material so hard that five or six hours were required to deepen the hole by only a few inches, and the drill bits were quickly dulled and deeply scored.[169] Barringer urged that an open bit of hardened steel be used to obtain cores of this material, but unfortunately this was not done.[170] Three weeks later at a depth of 1,376 feet, the tools stuck again—this time for the last time.[171]

All attempts to free them over the next two months proved futile, and U.S. Smelting officials, who had spent "upwards of $175,000"[172] now questioned the advisability of continuing. The evidence was strong that the metals in the material had been leached and redeposited, and even Barringer conceded that it was "the rottenest shale that Thomson and I have ever seen."[173] He nonetheless urged Jennings to try another bypass, and to Moore he pointed out that the drill must be in contact with a 100-foot thick mass at 1,376 feet, as the 1905–1908 drill holes indicated that the Supai sandstone, presumably below the mass, should be encountered at about 1,470 feet below the rim.[174] But Jennings now concluded that the mass had little or no commercial value, and on November 15 the drill crews were dismissed.[175] Although U.S. Smelting officials continued to consider their options and Crater Mining's lease would not be formally abandoned for another two years, no further operations at the crater were undertaken.

A few months later, in March 1923, Barringer wrote his fourth and final paper on the crater and sent it off to the Academy of Natural Sciences of Philadelphia for publication in its *Proceedings*.[176] "It will probably be of interest to the members of the Academy," he began, "to know that the meteoric mass— probably a large cluster of iron meteorites—which made this remarkable crater, has been located under the southern wall, as predicted by my paper read before the National Academy of Sciences at its autumn meeting in 1909." He continued:

> An exploratory drill hole . . . at a vertical depth of 1376 feet, has recently passed through about thirty feet of undoubted meteoric material,

that is to say, highly oxidized meteoric iron or so-called iron shale, cementing small fragments of sandstone and still smaller fragments of Variety A and Variety B of the metamorphosed sandstone previously described. The drill is immovably stuck at this depth, but so much iron shale has been brought up as to make the conclusion inescapable that the main mass of the meteorite is underneath the southern wall of the crater.

This material, he added, "represents the upper portion or outer shell of the iron mass, which, in my present opinion, was largely composed of this shale-ball variety of iron meteorites rather than what is known to the scientific world as ordinary Canyon Diablo meteorites.... From what has been disclosed by the drill, it is probable that a considerable portion of the impacting mass, on the theory that it was largely composed of shale-ball iron, has undergone oxidation and is now in the form of iron shale."

Ironically, however, Jennings now denied Barringer permission to publish his paper, fearing that tax officials might claim a discovery under federal income tax laws, and Barringer hastily notified the academy's editor to hold his manuscript "until further notice."[177] His paper, indeed, did not appear in print for another nineteen months, until November 1924, when Crater Mining finally abandoned its lease. At the same time, incidentally, and for the same reason, Jennings also withdrew his permission, given a few weeks earlier, for Campbell to publish, a decision that Barringer found more to his liking.[178] For Campbell, late in February, had outlined his ideas about the crater to Barringer, concluding that the bulk of the mass had been blown back out of the crater after impact and had since been oxidized and eroded away. "The probabilities seem strong for the following hypothesis," he had written:[179]

> A very great cluster of relatively small meteorites, such materials as may be thought of as composing the nucleus of a small comet, travelling through space in a compact mass, penetrated ... to a depth of 1,200 feet before the momentum of the descending group was overcome. The reaction of condensed atmosphere, meteoric vapor and rock vapor discharged the descending mass ... much as the short type of cannon known as a mortar discharges its contents.... We know that essentially all metallic meteorites contain chlorine, which is an excellent rusting agent. In the course of a few centuries or a few thousand years, the bulk of the meteoric matter would rust, disintegrate and for the most part be blown away. Only the relatively few meteorites remaining uncovered until recent decades or centuries would remain in the form discovered by the present generation.

In a series of letters Barringer immediately disputed Campbell's conclusion as to the disintegration and disappearance of the mass; one letter fills no less than thirty-nine pages of his letterbook, marshaling somewhat repetitiously all the arguments he had used over the past fifteen years against Merrill's similar views.[180] But Campbell was not persuaded, and early in April he reiterated his stand in what Barringer called "an astonishing letter."[181] In particular, Barringer complained to Thomson, Campbell had "missed the point" of his argument that the depth of penetration of the mass precluded its having been blown back out of the crater, that as the meteorite had struck at an angle of about 45 degrees, it had plowed through not 1,200, but more than 2,500 feet of rock before coming to rest. Campbell, reasoning from the circular form of the crater, "thinks the fall was vertical and ignores the hypotenuse of the triangle in measuring the penetration," he fumed, and "sneers" at the experiments with rifle bullets and shotgun charges that Barringer had cited to argue not only the mechanics of penetration, but that impact at such an oblique angle nevertheless produced a circular crater.[182]

Barringer, however, did not complain directly to Campbell himself; apparently he had concluded from the astronomer's uncompromising tone that further argument would be useless, and in any case he was preoccupied with other matters. Frustrated by U.S. Smelting's continuing indecision as to any future operations at the crater, he had decided to organize an exploration company of his own to sink a shaft into the south rim and mine the mass that now, more than ever, he was convinced was there. As U.S. Smelting officials "seem likely to throw it up," he wrote Bennitt, "I will take off my coat and hunt money for the shaft."[183]

THE LUNAR ANALOGY REVIVED

To this end, Barringer turned his attention once again to the craters on the moon, writing a lengthy paper arguing their impact origin and drawing frequent analogies to Meteor Crater, "the impact origin of which is now proved...."[184] This he planned to publish in *Scientific American*, explaining candidly to Thomson that he wanted "to draw the attention of people of wealth to the Crater" as "they don't read scientific journals, but they do read the 'Scientific American.'"[185] He had considered the *National Geographic Magazine*, he added, but had decided that it was "too close to the U.S.G.S." The purpose of his lunar paper, he later confessed to Magie, was simply "to advertise the Crater."[186]

In June, he sent copies of his paper to Swarthmore astronomer John A. Miller and to Thomson for criticisms and comments.[187] He also wrote to Walter S. Adams, director of the Mount Wilson Solar Observatory in California,

25 Aerial view of Meteor Crater.

26 W. W. Campbell

requesting and receiving photographs of the moon made with their 100-inch Hooker reflector to use as illustrations.[188] In August, he submitted the paper to J. Malcolm Bird, an associate editor of the *Scientific American* and a knowledgeable writer on topics ranging from Einstein's theory of relativity to psychic phenomena.[189] Bird's skepticism on certain points, however, required several revisions, and thus its publication was delayed for nearly a year. The paper finally appeared in two installments in July and August 1924 under the title "Volcanoes—or Cosmic Shell Holes?"[190] As the *Scientific American* would not furnish reprints in quantity, Barringer later reprinted this paper himself in a less sensational format and under the more prosaic title of "A Discussion of the Origin of the Craters and Other Features of the Lunar Surface."

Barringer's paper was in part a restatement of the arguments that he and Thomson before him had presented in their earlier papers on Meteor Crater vis-à-vis the lunar craters. But more interestingly, what was new in it came principally from a brief article on lunar crater origins that his twelve-year-old son, Richard, had written in January 1923 and which had been published that summer in *Popular Astronomy*.[191] Bird, indeed, had pointedly called attention to Richard's article when Barringer submitted his own paper, and Barringer conceded ruefully that the boy had "beat me to it."[192] Richard, who despite his youth had been an avid amateur astronomer for several years, had argued the impact origin of the lunar features primarily on the basis of the observed superposition of crater upon crater on the moon's surface. But he also dealt briefly with the problem of the circularity of the lunar craters, citing Barringer's rifle and shotgun experiments, and had even outlined a relative chronology, based on superposition, of the impact events that had brought the moon to its present condition.

Barringer, in turn, simply expanded on these same ideas, including Richard's theoretical model of lunar history, in his own words. Apparently, he considered these ideas as his own, and in fact he may well have been Richard's source for them in the first instance. In any case, while liberally bestowing credit on Thomson and Magie and citing Worthington and even Gilbert, he did not mention young Richard's prior publication.

Barringer's use of analogies to Meteor Crater in his paper shows up particularly well in his discussion of the generally circular configuration of the lunar craters. "It may seem to one who is not familiar with the behavior of projectiles and the sort of craters they make," he wrote, "that the lunar craters are too round to have been produced by impact, since the impacting body in all cases would have approached the moon vertically, which manifestly was impossible. In fact, this has usually been considered the fatal objection to the impact theory." However, he noted, Meteor Crater was "just as round as the lunar craters; and yet we have the strongest reasons to believe that the meteoric mass which made it approached at probably a low angle. . . ." Anyone who had not fired rifle bullets or shotgun charges into stiffish mud at close range and at an angle of 45 degrees, he declared, "does not know that the resultant crater is almost round and that the splashing distribution of the ejected material is exactly similar to these effects in the cases of the lunar craters and the Arizona crater. . . . Even in the lunar craters one in many instances can tell from what direction the impacting mass approached, for the simple reason that more ejected material is observed to be on the side of the crater opposite to that from which the mass approached, just as in the case of the Meteor Crater of Arizona."

Some distinguished astronomers, he added, and here he cited Moulton, although he might well have cited Campbell, "seem to have been under the erroneous impression that the impacting mass in order to make a round crater must of necessity have fallen straight toward the moon's center. Such experiments with a high powered rifle as I have mentioned would soon convince them of their error."

Like Richard, Barringer also based his exposition of the evolutionary history of the moon on the superposition of the lunar craters. "The fact that it is easy to pick out the lunar craters which are younger than others and others which are younger still is very important," he declared.

> It is easy to find on a photograph a vast number of craters which have been almost covered and obliterated by the material thrown out of a neighboring crater or craters. To my mind the seas show this effect very well, obliterating, as they seem to have done, or filling pre-existing craters with molten material. We note especially . . . that some craters and the areas surrounding them appear to be much brighter than others. . . . an effect which is best explained by the reflection of sunlight from the white or very light colored pulverized rock flour which has not yet had time to be covered with cosmic dust or the ejectamenta from neighboring craters. Certainly the younger craters, judged by other signs, have this characteristic. In not one of them do we find a smaller crater. It is an effect which is not observable in the manifestly older craters.

Thus the moon, he suggested, had passed through four successive stages "since it contracted, possibly into a liquid mass (although it may never have been wholly liquid. . .)." First, as its surface cooled to a solid state, it received "innumerable additions of solid extralunar material which made the typical lunar craters, often showing the central hill or nipple-like protuberance." Subsequently, it received "some very large additions which struck with force enough to melt, or more likely break through the crust with the result that molten matter welled up from below and flowed more or less radially over large areas of the surface . . . forming the roughly circular seas or plains and the mountains surrounding them." Possibly, he added here, the "terrific force" of impact was sufficient "to liquify the already hot but more deeply seated rock and cause it to erupt and flow radially outward with great violence." In the third stage, these areas, solidifying on exposure to the cold of space, received further impacts, some of which may have penetrated the crust and caused molten material beneath "to rise to the level of the surrounding sea or plain," and "a rim but no permanent central hill would be formed." Finally, when the crust became completely cold and thus more resistant, still further impacts would again produce craters with a central hill and, in addition, bright, "splash-like" ejecta, "but with no general fusion."

The first installment of Barringer's paper appeared in July 1924; and Bird, now chief editor, provided a skeptical comment on it in his editorial preface to the issue in which he quoted from a letter from "one of the best known of American astronomers."[193] Although Bird did not identify the writer, it was in fact Henry Norris Russell.[194]

"It has been for some time practically settled," Russell had written, "that Meteor Crater in Arizona is due to the impact of a large meteoric object. Barringer has done a very good job in furnishing data on the subject. When he attempts, however, to revive Proctor's old theory that the lunar craters are similarly produced, he goes beyond his evidence in the opinion of most authorities. Several geologists will agree with him, but the astronomers who have seen the moon under the most favorable circumstances are practically a unit against him."

On Meteor Crater, Bird himself agreed: "The editor might remark here, on his own responsibility, that Barringer has positive proof beyond anything that has been published," a reference to Barringer's still unpublished final paper on the Meteor Crater. As to the impact origin of the craters on the moon, however, the editor was polite, but unpersuaded. Barringer, he declared, "presents the argument for his theory better than we have ever seen it marshalled by his predecessors," and "brings up points which must be disposed of before the theory of the volcanic origin of the lunar craters can settle back into unquestioned dominance in the field."

Bird's prediction was inaccurate, however, and Barringer would see his theory vindicated over the next several decades.

EIGHT
THE NEW METEORIC HYPOTHESIS

> We grant at once that nearly all the impacts must have been decidedly oblique. But what of that? The crater is not the hollow made by the blow. . . .
>
> ALGERNON CHARLES GIFFORD, 1924

By 1924 the volcanic analogy for the origin of the craters on the moon was being sharply challenged; in fact, it would never again "settle back into the unquestioned dominance" that it had held over selenological thought for nearly three hundred years. Its influence began to wane progressively over the next few decades as the impact theory gained more and more adherents, until by the 1950s "the pendulum swung too far," as one modern selenologist would later note, and "some scientists tried to explain everything on the moon as the result of cosmic impacts."[1]

Barringer's demonstration of the impact origin of Meteor Crater certainly played a part in this gradual but nonetheless dramatic turn of events, for it bore directly on one of the main arguments that had been used historically against the impact hypothesis, i.e., that there was no evidence that such impacts had ever occurred on the earth. In a few years, indeed, and from criteria developed in the investigation of the Meteor Crater, a number of other terrestrial impact craters were identified near Odessa, Texas; on the island of Ösel in Estonia; and in isolated areas of the Arabian desert and the Australian "outback."[2] In a few years too, the first belated reports of a catastrophic meteoritic event that had occurred on June 30, 1908, in the remote Tunguska River region of central Siberia, provided a sensational demonstration that large extraterrestrial bodies could collide violently with the earth.[3]

But it was Algernon Charles Gifford, a little-known associate in astronomy at the Hector Observatory in Wellington, New Zealand, who made what surely was the most effective contribution to this quiet revolution in science. Gifford, in 1924, published two versions of a paper entitled "The Mountains of the Moon" in which he proposed what he called "the new meteoric hypothesis" to answer all the objections hitherto advanced against the impact origin of the lunar features.[4] In particular, he dealt with the key problem of their seemingly uniform circularity. It was not the mechanical effects of the impact itself which produced circular craters, he now declared, but the "terrific" explosive forces resulting from "the sudden transformation of the energy of motion into heat." These explosive forces, which would of course act radially, "must, in all cases of cosmical impact, completely obliterate the original wound. . . ." Moreover, these forces were more powerful than anything then known to human experience. A meteor striking the surface of the moon, he declared, "is converted, in a very small fraction of a second, into an explosive compared with which dynamite and T.N.T. [tri-nitro-toluene] are mild and harmless."

Gifford's basic idea, so simple in retrospect, was not new, and Gifford himself credited it to his long-time friend and fellow New Zealand scientist, Alexander William Bickerton, who had suggested it in discussions of lunar problems at two successive meetings of the British Astronomical Association in 1915.[5] And although Gifford was not aware of the fact, two other scientists—astronomer Ernst J. Öpik and physicist Herbert E. Ives—had developed it independently in 1916 and 1919, respectively.[6] However, Öpik's paper, written in Russian and published in a little-known Russian journal amid the disruptions of World War I, has remained quite obscure; and Ives's publication, which appeared in the *Astrophysical Journal,* was quickly criticized by no less an astronomical authority than William Wallace Campbell.[7]

But the roots of Gifford's idea go even further back in time, for a number of early investigators of the problem of the origin of the lunar features had touched upon one or another aspect of it without realizing its full implications. In 1874, it will be recalled, selenographers James Nasmyth and James Carpenter had noted that an explosion at some point beneath the moon's crust would form a circular scar on its surface. But as they could not explain how such explosions might occur, or their number, they did not pursue the point further and settled for their "expansion on solidification" version of the volcanic analogy.[8] Four years later Proctor, expanding on the "meteoric" hypothesis he had proposed in 1873, had remarked in passing that "almost every mass which thus strikes the moon must be vaporised by the intense heat excited as it impinges on the moon's surface."[9] He did not develop the idea, however; and clearly, he did not associate the process with the phenomena.

Gilbert, in 1892, had recognized that even his slow-moving "moonlets," falling freely on the moon from a terrestrial ring at a velocity determined by the force of lunar gravity alone, would raise the temperature at the point of impact 3500°, a temperature sufficient to melt portions of the surface as well as of the "moonlet," and thus sufficient to his purpose, which was to explain the apparent flooding of some of the larger lunar craters and of the maria by molten lava.[10] His consideration of the effect of the mass-velocity relation, however, was limited by the special conditions he had devised to account mechanically for the circularity of the lunar features and which allowed him simply to ignore the problem of the random impact of fast-moving meteoroids.

Shaler, on the other hand, ignored Gilbert's special conditions; although he rejected an impact origin for the lunar craters, or "vulcanoids," he nevertheless speculated at some length on the possibilities inherent in "meteoric" impact. In his 1903 paper, the Harvard geologist pointed out, for example, that such impacts not only would generate temperatures far in excess of Gilbert's estimate of 3500°, but would probably result in volatilization.[11] "Assuming that the impinging body came upon the surface of the Moon at planetary velocity, and that all the resulting heat was applied to its mass, the resulting temperature would exceed, according to my reckoning, 150,000 degrees. . . ." The heat generated by the impact of such a body "some miles in diameter" would, he thought, "convert the whole of the body into a liquid if not a gaseous state." But if volatilization occurs, it must be followed by condensation, he reasoned, and he could find no evidence for this, although he had looked for it specifically in the bright streaks or rays emanating from some of the larger lunar craters. "Yet the absence of any deposits of these temporarily volatilized materials," he concluded, "is indicated by the fact that the light streaks are not obscured."

Shaler had also understood that the impact of large bodies on the moon at planetary velocities might produce explosions. But he did not deduce radial forces or circular craters from this. Rather, he thought such explosions would be significant only in regard to the problem of the "apparent degradation of some of its older features. . . . On the supposition that the in-falling bodies penetrated deeply and were converted into a gaseous state so that they produced explosions," he noted, "we gain an agency to break down reliefs in the manner in which many of the ancient features seem to have been mined."

Following Shaler, the debate over the impact hypothesis vis-à-vis the volcanic analogy turned briefly from the moon to the earth, and to the specific consideration of the claims that Barringer and Tilghman had made in 1906 for the meteoritic impact origin of Coon Mountain's crater. In this narrower controversy, ironically, it was the believers in a volcanic origin who argued that the Arizona crater had been formed by an explosion, as Gilbert had originally concluded, and Barringer, the vociferous advocate of impact, who stubbornly refused to concede that impact might result in the production of explosive forces. Only Merrill, as we have seen, suggested that a meteorite large enough to form Meteor Crater might produce at least partial volatilization and a consequent explosion; but even he, then or later, did not generalize his reasoning beyond the particular case.

DR. THOMAS JEFFERSON JACKSON SEE

A quite bizarre and far more imaginative version of the impact hypothesis was offered in 1909 and thereafter by astronomer Dr. Thomas Jefferson Jackson See, who is surely one of the strangest figures in the history of twentieth-century science.[12] In papers in both European and American journals and in his magnum opus, *Researches on the Evolution of Stellar Systems,* See proposed that impact was a fundamental process in the origin and evolution of the solar system.[13] Not only was the moon "a battered planet," he wrote, as was evident from the thousands of craters on its surface, but "all the smaller planetary bodies, such as Mercury and the satellites, have battered surfaces essentially analogous to that of the terrestrial satellite, which alone admits of minute telescopic investigation."[14]

But See's impact hypothesis, like Gilbert's, was also subject to special conditions, in this case those imposed by his grandiose and far more controversial "Capture Theory" of solar system origins to which it was, he noted, "a necessary corollary." This larger theory postulated the formation of the sun by in-falling bodies that had accreted in the outer reaches of a vast primordial nebula. Most of these bodies, their highly eccentric orbits reduced by what he called "the resisting medium," were "captured and absorbed" by the growing sun. As the sun gradually cleared the inner portions of the nebula, some of them remained in permanent solar orbits to become the planets. These, in turn, "captured and absorbed" other in-falling bodies as satellites which, spiraling in under the forces of gravity, fell on the planet's surface and produced what See called "satellite indentations." Earth's moon, he suggested, had once been such a planet, orbiting the sun in the region of the asteroids between Jupiter and Mars; and thus the lunar craters and maria were the "satellite indentations" it had acquired during this earlier period in its history before it had been "captured" itself by the earth.

See's "Capture Theory," of course, is more complex than this, but this brief summary should suffice not only to indicate its unorthodox character but the basis for his postulate of an impact origin for the lunar features. To See, these were "*simply Satellite Indentations in a surface of loose and largely uncemented cosmical dust and fragmentary rock, and not volcanic at all* [See's emphasis]." As there was no water or other chemical agent on the moon to cement this dust and rock into a coherent mass, he argued, his satellites which, like Gilbert's "moonlets," were falling in at the relatively slow velocity induced by lunar gravity, "could therefore easily indent the surface by compression of this uncemented material under the force of impact."

The mechanics of this process are perhaps vaguely reminiscent of Gruithuisen's concept of impact. "Accordingly, if a satellite fifty miles in diameter collided with the Moon, it would sink down into the soft and uncompacted surface, and at the same time be flattened and spread out at the base.... In flattening and spreading at the base, the satellite would force the walls of the crater outward, and itself would be reduced to fragments, resting on a fluid base, with the highest peak in the centre.... These central peaks," he added, "are conspicuous in all large craters, except when they have been made flat-bottomed by the fusion of the satellite, or largely filled up by a deposit of cosmical dust." In the smaller craters, he thought, the central peak had been covered by loose debris sliding down the crater's steep inner walls.

Like other theorizers before him, See also drew morphological comparisons between the lunar features and the crateriform structures formed by the fall of raindrops in mud, or by bullets fired into lead targets, citing "the large experience in various kinds of target practice" of a colleague, U.S. Navy Captain A. W. Dodd. See was, in fact, quite conversant with the early literature on impact back to Proctor, and in particular with Gilbert's paper in 1893 which, while he considered Gilbert's postulate of a terrestrial ring of "moonlets" inadmissible, nevertheless offered "many excellent features." Like Gilbert, See used the mass-velocity relation to compute the heat generated by a satellite spiraling in and impacting on the moon at a velocity of 2.27 kilometers per second [1.5 miles per second], finding that it would be 3367°, essentially Gilbert's figure. But unlike Gilbert, he recognized that some bodies "would collide with the lunar surface with a velocity even greater than this, and perhaps as high as the parabolic velocity around the sun, 42 kms per second [26 miles per second], giving . . . a temperature of about a million degrees." Consequently, "many of the satellites colliding with the Moon would be more or less melted and vaporized," a circumstance which to See, however, was significant only because on the atmosphereless moon "the dust arising from such a conflagration would rapidly fall as metallic and lithic rain, and tend to cover up the ancient craters."

See then proceeded to list twenty-four "facts" which the impact theory explained, including the superposition of craters, their distribution over the lunar surface, the depression of their floors below the surrounding surface, the deficiency of material in their rims vis-à-vis the volume of the crater, the streaks or rays emanating from some craters, and even the clefts and rills, the former being "paths cut by glancing satellites," and the latter being due to the settlement of loose, post-impact material along cracks in the lunar surface.

As to the circularity of the lunar features, he declared:

> If it be thought that more larger craters ought to be elliptical than are observed, it may be recalled that even if the first contact with the Moon produced such an outline, the impact of a large satellite would generate enough heat and underlying flow to force out the walls about symmetrically all around, and the final figure would be circular like the globular figure of the satellite. Thus craters which are, say ten times as wide as they are deep, ought to be almost circular; while smaller craters would be more irregular and elliptical, as found by observation. This is because the forcing out of the material beneath the small craters is less effective than in the case of large craters, and they retain more nearly their original shape of first contact.

See, incidentally, mentioned Meteor Crater only once in his book, in a footnote to his summary of his ideas some 350 pages after his presentation of his impact hypothesis, and then he seems to have been under the impression that Meteor Crater and Coon Butte were two different places.[15]

Reaction within astronomy to See's papers and his massive treatise was concentrated largely on his all-embracing "Capture Theory" and ranged from uncritical praise, through polite skepticism, to devastating denunciation. In particular, the theoretical astronomer, Forest Ray Moulton, coauthor of the "planetesimal" hypothesis, attacked See's claims and conclusions in a bitterly sarcastic review entitled "Capture Theory and Capture Practice," in which he bluntly accused See of prevarication and plagiarism.[16] Many of the equations and explanatory passages in See's book had been lifted entire from his *Introduction to Celestial Mechanics,* published in 1902, Moulton charged, and yet "in no place whatever is there a hint that the Capture Theory is indebted to the Celestial Mechanics for the material. . . . The explanations of the derivations and meanings of these equations, so far as the form is concerned, have also been captured from my book with only slight, but for the actual meaning, often fatal alterations." Moulton juxtaposed portions of the two texts side by side "to show with what success the form of my exposition, even to its minute details of printing, has been captured," while "the meanings of the equations and figures, unfortunately, have not been so fully captured. It would have been much better from the standpoint of accuracy if absolutely literal appropriation of the material had been made." See's treatise, Moulton fumed, was hardly the "Magnificent Work Indispensable to Libraries" or "The Only Great Treatise on Cosmogony Ever Published," as its author pretentiously claimed it to be.

Moulton did not comment specifically on See's impact theory, but in England several reviewers mentioned it in passing in discussing his more general "Capture Theory." English astronomer F.J.M. Stratton, for example, thought that See "gives a very interesting statement on the origin of the lunar formations: and that "a very good case is made out for the theory of impact."[17] Another English astronomer, F. W. Henkel, found that See's "wonderful work" provided in general "the most consistent explanation of a wide range of phenomena as yet developed," and that his "researches mark a new era in the history of theories of Cosmogony," a statement, perhaps, that even See's severest critics could accept, although not in the sense that Henkel meant it. In specific reference to See's impact hypothesis of lunar origins, however, Henkel simply noted briefly that "So different a theory from the ordinary volcanic one will not be easily accepted by selenologists, though a not dissimilar idea has occurred to others."[18]

And indeed, selenologists and selenographers, at least those in the British Astronomical Association, which is to say a majority of the active ones, did not accept See's impact theory. As noted earlier, they agreed with what, in effect, was a summary restatement of the volcanic analogy offered to refute See's ideas at the association's October 1910 meeting. The statement was delivered by Walter Goodacre, chairman of the association's Lunar Section and one of the leading selenographers of that day.[19]

Goodacre began with an appeal to collective authority. "This impact theory has been so little countenanced in the past by selenographers that when we find it advocated by an eminent astronomer like Dr. See we are bound to carefully examine the ground on which he claims our adhesion to his views. . . . It is, I think, important at the outset to state that, so far as I am aware, no selenographer of repute has held similar views to those set forth by Dr. See."

Goodacre then took up a number of the twenty-four "facts" relating to the lunar features that See contended his impact theory explained, noting that "it is not necessary that we who stand by the volcanic theory should be able to demolish every one of the arguments Dr. See puts forward. It will be sufficient if in one or two instances only we can show that his theory is untenable and contradicted by facts."

The superposition of lunar craters and the presence of small craters on the rims of larger craters, for example, could be explained equally well by the volcanic analogy, he declared, and here he pointed to the secondary craters on the outer slopes of the Sicilian volcano, Etna. And the reason that so few craters appeared in the lunar maria, he argued, was not that their surfaces had been fused by "terrible impacts" as See believed, but simply that after the outpouring of lava that created these level plains "the internal energy was too feeble to form anything but the smallest of these objects." To See's argument that terrestrial volcanoes "generally follow the mountain ranges along sea-coasts" while the lunar craters were randomly distributed, Goodacre noted that the "chains" of craters or craterlets seen on the moon could not be explained adequately by the impact theory. He also dismissed See's comparison between the lunar craters and bullet scars on leaden targets, referring specifically to a photograph of one of Captain Dodd's pockmarked targets that See had published to illustrate his point. "It may be admitted that at first glance they have a striking resemblance to the lunar craters," Goodacre conceded, "but the central peaks upon which Dr. See lays stress, and which are so necessary to complete the analogy, are not there."

Goodacre's most telling point, however, dealt with the circularity of the lunar features. See, he noted, had contended that under his impact theory the larger craters would be circular, while the smaller ones would tend to be more irregular "as found by observation." But, Goodacre pointed out, "it appears that observation shows conditions the exact opposite of this. We find that the largest rings depart most from the circular form, and that the smaller ones are the most symmetrical; indeed the smaller the crater the more does its shape approach that of a perfect circle. This," he added, "is precisely what we would expect as a result of volcanic action." There were no dissenters to Goodacre's thesis at the meeting.[20]

In 1911 still another idea was advanced to explain the mechanics of impact, this time by meteorites rather than "satellites," by M. E. Mulder, a professor at Groningen, in an article in the Dutch journal *Ingenieur*.[21] Mulder suggested that in their swift flight through the earth's atmosphere, meteors and meteorites became, in effect, shaped charges; and he applied this concept specifically to the phenomena at Meteor Crater, as described by Barringer in his 1909 National Academy of Sciences paper.

Mulder proposed that the leading surface of a meteor or meteorite would become hollowed out, like the concave bottom of a brandy bottle, by a combination of frictional heat and aerodynamic forces during its fall through the earth's atmosphere. For small meteors, the compressed air and gases in this hollow, or cavity, would cause it to explode in the air, as many meteors had been observed to do. In such a case, he pointed out, the fragments of the meteor would fall on the earth in an oval pattern because of the motion of both the meteor and the earth itself. A particularly dramatic illustration of this, indeed, came within a year when, on July 19, 1912, a large bolide flashed across the sky over Meteor Crater and exploded "one or two miles above the earth" near Holbrook, Arizona, some fifty miles to the east, showering more than 15,000 "stones" over an "elliptical" area three miles long and half-a-mile wide.[22]

A larger meteor, however, might reach the earth before being thus shattered, and the entrapped air and gases in its hollow, or cavity, being violently compressed on impact, would then explode, much in the manner of a military mortar, blasting rock and the fragments of the meteorite itself back out of the crater to fall in a circular pattern and form a rim. At Meteor Crater, he thought, "The meteor penetrated, with cavity pointed down, to a no-longer definable depth. Thereby, the pressure in the interior [of the cavity] was further increased so that . . . the gas contained therein was suddenly compressed. Thus, through this shock, the meteor bursts with great force, flinging its pieces in all directions, exactly as in an explosion in the air, but now no longer in an oval but in a roughly circular ring."

As a result, Mulder also concluded, very little if any of the original meteoritic mass would remain in the crater itself, a circumstance which, he suggested, Barringer and his associates might well consider. "I am therefore afraid," he declared, "that the Standard Iron Company, which wants to mine the crater in order to extract from it great treasure, will come away empty-handed."

Mulder did not apply his concept of impact directly to the lunar features. He did, however, quote with approval a brief passage from the many-editioned book, *Kometen und Meteoren*, by the German popularizer of astronomy, M. Wilhelm Meyer, who had referred to Meteor Crater not only as an "immense mortar," but as a "moon crater-like hole we at least cannot explain in any other way except by the impact of a cosmic mass."[23]

BRITISH ASTRONOMERS REACT

Mulder's paper apparently passed unnoticed within science; Barringer, who certainly would have objected to its conclusions, was not aware of it. But See's voluminous writings, of which Barringer *was* aware,[24] continued to stir controversy and, early in 1915, again became the subject of extended discussions of lunar problems in the British Astronomical Association—discussions which, incidentally, provide insights into the state of selenological thinking at the time. These were precipitated by a typically rambling paper by Harvard astronomer William Henry Pickering, published in November 1914, who reported his observations of a Mars-like "double canal" in the lunar crater Aristillus and suggested that an apparent darkening of a "field" near the crater at full moon was due to "a covering of vegetation."[25]

Pickering's suggestion of lunar vegetation was greeted with polite, but firm skepticism at the association's January 27, 1915, meeting, largely because it presumed the existence of a lunar atmosphere of sorts.[26] The point was stressed by the Reverend M. Davidson, who pointed out that a lunar atmosphere "would rather seem to contradict See's numerous theories," including his postulate that the lunar craters were "due to bombardment by planetoids." This remark, though made in passing, inspired S. L. Fletcher to comment that there was "one simple objection" to impact theory as it related to the craters on the moon. "If the Moon were struck by meteorites," the report of the meeting paraphrased his remarks, "they would be moving chiefly with their original velocity, and not merely with the velocity imparted by the Moon's gravitation. . . . Hence there was no reason why such meteorites should approach the Moon directly towards its centre. They might strike its surface at any angle, and most of the impressions they made would be elongated rather than round. The fact that the lunar craters were, on the whole, very truly round, seemed to point to their having been formed by some force acting directly towards, or from, the Moon's centre, and this seemed to bring us back to the volcanic theory."

When the report of this discussion appeared in the association's *Journal*, See immediately dispatched a brief paper defending his theories in general and expressing his "regret" that the results of his researches had been "imperfectly represented. . . . In regard to the Lunar craters," he added, "I have shown that they may be accounted for by impact; in fact, the approach of the Moon to the Earth since its capture and the similar approach of the other satellites to their several planets can come about only by increase of mass, due to collisions against smaller bodies. . . ."[27]

See's paper, published in the association's *Journal* in April, in turn brought a quick and apologetic reply from Davidson in the form of a letter read before the association's May 26 meeting.[28] In this, Davidson explained that the discussion of See's impact theory "arose out of the question of lunar vegetation," and that he had merely mentioned See's various theories "without any comments; the chief point emphasised was that the Moon had no sensible atmosphere either before or after its capture by the Earth." Apropos See's impact theory, however, Davidson added that there were "difficulties in accepting the theory of the lunar craters having been produced by satellite indentations in the surface of loose cosmical dust and fragmentary rock. It is true that the illustrations of imprints of raindrops and of bullets on leaden targets seem to corroborate the theory of impact of cosmical masses," he conceded, "but an objection, for which I have been unable to find an adequate solution, is that the lunar craters are generally round in appearance. Is there any explanation of this if we are to adopt Prof. See's theory?" Only one member at the May 26 meeting, A. W. Bickerton, offered an answer to this question.

Bickerton, in some ways, is almost as interesting a figure as See himself, although he is far more obscure in the history of science. Born in England in 1842, he went briefly into "railway engineering" and then had "a very distinguished career" at the Royal School of Mines and the Royal College of Chemistry, according to an obituary "appreciation" written shortly after his death in 1929 by his close friend, Gifford.[29] In 1874 he took the chair of chemistry at Canterbury College in Christchurch, New Zealand, where for nearly thirty years he taught not only chemistry but physics, and actively pursued—too actively as it turned out—an avid interest in astronomy. Intrigued by the appearance of a nova, or new star, in the constellation Cygnus in November 1876, Bickerton concluded, "in a flash of genius" to Gifford's mind, that the nova had been formed as a result of the partial, grazing collision of two stars. Subsequently, he applied this concept of partial collision to a wide variety of astronomical phenomena. "Celestial encounters, he saw, may take place between all sorts of celestial bodies," Gifford recalled, and could account for such observed phenomena as variable stars, double and multiple stars, and the planetary nebulae. Bickerton also seems to have anticipated a key postulate of Chamberlin and Moulton's "planetesimal" hypothesis. "A consideration of a rather more direct type of stellar encounter," Gifford noted, "gave him a picture of the origin of the Solar System that fits in with modern knowledge far more perfectly than the celebrated nebular hypothesis of Laplace."

Unfortunately, "troubles arose" for Bickerton in the mid-1890s for, as Gifford put it, "Some thought that a man who was making discoveries in astronomy could not at the same time be fulfilling all the duties of a Professor of Chemistry. The protracted inquiries that followed, although they emphasized the excellence of his work for the university, seriously undermined his health, and at last, in 1903, led to the loss of his professorship." Not for twenty-five years and until shortly before his death in 1929 did Canterbury University confer on him "the rare honour of Professor Emeritus," or did the New Zealand government award him a "well-earned pension."

Bickerton, addressing the Reverend Davidson's question, noted that he had recently published an article in the English journal *Knowledge* in which he had pointed out that a meteor, "striking the sun at critical velocity, would have something like 100,000 times the intensity of fulminate [of mercury], and that this would liberate perhaps 1,000 or even 100,000 times that amount of solar energy. So," he declared, "in the same way with the Moon."[30] And he explained that "in all probability a meteor, striking the Moon and going in for some distance, would produce volcanic action that would be intensified by the internal heat of the Moon. Thus, although the primary cause of the craters in the Moon might be due to impact, the real force that produced the crater was due to volcanic action," and this, he suggested, was "a possible interpretation of their circular form."

The ensuing discussion, however, ignored Bickerton's idea and centered on the contention of member J. E. Maxwell that the circularity of the lunar craters was really not a serious obstacle to the impact theory at all. He thought that "one would expect that most meteors would hit almost straight, and that the number of round craters would be a bigger percentage than might at first sight be supposed." But Maxwell himself was not too sure on the point, and he finally agreed with the association's president, the Reverend T.E.R. Phillips, who, in closing the discussion, declared that "Of course there ought, if the [impact] theory is correct, to be all over the Moon a certain proportion of elliptical craters that I do not think we see."

Maxwell's contention, nonetheless, led Davidson to make a geometrical investigation of the effects of random impact on the moon, even as Gilbert had done in 1892, and to report his results at the association's next meeting on June 30. In this brief paper, he showed that there would be just as many impacts at angles of incidence of 60 and 75 degrees from the vertical as at 15 or 30 degrees.[31] "We should thus expect to find a larger number of craters with high eccentricities," he declared.

In response, Bickerton again noted that "the normal speed of a meteor in space . . . would produce an explosive action by impact, and that the action would be augmented by the internal energy of the Moon; consequently oblique impact would produce roughly circular volcanic rings, the explosive action being from the centre outwards owing to the actual energy of the explosion of the hitting of the meteor becoming explosive heat. . . . Therefore, oblique impacts would, in these circumstances, become circular volcanoes."[32]

Davidson, however, protested that, while this explanation might be considered for large bodies impacting on the moon, it "would scarcely hold in the case of small craters produced by small bodies." To this, Bickerton answered that "the injury would be in proportion to the mass, just as great with small meteors as with big meteors as regarded the velocity," the meeting's record notes. "It would develop energy greater than any explosive with which we are familiar. . . . So the meteoric energy added to the internal explosion would produce a pretty fair bang even with small meteors."

Bickerton's suggestion, however, did not stimulate further discussion by association members and, despite the fact that it was reported in both the association's *Journal* and in *The Observatory*,[33] does not seem to have attracted any interest on the part of other astronomers or selenologists. Only Gifford among later impact theorists apparently saw any significance in it and referred to it in his own later papers.

In 1916, the idea that a meteorite impacting on the moon would produce "a pretty fair bang" even without the assistance of internal volcanic forces was spelled out in some detail by a young Estonian astronomer, Ernst J. Öpik, who would later become renowned for his theoretical studies of meteors, meteorites, and comets. Öpik published a paper in the *Bulletin de la Société Russe des Amis de l'Etude de l'Univers* in which he analyzed the mechanics and dynamics of meteoric impact, without recourse to any volcanic forces, in terms of the relation, $e = \frac{1}{2}mv^2$.[34] In this, he showed mathematically that such impacts of massive bodies on the moon at cosmic velocities would release enormous amounts of energy simultaneously, and thus explosively, and must result in circular craters regardless of the angle of incidence of impact on the lunar surface.

Öpik also noted that only a small fraction of this energy would be used up in the production of the mechanical effects of impact, i.e., excavating a crater, raising its peripheral walls, and shattering and pulverizing the material at and around the point of impact. Thus, he suggested that by setting the kinetic energy of the meteorite equal to the amount of mechanical work done on impact, it was possible to derive an estimate of at least the minimum mass of the projectile, and he then proceeded to work out the mathematics of the case. Some years later, Öpik would apply the method in an attempt to estimate the mass of the Meteor Crater meteorite.[35]

Unfortunately, from the standpoint of its influence on the subsequent development of impact theory in general, Öpik's 1916 paper seems to have been no more effective than was Bickerton's suggestion to the British Astronomical Association the year before. Those few who over the next ten years or so seriously considered the problem of meteoritic impact apparently were not aware of it, and in fact it has been cited only rarely in the later literature on the subject, even by Öpik himself, whose subsequent contributions to what came to be called "meteoritics" were not only extensive but particularly fruitful and, perhaps, have overshadowed his early work. Moreover, in 1916 Öpik, at age twenty-three, was only beginning his long and distinguished career and was not yet known in the international astronomical community. This same year, indeed, he had just obtained his first professional position on the staff of the Tashkent Observatory in remote Russian Turkistan. Given this, and the fact that his paper was written in Russian and published in a Russia already devastated by the catastrophe of a great world war, it is hardly surprising that Öpik's work did not come to the immediate attention of other astronomers or selenologists.

A CONTINUED SPATE OF PAPERS

World War I, however, was directly responsible for the next important statement in the development of impact theory, made in 1919 by the American physicist Herbert Eugene Ives. Ives himself would note that his researches had been undertaken "not in reference to lunar theories, but were incidental to the development of munitions" and were "therefore a scientific by-product of the Great War."[36]

During the war Ives had been assigned to Langley Field, Virginia, to study the craters made by the explosion of experimental aerial bombs, and had immediately recognized "points of resemblance between these bomb craters and the lunar configurations." Consequently, he had expanded his study to include the crateriform features on the moon, reporting his conclusions after the war in a paper in the *Astrophysical Journal*. In this, he quickly dismissed the reigning volcanic analogy of lunar origins in a brief opening paragraph. "The explanation readiest to hand, that the rings, pits and peaks are the result of volcanic action, does not appear to be adequate when closely studied," he declared. "While superficially similar to terrestrial volcanoes, the lunar 'craters' exhibit significant differences in structure. . . ."

On the other hand, aerial photographs of the craterpocked bombing range at Langley Field, which he published with his paper, "largely speak for themselves" in showing a "very striking similarity between the craters made by the explosion of bombs and the craters of the moon." In particular, they confirmed, on an "enormously greater" scale, the familiar laboratory experiments made over the years by many investigators who duplicated lunar appearances by shooting clay pellets or lead bullets into targets of similar materials. It was "a point of considerable importance," he noted, that the central elevation or peak, "formed apparently by a species of rebound," occurred both in these small-scale experiments and in some of the larger Langley Field bomb craters—"A most conclusive demonstration of the ability of a body (of the proper sort) striking a surface to produce an elevation."

The main thrust of Ives's paper, however, was not on such morphological comparisons, but on the dynamics of explosion as it related to impact. "It may at first seem farfetched to liken meteors to explosive bombs," he conceded, but it was well known that "meteors striking the earth's atmosphere not only flash into incandescence, but do frequently burst with terrifying reports." Then, noting that the velocity of meteors entering the atmosphere ranged from 16 to 64 kilometers per second, he applied the mass-velocity relation to the problem of meteoritic impact, using the formula $\frac{1}{2}mv^2 = ms(T-T_0)J$, in which s was the specific heat of the body, T the temperature to which it would be raised, T_0 its original temperature, and J the mechanical equivalent of heat. At a velocity of 16 km / sec, and assuming 0.2 for the specific heat, zero for the original temperature, and 41.8×10^6 as the mechanical equivalent of heat, "this equation gives for T the figure of 150,000 degrees Centigrade! Even if we assume that nine-tenths of this heat is given up to the surroundings, we still have in the 15,000°C a temperature amply sufficient to gasefy any known material, that is, *to produce an explosion* [Ives's emphasis]."

> Thus our calculation leads to the conclusion that a meteor striking the moon, even with the lowest velocity at which these are observed, would become a very efficient bomb, and should therefore produce the kind of crater we can imitate on earth only by filling our slowly moving military aerial bombs with explosive material. And not only does this explanation take care of the general appearance of the craters, but it affords an answer to the perplexing question presented by the almost uniformly circular shape of the lunar craters; for it is clear that the shape of the cavity has no reference to the angle at which the bomb strikes, but takes its form from the symmetrical explosive forces. Moreover, the available energy is so great that even if the meteor strikes at very great angles to the vertical the result will be an explosion.

The idea of explosion on impact also explained some of the other objections that had been raised to the "meteoric" theory, he pointed out. "One is that the heat generated by the impact of a meteor would be so great as to melt the crater walls. Obviously this criticism errs only in not going far enough into the matter and finding that the generation of heat is so much beyond that necessary for the melting of rock as to put an entirely new face on the problem, leading, as we have seen, to the conception of a meteor as an explosive bomb."

To the objection that the earth itself should show the effects of meteoritic bombardment, Ives pointed out in passing "that the Canyon Diablo [crater], the most perfect imitation we have of a lunar crater, bears numerous evidences of having been caused by an explosion of other than subterranean origin. But the most complete answer to this criticism is found by noting, first, that the earth is surrounded by an atmosphere which in previous ages must have been denser than now and so would dissipate the energy of falling meteors, as indeed we see it doing now; and second, that the earth's surface has been undergoing the processes of upheaval and weathering for perhaps countless ages since the collision with the giant meteor swarms which permanently marked the dead and atmosphere-less lunar surface."

A few months after Ives's paper appeared, however, three eminent astronomers—George Ellery Hale, director of the Mount Wilson Solar Observatory; William Henry Pickering of Harvard; and William Wallace Campbell, director of the Lick Observatory—strongly reaffirmed the volcanic origin of the lunar features in a series of papers in the *Publications of the Astronomical Society of the Pacific*,[37] inspired by the first photographs of the moon taken by the newly operational 100-inch Hooker reflector on Mount Wilson, then and for many years the largest telescope in the world.

Hale led off the series by noting diffidently that "I cannot pretend to have made any serious study of the Moon, and therefore my comments on lunar research are entitled but to little weight." Nonetheless, he voiced his opinion on the origin of the lunar features:

> In fact, altho many lunar craters are depressions below the surrounding country, and altho traces of lava streams, so characteristic of our volcanoes, are usually lacking on the Moon, it nevertheless seems to one who has considered the subject only superficially that the present photographs favor the view that the lunar craters are more probably due to volcanic or other internal phenomena than to the fall of meteorites. The fact, that minute craters are much more numerous on the slopes and in the vicinity of large craters than in open plains . . . and the almost complete absence of evidence indicating impact at angles differing greatly from ninety degrees, are points to be remembered in this connection.

Pickering's arguments against impact theory were somewhat longer. He began by remarking that "a crater probably formed in this manner actually exists upon the Earth in Arizona. . . . But why there are not more of them, or at least some evidence of their remains, since the Earth is so much more massive than the Moon, has not been explained." For Pickering there were other objections to the impact concept as well. First, "the lunar craters are nearly circular, whereas on the meteoric theory they should in the great majority of cases be ellipses." In particular, the postulate of a terrestrial ring of "meteors," which "simply dropped into the Moon by their own gravity," could not be used to account for this circularity as it was otherwise flawed. On the moon, he pointed out, "smaller craters impinge on the larger, as they should if they were of volcanic origin." But if caused by "meteors" falling from a terrestrial ring, "a large proportion of the smaller meteors would fall early, and their craters would be impinged on by larger ones."

Pickering did not refer specifically to Ives's paper on aerial bomb craters, but he did contend that there was a difference between the lunar craters and those formed by small-scale experiments involving clay projectiles. In the small laboratory craters, "the outer walls are steep or perpendicular, while the inner walls are shelving. This is the reverse of what we find on the Moon...." In craters formed by solidifying paraffin or iron slag, as in his own experiments, "the inner walls are steep and the outer ones shelving, as on the Moon." Then again, there was the problem of the terraced interior walls of some of the craters, and here he singled out the lunar craters Tycho and Moretus as shown in the 100-inch photographs. "Tidal action on a liquid floor would produce just such an effect," he noted. For his last, but not his least objection to impact theory, Pickering cited a lunar feature that apparently had not been commented on before, noting that "many of the central peaks exhibit small craters on their very summits.... This is precisely the phenomenon we should expect to find if the craters were due to volcanic action."[38]

Campbell, in his turn, combined Pickering's points with some of his own to fashion an even more comprehensive assault against the impact origin of the lunar features. He introduced his discussion by citing the parallel controversy over the origin of "the so-called Meteor Crater" and deploring the fact that astronomers had largely ignored the significance of this unusual terrestrial feature. "The literature of the subject," he declared, "is due almost wholly to those whose chief interests are geologic.... Astronomers have not, in my opinion, given the crater the attention it deserves.... It is to be hoped that this neglect may soon be remedied." As to the origin of Meteor Crater, however, Campbell would not go beyond the statement that "it seems to admit the possibility of a meteoric origin." Privately, of course, as we have seen, he was not so circumspect.

At the outset, too, Campbell delivered a pronouncement on a point that Ives had cited and that would be hotly debated in the subsequent denouement of the Coon Mountain controversies—the role that atmospheric retardation played in decelerating a meteoric mass impacting on the earth. "The scarcity of great meteoric craters on the Earth is not to be explained by the resistance of the atmosphere to meteoric bombardment," he declared; "meteors massive enough and speedy enough to create great craters on the Moon would pay no appreciable attention to the Earth's atmosphere. They would come thru the atmosphere, to the rock or water surface of the Earth, with speed and mass not appreciably reduced." If such craters had ever existed on earth, he added, their present scarcity would best be explained by "erosion, sedimentation, flows of molten rock, or other meteorologic and geologic forces."

Campbell also wondered briefly that if the impacts of meteors on the moon had been as numerous as the existing craters, "should we not expect abundant evidence of impacts at all angles of incidence, varying from the tangential to the vertical?" Yet he knew of only a single feature, the Valley of the Alps, that might be explained by the oblique impact of a meteor. "It is as if a great meteorite, travelling northerly over Italy, had struck the southern flank of the Alps at Como and ploughed its way horizontally thru the alps as far as Lucerne." But there might be a "lingering doubt" here, he added, "as to whether a great metallic meteorite, travelling with a relative speed of forty miles, more or less, per second could save itself from volatilization or other destructive consequences while penetrating and overcoming the resistance of a great mountain range on a course of 83 miles. . . . There appear to be no lunar features clearly illustrating the effects of collisions by massive meteoric bodies whose angles of incidence lie between that assumed to have attended the formation of the Valley of the Alps and those postulated as the sources of the lunar craters. This is a fact in strong opposition to the impact theory of crater origin."

Campbell discussed at some length the experiments by Ives and earlier investigators that duplicated lunar craters in the laboratory. These, he conceded, gave "interesting results, but I do not believe there is any chance that they can afford serious evidence as to the origin of any lunar craters large enough to be seen with telescopes. . . . A laboratory crater, whose diameter is only a few inches or a foot or two, must not be expected to tell us much about the origin of a crater 250,000 feet (50 miles) in diameter. A laboratory crater may have a central peak an inch or two or three inches high; will that tell us anything of value as to the origin of central peaks 6,000 feet high in lunar craters?" Even Ives's aerial bomb craters, which provided "the most interesting laboratory experiments of this nature conducted to date," were inadequate, he argued. "The largest craters developed by this means were of the order of one hundred feet in diameter, and possibly thirty feet in depth. Would we be justified in reasoning from these craters to the lunar craters, whose diameters are forty, fifty and sixty miles, and whose depths are four, five and six thousand feet? I think such comparisons must be of very limited value."

The central peaks were of particular importance, the Lick director stressed, as impact theorists ascribed them to reactions from impacts "such as are illustrated by the central peaks in some of the small laboratory craters." However, he pointed out, citing Pickering as his selenographic authority, the "vast majority" of lunar craters contained no central peaks. "Do the central peaks in the little laboratory craters then mean anything in the lunar problem? If these central peaks on the Moon are the reactionary results of impacts, why should they be so relatively few?" Moreover, if the "throwing up of crater walls" was a general effect of impacts, "why should they not have the common effect of central peaks?" Finally, he argued, in the case of a large meteorite impacting on the moon, the "materials left in the crater could scarcely avoid being reduced to liquid form. . . . A central peak was thrown up suddenly and, notwithstanding its probable liquidity or viscosity, remained standing in opposition to the force of gravity. Are not these results highly improbable and essentially unthinkable?" he asked. "To me, a vastly more probable description of the origin of a lunar crater and its central peak is the well known one held by those who look with favor upon the hypothesis of volcanic origins."

Like Pickering, Campbell cited the small craters or craterlets atop some of the central peaks. "Must we not agree that the unquestioned existence of craterlets in the summits of central crater peaks is absolutely fatal to the impact theory . . . and at the same time in full and complete harmony with the volcanic upbuilding of those peaks?"

This spate of papers, however, stirred active interest only among a few members of the British Astronomical Association, which had published abstracts of Ives's and Pickering's papers in its *Journal*.[39] E. O. Fountain, a physician, for example, contended in connection with Ives's paper that bomb explosions involved forces acting outward from a center, while in the case of meteorites, "the explosion commences where it is heated by friction, that is to say on the outside."[40] This, he added, "should be pertinent to the relative dimensions of central peaks and of craters themselves." Fountain, incidentally, also suggested that because it had been formed in rock, the "Canon Diabolo" crater "probably required a larger meteor than would be necessitated by the loose fragments on the Moon." One A. C. Curtis wondered that if meteors large enough to form the larger lunar craters had once existed, "why should they not still exist?" and asked whether a minor planet might "suffer destruction on the same principle of the 'collision theory' . . . so that the debris would fall on the Earth and the Moon with the same effect as a shoal of meteors?"[41] Fountain's brief answer to this was that meteors "of large size have long since been used up."[42]

Pickering's defense of the volcanic analogy won support from a British Army captain, G. H. Lepper, who declared that he was "glad to find that so well known an authority has expressed his disbelief in the meteoric bombardment idea, as this confirms the impression I had formed as a result of a fairly intimate acquaintance with all the worst crater fields of the British portion of the Western Front. There may be a superficial resemblance between a photograph of a heavily shelled region and a photograph of the Moon, but a close examination of the actual *terrain* [Lepper's emphasis], with the eye in the former case and the telescope in the latter, soon shows the essential dissimilarity between the two."[43] To this, Fountain replied that shell craters "could not be expected to entirely resemble large meteorite holes . . . because the results would not be 'to scale.' Shell holes are only to be compared with the very smallest crater-pits on the Moon, and these have no central cones."[44]

In 1920, and thereafter, the impact theory received support from the German meteorologist Alfred Wegener, who in 1912 had advanced the highly controversial theory of continental drift, another fruitful concept that was widely ridiculed within the scientific establishment for some fifty years before generating a not-so-quiet revolution in geology in the 1960s.[45] Wegener, however, argued the impact origin of the lunar features primarily on morphological grounds, citing his own extensive experiments with mud, powder, plaster and cement projectiles and the appearance of the moon itself.[46] He concluded that the shapes and proportional dimensions of his experimental craters were closely comparable to those of the lunar craters, and thus that the impact hypothesis, "which is the one most rarely represented by the experts, upon close examination proves to be the most promising."

Wegener focused his exposition largely on what he considered the "trivial [stickhaltig]" objections to the impact theory relating to the absence of impact craters on the earth. Some contended, he noted, that the much larger earth should "very much show the traces of such cosmic bombardments, and that this is not the case because the meteorite crater in Arizona, as decisively as it appears to prove the possibility of the origin of the large [lunar] craters by this method, is only a single case, and it would still remain that if some of the craters that have been hitherto called volcanic in other locations would, upon closer scrutiny, turn out to be impact craters." Several schemes had been suggested to account for this deficiency in terrestrial impact craters, he pointed out. For example, if the moon had been captured by the earth, "then, obviously, we cannot expect that the Earth has had the same experience as the Moon." Others "assume that the lunar mass earlier surrounded the Earth in the form of a ring and only later combined into a single body, and this is supposed to be exactly the process whose decay has left us the visible marks on the Moon."

But such schemes were unnecessary, he declared, for the moon's appearance alone could explain the dearth of impact craters on earth. In particular, the lunar maria and the "older craters" showed clear evidence of "melting" and seemed to be all of an age, "while the freshest, youngest craters show no melting, but instead ray systems." This, he felt, suggested that the moon had undergone "very rapid cooling during the time of the formation of the craters that are visible today. Nothing is more reasonable than to assume that an initial high temperature was produced through the collision of the masses themselves and that it was already being lost due to radiation" during the more recent impacts. The much larger and more massive earth would retain its initial high temperature for a longer period of time, and thus the final phase of the meteoritic bombardment of the earth-moon system "had fallen into the still molten earth and no traces remain. . . . Even with equality of processes, one could not demand the same traces that are visible on the Moon should be visible on the Earth."

Four years later, Gifford independently employed this idea of a molten earth to explain the absence of terrestrial impact craters. "Some say it is because denudation has obliterated the ancient craters on its surface," Gifford wrote.[47] "But there is a far stronger reason than this. As long as both bodies were in a liquid state, not even the greatest meteorite left any permanent scar. The Moon being so small in comparison with the Earth cooled much more quickly. As soon as a crust began to form every collision left some mark. The Earth, still liquid at the time, though receiving more and slightly heavier blows, remained unscarred. The bombardment simply aided it in its growth, and rendered its orbit more nearly circular." Another reason, he thought, was that then, as now, the earth was "probably protected by its surrounding atmosphere."[48]

But Gifford's thesis went far beyond Wegener's analysis of impact theory. While Wegener felt that it was premature to consider the nature of the impacting bodies or the impact process itself, Gifford saw these aspects as the crux of the matter. And although there were many features on the moon to be explained—the maria, the ray systems, the cracks, rills, and isolated mountain peaks—the key to the problem lay in the craters. These, he pointed out, were "undoubtedly the most characteristic of all lunar forms. If their origin can be determined, it seems probable that a flood of light will immediately be thrown upon other dark parts of lunar geological history."

GIFFORD'S HYPOTHESIS

Gifford, perhaps, is an even more interesting figure than his friend and colleague, Bickerton. His ideas, like Bickerton's, more often than not were outside the mainstream of contemporary scientific thought, and in his later years, indeed, he prided himself on his unorthodoxy. Born in 1861 aboard a ship enroute to New Zealand, he attended Denstone College in England, where he gained a sizarship to St. John's College at Cambridge University. There, he took his M.A. degree in 1880, graduating as fourteenth wrangler in the mathematical tripos and winning the Herschel Prize in mathematical astronomy. Returning to New Zealand, he taught science and mathematics at Waitiki High School, then at Christ's College in Christchurch, and finally from 1895 to 1927 at Wellington College.[49]

During these years, he became affectionately known as "Uncle Charlie" to two generations of New Zealanders through his varied interests and activities and his articulate writing on scientific subjects in New Zealand periodicals and newspapers. In his later years, after formal retirement, he was no less active. "As a gardener on a scale which would have deterred most young men, an amateur astronomer and with an extremely live interest on the question of the authorship of Shakespeare and economic questions of the day, he was never idle," his obituary notes. In his early days, he was also "a keen geologist, pursuing the subject in the field and exerting himself in extensive explorations of little-known areas."

Astronomically, Gifford's professional credentials were sound and allowed him to keep abreast of developments in the field elsewhere in the world despite his remote situation. He was a Fellow of the Royal Astronomical Society, and a member of the British Astronomical Association, the Société Astronomique de France, the Royal Astronomical Society of Canada, and the Astronomical Society of the Pacific in the United States. At home, he helped organize the New Zealand Philosophical Society, served as president of the New Zealand Astronomical Society, founded the Wellington College Observatory, and was for a time an associate of the Dominion Observatory in Wellington. With his death in 1948, his obituarist declared, "New Zealand lost its most enthusiastic and inspiring astronomer," and one of the few who "may well be said to have established whatever astronomical foundation this small country, just past the pioneering stage, possesses."

Gifford's career, however, was dominated by his relationship with Bickerton, for although he was "by no means unaware of the failings of Bickerton," he had "an intense admiration for the old professor." Bickerton's theory of partial impact for the origin of novae, and of the solar system, "should be labelled the Bickerton-Gifford Theory," his obituarist suggested. "Bickerton . . . supplied the general idea, and Gifford polished it to a state where it might receive some consideration." Gifford not only "pushed all aspects of impact to their limit," but the "meteoric theory of the origin of the lunar craters found in him a staunch champion. . . ."

The "new meteoric hypothesis" of lunar origins that Gifford proposed in

1924 was, of course, one of Bickerton's general ideas. Gifford, however, jettisoned Bickerton's belief that supplemental volcanic action was required to produce the observed lunar surface forms. Impact alone, he found, was sufficient.

Gifford first summarized the history of the problem, the various theories that had been advanced to meet it, from Robert Hooke to Thomas Jefferson Jackson See; and contemporary opinion regarding these theories, concluding from the latter that the volcanic analogy enjoyed "wide acceptance." But, he declared, "there does not appear to be much—beyond a slight superficial resemblance between lunar rings and certain terrestrial craters—that can be said in favor of the volcanic theory." For one thing, the lunar craters were so numerous and of such vast dimensions that there was "no room for all of them on the lunar surface," and they "overlap and obliterate one another. We cannot imagine that, at any stage of its evolution, the Moon required such a multitude of blowholes." There were also no terrestrial forms that closely resembled the lunar features, and here he cited Gilbert's "impressive list of contrasts" between volcanoes on the earth and craters on the moon. Nor was there any system to the distribution of the lunar craters while "volcanic action on the Earth appears to take place along lines of weakness or fracture." Moreover, the "universal circularity of form" of the lunar craters was difficult to explain on the volcanic analogy; and while terrestrial volcanoes "appeared to be due to imprisoned steam," the absence of water on the moon makes it difficult to account for the presence of vapour in sufficient quantity to produce eruptions capable of forming such vast craters." The lack of evidence of lava flows from craters and the superposition of craters were additional points, he noted, and finally, any theory that explained the craters must also explain the maria or "seas" and the lunar mountain ranges which were "portions of the curved rims of the seas," and the "volcanic theory cannot do so."

The only serious objection for Gifford to the impact hypothesis, on the other hand, was that "Meteors colliding with the Moon would be certain to strike the surface obliquely, but the universally circular form of the lunar craters implies that, if they are due to impact, every meteor must have fallen vertically." This objection, he thought, was "strong enough by itself to lead to the rejection of the meteoric hypothesis, as it is usually presented. But in all discussions hitherto," he added, "one all-important factor has been overlooked." This—the explosive effect of violent impact—he now made the basis of his "new meteoric hypothesis."

Gifford provided a quite vivid description of what would happen when a meteor traveling at forty miles a second impacted on the moon. Its velocity, he declared, "is destroyed in one-tenth of a second, by a constant resistance," and whilst forcing its way into the Moon's crust, its entire store of energy of translational motion is transformed into molecular heat agitation. When this transformation is complete every gramme of the meteor's mass is endowed with 474,700 calories of heat energy.

Now, in a tenth of a second, it is changed into a gas ever so far above the critical temperature of its most refractory constituent, but confined temporarily in a severely limited space. The pressure is so intense, the expansion so rapid, that there is no time for the superimposed or surrounding material to be liquified. It is instantaneously shattered and ejected in an explosion five hundred times as powerful, as much more rapid, than that which would result if a quantity of dynamite equal in mass to the meteorite were detonated in a cavity within the lunar crust. The explosion commences before the onward motion of the meteorite is stopped, and so the initial cavity widens out as it deepens. The subjacent materials are forcibly compressed, whilst those above and around are shattered and pulverized and hurled aloft. In a few seconds a vast structure is fashioned which may preserve for untold geological ages a record of its birth.

The meteoric hypothesis, Gifford contended, had not yet received due consideration "because the explosive character of the missiles has not been recognized." Moreover, "the form produced by an explosion is very different from that produced by the fall of an inert mass." Gilbert's idea that "the meteorites were originally satellites of the Earth, but, finding their way into a neutral region between Earth and Moon, dropped vertically on to the surface of the latter, is not at all convincing."

In the case of craters produced by exploding meteorites, the forces generated would be equal in all directions, but the effects of these forces would vary greatly. "The downward pressure results in a considerable compression of the imperfectly consolidated materials below. The lateral pressure produces both compression and ejection of matter. The upward pressure, assisted by the reflection of gas from the consolidated surface below, carries all before it, and the disintegrated material is scattered with some approximation towards equal distribution in all directions. . . . The reflected lateral pressure causes some excess of material to be sent in directions near the horizon. The resulting cavity appears to be essentially saucer-shaped."

Gifford then proceeded to compute the density distribution of the ejected material falling back on the lunar surface after an explosive impact, finding that, in an "ideal case" it produced a crateriform structure with vertical outer walls. But this, he noted, required modification as the material of the wall, "instead of forming vertical cliffs, would roll down and assume the natural maximum slope." When this modification was made, he added, "the resemblance to typical ringed plains on the Moon is very striking." To illustrate this, he presented cross-sectional drawings of the structure formed by the "ideal" distribution of the ejected material, of the same structure modified for lunar gravitational effects, and of the actual lunar craters Copernicus and Theophilus. "The typical form of the lunar crater then," he declared, "agrees exactly with what might be expected as the result of an explosion."

Gifford then showed that the available energy of an impacting meteorite would be sufficient to do the work of producing such craters, computing tables of the energy produced in calories per gram by six of the most powerful explosives known, ranging from picric acid to nitroglycerine, and by meteorites striking with velocities ranging from one to 400 miles per second. These tables demonstrated, he pointed out, "that when a meteorite strikes a surface with a velocity of many miles a second it becomes an explosive compared with whose violence that of dynamite is insignificant." He also noted that in the case of a meteorite, probably 1,000 calories per gram would be required to raise its temperature from absolute zero and to vaporize it, "but the rest of its vast store of energy is available for pulverizing and scattering the materials into which it has penetrated. . . . The mass of the ejected material may exceed that of the original meteor thousands of times."

The diameter of a crater produced by an exploding meteorite, moreover, depended on the velocity at which the materials were ejected, he contended, "and it is surprising how small are the velocities of the projected materials required to produce the craters of the Moon. . . . Very few even of the vast lunar craters require a velocity of ejection of more than a quarter of a mile per second. Now, a velocity of forty miles per second implies energy 25,600 [times] that due to a quarter of a mile per second. Thus, a rapidly moving meteor may eject more than 25,000 times its own mass." Here he provided a table to show that velocities of ejection ranging from 0.01 to 0.32 miles per second would result in craters ranging from 0.197 to 202 miles in diameter, and then listed the diameters of fourteen lunar craters, from Hyginus (4 miles) to Grimaldi (147 miles), for comparison.

Gifford's "new meteoric hypothesis" was distributed quite widely in the United States, Europe, and England in the form of a Hector Observatory *Bulletin*. But not surprisingly, given the indifference of most astronomers to lunar problems, it was seriously considered only in the British Astronomical Association, where it figured prominently in still another extended debate over lunar origins that occupied the members at its May and June 1925 meetings and continued thereafter in its *Journal* for nearly a year.

THE ENSUING DEBATE

The stage for this debate was set two months earlier when, at the end of the March 1925 meeting, J. W. Wilson read a short paper on the "meteoric" origin of the lunar craters.[50] In this, he calculated that if all the kinetic energy of impact were used in creating a crater, it would require a meteor of "*at least* 220 yards in diameter when moving at 20 miles per second [Wilson's emphasis]" to produce the lunar crater Copernicus. While "no known meteor has been as large as this," he conceded, the great size of some lunar craters nonetheless might be explained if the earth-moon system had encountered a swarm of such large meteors at some earlier period in its history or if meteors "acted merely as a trigger for the release of internal volcanic energy. . . ." As the hour was late, discussion of Wilson's paper was postponed until the association's next regular meeting.[51] As it turned out, when the members again convened in formal session late in May, they were prepared to discuss not only Wilson's paper, but to mount a full-scale debate on the whole question of how the lunar features originated.[52]

W. H. Steavenson, a well-known lunar and planetary observer, began with a sharp attack on the volcanic analogy and a plea for the consideration of alternatives, particularly the impact hypothesis. "Really," he explained at the outset, "I am not bound to the impact theory as the only correct one, but what I do want to suggest to you is that, whatever theory may be correct, the volcanic one is certainly untenable, and that in supporting it, Mr. Goodacre and his followers of the opposition are wrong. At least I hope you will end by being convinced that the volcanic theory has been too readily accepted, and that the impact theory has been too lightly set aside." He then offered seven "chief objections" to the volcanic analogy, including the "enormous" number of lunar craters, the great size of many of them, their structure, their lack of special distribution, and the frequent absence of central cones or peaks.

As to the impact theory, he noted, "There have always been three main objections to it, which ought now, I think, to be revised." Of the first, the great size of some lunar craters, he pointed out that "Until lately it has not been realised that an explosion is involved. . . . But if a meteor falls at 40 or 30 miles per second, it is more potentially explosive than hundreds of times its weight of dynamite. If it is brought up sharply by a hard surface it seems that some sort of explosion *must* occur [Steavenson's emphasis]. The only question is, how much of the available kinetic energy is used in raising and spreading the portion of the surface struck," and here, citing Wilson's calculations, he noted "the rather striking result" that "with an expenditure of only 1/100th of the kinetic energy, a meteor travelling at 25 miles per second may make a crater one hundred times its own diameter." To the second objection, the circularity of craters, he simply pointed out that "the example of craters produced by low-angle shells in the War showed that the result of an explosion is always a roughly circular cavity, unless the angle of impact is extremely small." The third objection, the absence of impact craters on earth, could be met by the "reasonable assumption that the earth's crust had not yet formed (when the bombardment was in progress. . . .)." And finally, he thought that the idea that "the large craters were clearly formed first, and that this favors the volcanic theory" was based on a false assumption. "Whatever was the order in which different sized bodies impinged on the Moon, the result would be the same. The larger would obliterate any smaller craters that were formed earlier, and only the small ones which fell later would be observable."

Steavenson, although he used Gifford's terminology, did not refer specifically to Gifford's paper, but Goodacre, in a prepared response, not only cited it but, initially at least, misstated its primary thrust. Gifford, he declared, "discusses the *pros* and *cons* of the volcanic and impact theories, and concludes that neither fully meet the difficulties of the case." The eminent selenographer then proceeded to defend the volcanic analogy in a largely rhetorical attack on the impact hypothesis, focusing on what he considered to be a discrepancy between Gifford's and Wilson's calculations regarding the size of the craters that impacting meteors might produce.

Both Gifford and Wilson, he noted, "hold the view that the impact causes an instantaneous and tremendous explosion, which explosion is capable of ejecting material exceeding in mass that of the projectile many thousands of times. . . . Mr. Wilson commits himself to the statement that a meteorite 220 yards in diameter, moving at 20 miles per second, would produce a crater the size of Copernicus. Now Copernicus is 56 miles in diameter and the meteorite one-eighth of a mile in diameter; it follows that a projectile of this size is capable of displacing material 448 times greater than itself. This is quite modest in comparison with Mr. Gifford's statement about the meteorite displacing 25,000 times its own mass."

At this point, Fountain, who apparently had lost his earlier interest in Ives's bomb craters and in impact theory in general, proposed the revival of the glacial hypothesis.[53] This he based on the nonuniform freezing of "lunar oceans" as the moon lost its primordial heat, leaving "patches of water" which froze at their outer edges and, abetted by the condensation of water vapor, formed circular walls of ice. "A deposit of an inch a day would produce 3,000 feet in 100 years," he opined. Of all the theories, "the glacial requires the least expenditure of power. . . . It, therefore, deserves a claim for consideration by the side of other theories."

Fountain's plea, however, drew no immediate response from his fellow members, and Major A. E. Levin quickly turned the debate back to Gifford's and Wilson's papers, finding specific difficulties with both on minor points, but not disagreeing with their major conclusion of explosive impact. From an examination of Wilson's calculations relating to the crater Copernicus, he noted, "it appears that there is difficulty, on the Meteoric theory, to account for the great depth below the surrounding surface of some of the crater floors. Thus, taking the figures given for Copernicus, the material necessary to build up the walls would only require the floor to be about 1,500 feet below the original surface."* Yet he thought it "most likely" that only one hundredth of the available energy of impact was used "for raising the walls," while the rest was transmitted to other parts of the moon in the form of a "moonquake," or dissipated as heat. Levin also noted, as had Gifford, that "instead of simple penetration with final explosion, as in the case of a shell, the explosive action would be going on throughout the period of penetration." As to Gifford's calculations of crater form, Levin questioned the New Zealander's assumption that impact would scatter matter uniformly in all directions with equal velocity from a central point, noting that this involved the "impossible assumption of infinite density. As a matter of fact, the conditions of differing velocity and directions of ejection are probably so complicated that it is better to rely on the results of experience collected on such a vast scale in the years 1914–1918. . . ."

A final word before adjournment came from Captain Maurice Ainslie, who urged his fellow members "to give up terrestrial analogies altogether. Whenever I look at the Moon I am struck by the absence of anything in the least suggestive of a terrestrial volcano," he declared. "I see certain difficulties about accepting the Meteoric Impact theory, but the difficulties in the way of the Volcanic theory seem to me incomparably greater. Let us face the Lunar Mountains as they really are."

When the debate resumed at the June meeting, the proponents of the volcanic analogy were clearly on the defensive.[54] Their chief spokesman now was H. G. Tomkins who, in two years, would propose an "igneous explanation" of the lunar craters involving the gradual breaking up of laccoliths on the moon.[55] His objection to impact theory, he noted now, was fundamental. "It seems to me to be nothing but an imaginary conception, completely outside our experience, and based on assumptions which have no support either from what we see going on around us or from anything which history or scientific investigation tells us has happened in the past." Along with such rhetoric, satire was another Tomkins device, as witness his review of the development of the impact theory:

---

*Copernicus is about 11,000 feet deep.

Everyone who has shown the Moon through his telescope to friends is familiar with their remarks that it looks like a cheese with holes in it, or else as if it had been shot at. The latter was the origin of the meteoric theory. Then we had people firing revolvers into beds of mud, lead or other substances, and when they examined the holes so made, they declared them to be miniature craters; and after that they looked for something to represent the bullet for the Moon and seized on our meteorites. . . . The thing has been repeated so often in one form or another, that it has reached the stage of a probable explanation. It seems to me utterly fantastic, and a similar theory might be put forward about any kind of hole or pit in any place which could not be otherwise explained.

More specifically, Tomkins wondered where were all the impact craters on earth that must result from the postulated earth-moon bombardment. Wilson, he contended, found "that such large meteors as would be required could not have existed, or, if the features *were* due to meteors [Tomkins's emphasis], the latter could have only pulled the trigger for internal volcanic forces to act." In Gifford's paper, he added, "the heat generated by such impacts seems to have been neglected," and here he cited Shaler's conclusion that the heat generated by an impacting bolide "would be sufficient to liquify an area many times its diameter. . . . It is certain that no explosive theory could produce the terraces such as we see in Copernicus. But from the start, the bolide is a mere speculative assumption and my view is that there is nothing to support it."

Wilson quickly responded to Tomkins's "lively attack," noting that "it is rather tempting to turn some of his words back on him. Surely," he pointed out, "the idea of a meteorite larger than any in our experience is no stranger or more unlikely than the idea of a lunar volcano many, many times larger than any known terrestrial one." He then clarified a point on which "Mr. Goodacre and others of his followers seem to be in some doubt. . . . The actual extent of the explosion only depends in any given case on the proportion of lost energy that remains kinetic, i.e., is not converted into heat. The trouble arises, I think, because we are accustomed to associate an explosion with a sudden transformation of energy from still a third form, namely chemical energy, without realising that the same effect is produced if the change is simply one from kinetic energy of one body (the meteorite) to kinetic energy of other bodies (in our case, the fragments of the lunar surface)."

Finally, the noted astronomer A. C. D. Crommelin pointed out that "as regards the argument that bodies of a size adequate to produce lunar craters are absent from our system, I think we ought to remember that the known minor planets range in diameter from 450 to 3 miles," and that according to the Chamberlin-Moulton planetesimal hypothesis, "such small bodies were more widely distributed in the early days of the solar system." As to the absence of impact craters on earth, he also noted that "There is one instance—the crater Coon Butte in the United States of America—where we have reason to suppose that a crater was formed by a meteor. . . . I do not wish these remarks to be taken as an endorsement of the impact theory of lunar craters," he added. "I still have an open mind on that question; but I think that some of the objections to the meteor hypothesis can be surmounted." At this point, president C. D. P. Davis closed the discussion, declaring that "I am, myself, a crater man, but I see that there is more to be said for the meteoric theory than I had thought."

This was not the end of it, however. In subsequent months various members voiced their opinions on all three theories—impact, volcanic, and glacial—in the association's *Journal,* the sharpest exchanges coming, oddly enough, over the glacial theory. Gifford himself participated in this phase of the debate, sending "communications" which appeared in October and December to answer points raised against his "new meteoric hypothesis" at the May and June meetings.[56] To Levin's criticism of his assumption of infinite density, for example, he replied that his distribution curve for the ejected material had been "an ideal not an actual one." To Tomkins's statement that there was no trace of any meteorite having struck the earth, he pointed to Meteor Crater as showing "what can be done by a meteorite, even after it has penetrated the Earth's atmosphere." Again, the objection that he had neglected the heating and melting effects of impact "does not appear sound," he declared. "Even T.N.T., nitroglycerine, and dynamite act too quickly to liquify the surrounding materials." And as to terraces in Copernicus: "Calculation shows that an explosion can produce a mountain ring and central peaks; the terraces appear just what one might expect to be the result of landslips on the material so placed." Wilson, he added, had made the "most useful contribution towards the solution of the problem by calculating the exact quantity of energy required to produce certain definite results."

Comments pro or con for one or another theory continued to appear sporadically in the *Journal* through May 1926. A tally of the opinions of the active participants in this long debate indicates that only six stood by the volcanic analogy, six others either favored or leaned toward the impact theory, and another three, rejecting both of these hypotheses, opted for the alternative glacial explanation.[57] Clearly, in the one astronomical forum where lunar problems were still being regularly discussed, the long-dominant volcanic analogy was now barely holding its own.

NINE

BLUE SKY LAWS AND BUREAUCRATS

> Aside from the assured scientific value of the exploration, this expenditure, or as much thereof as may be necessary, will be purely a gamble, and any subscription to the stock of the Meteor Crater Exploration and Mining Company should be made with that fact clearly in mind. . . .
>
> G. M. COLVOCORESSES AND D. M. BARRINGER, 1926

By the spring of 1925, Barringer was immersed in plans for the organization of what soon would become the Meteor Crater Exploration and Mining Company, through which he hoped to exploit the nickel and platinoid metals in the meteoritic mass he believed was buried beneath the southern wall of the crater. Certain now that the upper part of the mass had been probed by U. S. Smelting's drill, he proposed to sink a 1,500-foot shaft some 1,000 feet south of the U. S. Smelting drill hole and then drift back horizontally to a point directly under the hole, where he was confident that the main mass would be encountered. As a mining operation, this would be quite routine, he felt, for the strata under the surrounding plain should be in place and unshattered, and outside the rim water should not be the problem it had proved to be in earlier attempts to sink shafts within the crater itself.[1]

To finance this new enterprise, he planned to form a company, to be capitalized at $1 million, which would undertake the project under a ninety-nine year lease, with the Standard Iron Company receiving a 50-percent share of the eventual profits as rent for the crater property. Initially, $500,000 would be raised through stock subscriptions, of which only half would be required for sinking the shaft and running the drift, with the balance being available once the mass had been reached and mining and smelting operations could begin. Then, if additional monies were needed, subscriptions to the remaining stock should be easy to obtain.[2] Once the profits began to accumulate, he also planned to build a smelter at the crater to further reduce costs; and even as he began to formulate his plan, he had informed his old friend H. H. Alexander that he wanted him to be "Chief Consulting Metallurgist" of the facility.[3]

Barringer now envisioned an unprecedented mining bonanza: Ten million tons of meteoritic material with a gross value, estimated conservatively at $50 a ton, of $500,000,000 which, at $25 a ton for mining and smelting, would yield a net profit of $250,000,000 or more on a $500,000 investment![4] It is hardly surprising that many men of capital were skeptical, even incredulous; what is more surprising, perhaps, is that enough of them took him up on his proposition to bring it, in time, to fruition.

In the two years since U. S. Smelting had suspended drilling at the crater, however, his attempts to persuade wealthy individuals to join him in his scheme had been unproductive. In June 1923 he had contacted Percy Matthey, a principal stockholder in the London firm of Johnson Matthey, refiners and traders in precious metals, subsequently conferring with Matthey and his associates and later sending him some 200 pounds of meteoritic material from the crater for his chemists and metallurgists to analyze. But after eight months of sporadic negotiations, Matthey turned him down, concerned that the mass might prove to be a single, solid piece of iron and thus impractical to mine or, if mineable, that it would prove too costly to extract the platinoid metals.[5]

In April 1924 Barringer had turned again to James S. Douglas, entrepreneur of the rich United Verde Extension mine at Jerome, Arizona, offering him a "chance of making an even greater fortune" and sending him an early draft of a prospectus for his proposed company.[6] In May his second oldest son, Daniel Moreau Barringer Jr., now a Princeton graduate and a geologist at the Inspiration Copper Mine in Arizona, talked with Douglas about the crater. Later that month Barringer himself met with Douglas at Flagstaff to discuss his proposal in detail.[7] Over the next few months, he kept Douglas well supplied with literature, forwarding his *Scientific American* articles on the impact origin of the lunar craters, along with copies of his papers and letters solicited from Merrill, Russell, and the Carnegie Institution's Arthur L. Day attesting to the impact origin of Meteor Crater and the meteoritic nature of the iron and iron shale there.[8] Despite his initial interest, however, Douglas remained unconvinced, citing among other problems the possibility that the mass might have volatilized.[9] When, early in August, Barringer pressed him for a decision, Douglas quickly obliged by rejecting his proposition.[10]

A NEW GAMBLER

But Barringer's trip to Arizona to meet Douglas did have one important consequence. For during a stop at Prescott, he had met an able and experienced mining engineer, George M. Colvocoresses, who had expressed interest in what he had to say about Meteor Crater.[11] Thus began an association on which Barringer would rely heavily in the years immediately ahead as the Coon Mountain controversies moved toward their bitter denouement.

Colvocoresses, a 1900 Yale University graduate, was at this time president of the Association of Copper Producers of Arizona. Before coming to Arizona, he had been employed for twelve years by the International Nickel Company and had worked as an engineer for mines and smelters in New Caledonia and in Canada's metals-rich Sudbury district. More important from Barringer's standpoint, he was now general manager of the Southwest Metals Company smelter at Humboldt, Arizona, in which he was associated with Colonel Robert E. Thompson. Thompson's brother, Barringer knew, was financier and philanthropist William Boyce Thompson, a multimillionaire whose fortune derived from the lucrative operations of the Newmont Mining Company and Arizona's highly productive Magma Copper Mine. He suspected, and his brother-in-law Bennitt in Phoenix agreed with him, that William Boyce Thompson, known for his endowments of arboretums, might be more interested in the scientific than the commercial aspects of the new venture at Meteor Crater.[12]

Barringer's association with Colvocoresses, however, got off to a slow start, for through the summer he was preoccupied with his negotiations with Douglas and with debating the conclusions in his *Scientific American* articles with Russell, Merrill, and Pickering, among others.[13] He was now also campaigning vigorously in his correspondence to convince geologists Frederick E. Wright of the Carnegie Institution[14] and the aging but still influential Thomas C. Chamberlin of the impact origin of both Meteor Crater and the craters on the moon, urging Chamberlin, as he had nearly twenty years before, to visit the Arizona crater and pointing out the pertinence of impact to the Chamberlin-Moulton planetesimal hypothesis.[15] He failed to sway Chamberlin, but he could in time count both Wright and Moulton as believers in the impact origin of Meteor Crater and Moulton, at least, as an adherent of the meteoritic origin of the lunar craters.[16]

By midsummer, his other prospects having vanished, Barringer sought Bennitt's opinion as to whether Colvocoresses might be a "desirable associate," noting that he appeared "very much interested" in the crater.[17] Bennitt was impressed, and Barringer now tried unsuccessfully to arrange a meeting with Colvocoresses in New York City early in October.[18] Not until later that month, however, and only after wealthy chemist Dr. Leo H. Baekeland had again expressed disinterest in his proposition,[19] did Barringer commit himself. Now he offered Colvocoresses, and through him Colonel Thompson, an "opportunity to cooperate" in the exploration and exploitation of Meteor Crater, sending him his preliminary draft of the prospectus for his proposed company along with copies of his papers on the crater, and assuring him that he would approach no one else until Colvocoresses had made up his mind.[20] He had decided on Colvocoresses as an associate in the venture, he advised Bennitt, because of his experience, his "closeness" to Colonel Thompson, and because he was "one of the few who are interested."[21]

Colvocoresses, clearly, was not a man to act on an impulse; he wanted time to study Barringer's proposition carefully and to investigate the crater and its commercial potential for himself. Barringer encouraged him and urged Colvocoresses to "try out any plan that occurs to you for prospecting the deposit," offering to have his son Reau accompany him to the crater to describe in detail the exploratory work that had already been done there. "It may be," he wrote, "that between us we can get together a syndicate," adding that he was amenable to any changes in his proposal that did not affect Standard Iron's title to the crater. He was, he explained, "anxious not only as a matter of security, but of pride, that the property should descend to my children."[22]

Through the late autumn and winter Colvocoresses sent a steady stream of letters to Philadelphia, asking questions, discussing various points, and making suggestions. In December, for example, he inquired about the possibility of confirming the location of the mass by electrical methods developed by geophysicist Hans Lundberg of the Swedish-American Prospecting Company, and Barringer was agreeable "if you realize that negative results won't mean too much."[23] The mass was too deep, he thought, and sought Thomson's opinion; when Thomson concurred, Barringer forwarded his letter to Colvocoresses.[24] Colvocoresses also suggested that the strata south of U. S. Smelting's drill hole should be tested with a diamond drill prior to sinking the shaft and this, in fact, was later done.[25] A few weeks later Barringer could exclaim to Magie that "I have at last found a man intelligent enough to ponder deeply over our arguments."[26]

Late in February Barringer acknowledged Colvocoresses's "substantial agreement" with his ideas on the crater and on the general plan for its further development, as well as his willingness "to do everything you can to cooperate with me in raising the necessary sum."[27] Once northern Arizona's winter snows had melted, Colvocoresses finally journeyed to the crater, spending several days there studying the ground, collecting samples of meteoritic material, and talking with O. A. Hart, who had been the caretaker there during U. S. Smelting's operations, and his successor, William V. Smith. "Now that you have got the facts clearly in mind," Barringer wrote him early in March after his return, "we can thrash out everything to our hearts' content when you come east. . . . Needless to say, I agree with all you have written in regard to the work being of great scientific value, but from the evidence at hand, I cannot escape the conclusion that it possesses great commercial possibilities as well. . . . I note that you feel as I feel, that the combination will prove attractive to the right individuals, and I assure you that it will be a pleasure to cooperate with you in endeavoring to raise the money for the proposed exploration company."[28]

Yet Colvocoresses still had doubts, as Barringer reported to both Thomson and Magie in alerting them to Colvocoresses's impending trip east early in April. For one, he questioned Barringer's insistence that the crater was less than 3,000 years old, estimating its age on the basis of the erosion he had observed as more probably 100,000 years. Aside from this, Barringer confided to Thomson, "I think Mr. Colvo is already won over, but I would like you and Magie to put the finishing touches on him, so as to get him as wholly converted as we three are."[29] Colvocoresses, he added to Magie, was worried that the nickel and platinoid metals might have been leached out of the mass over 100,000 years, but "I think I can convince him, as Branner, Merrill, Fairchild and Hovey were convinced, that the crater is of very recent origin. . . . Personally, I do not see the relevancy of his criticism as the shale ball iron would oxidize just as thoroughly in 2,000 years as in 100,000 years."[30]

Colvocoresses spent the first week of April 1925 in Philadelphia, discussing the crater not only with Barringer, Thomson and Magie, but with Alexander, and Thomas Barbour, a noted naturalist associated with the Museum of Comparative Zoology and the Peabody Museum at Harvard University. Barringer had already sounded out Barbour as a potential investor in the new venture.[31] Before returning to Arizona via the crater, Colvocoresses joined with Barringer in drawing up a "syndicate agreement" under which Colvocoresses would organize the Meteor Crater Exploration and Mining Company and help obtain subscriptions to its stock to the amount of $500,000, at which time the ninety-nine-year lease on the crater property that Barringer agreed to execute on behalf of the Standard Iron Company would become effective. The company would be authorized to issue 10,000 shares of preferred stock with a par value of $100 per share and 12,000 shares of common stock with no par value, 2,000 shares of which would go to Barringer and his associates in return for the assignment of the lease.[32]

Neither Barringer nor Colvocoresses was under any illusions about the highly speculative nature of the venture. "If the mass is found in the condition we expect," Barringer wrote Alexander the same day of the agreement, "the gamble would be repaid perhaps 200 to 1. . . ."[33] And to Thomson, he noted that Colvocoresses "has said over and over again that it is the greatest gamble on earth, and should be regarded by every investor as a gamble."[34]

A few days later Thomson sent Barringer copies of Gifford's papers. Barringer could not accept Gifford's "new meteoric hypothesis" any more than he could accept Merrill's suggestion that the Meteor Crater meteorite had been vaporized and had exploded. Gifford's idea was not only persistent, he had found, but lately it seemed to be gaining ground. As recently as January he had argued the point with geologist Benjamin L. Miller of Lehigh University, who proposed that all of the meteoritic mass had been blown out onto the surrounding plain and had subsequently disappeared through oxidation.[35]

Barringer now voiced two "principal objections" to Gifford's thesis. "First," he advised Thomson, "that he has the same idea that we have not been able to get out of the minds of Merrill and others that a meteorite entering the earth must of necessity explode. This has always seemed silly to me and I can see no more reason why it should explode or do otherwise than keep penetrating until it comes to rest."[36] No meteorite had ever been known to explode at impact with the earth, he pointed out. Moreover, it was "difficult to see how iron meteorites, or a cluster of meteorites, could explode," and here he cited what he considered a "fatal objection . . . our shale ball and Canyon Diablo meteorites when sawed in two and etched show the Widmanstätten figures which, as I understand it, utterly disappear when heated to 800°C."

Second, he disputed what he thought was Gifford's "really ridiculous idea that the meteorites penetrating the moon, say to a depth of a mile, explode and the material ejected by the explosion comes out of a small central vent and is perfectly equally distributed in a circular ring. This to my mind is absurd." Shell craters in the war provided "strong evidence" against this idea, as did the craters produced by mining explosions in the nitrate fields of Chile. "The result of the explosion," he declared, "is that the dislodged material is thrown high in the air and upon falling is scattered around about in no very definite order, and the hole is usually irregular, not perfectly round, and there is certainly no central hill left."

Men like Gifford, he fumed, "soon become befogged in their own reasoning," and he had "no patience with certain scientific men who spend days and nights in tremendous efforts to prove conjured up theories, mathematics and whatnot, of the causes of effects which to common sense thinkers are obvious. So very rarely do we find their reasoning to be clear and crisp and easy to follow, simply because it is not based on *common sense* [Barringer's emphasis]. . . ."

Barringer also attacked Gifford's thesis in a long letter to Charles P. Olivier, an astronomer at the University of Virginia's Leander McCormick Observatory, adding two more arguments that dealt specifically with Meteor Crater. Olivier, an authority on meteors and meteorites who had just published a book on the subject in which he opted cautiously for an impact origin of the Arizona crater,[37] would eventually become a modest investor in the Meteor Crater Exploration and Mining Company and an important Barringer ally in the Coon Mountain controversies.

"The best physicists whom I have consulted," he advised Olivier, "tell me that such an iron mass as made the crater could not be volatilized and could not explode."[38] Had volatilization occurred, he argued, "this gas would immediately unite with the oxygen in the atmosphere to form an incredible amount of iron oxide which would have certainly discolored the millions of tons of snow white silica and covered the plain around for quite a distance with iron oxide." Also, he pointed out, "the shape of the hole absolutely precludes the possibility of any such explosion as Dr. Merrill seems to have imagined. We have found meteoric material 1360 feet beneath the surface of the southern wall. In the event of such an explosion as he imagines, there certainly would be no southern wall, with the peculiar uplifted and arched portion of the strata, as exists today."

Later, answering a query about Gifford's thesis from a physician friend in Philadelphia, Barringer noted that "while he accepts the impact theory of the origin of *typical* craters [Barringer's emphasis], his visualization of how they were made is absurd. . . . [T]he impact alone is sufficient to explain every feature of the craters and their rings of ejected material. Rifle bullet experimentation is convincing as to this. . . . The trouble with so many of these scientists like Gifford is that they are not practically minded."[39] And to Thomson, he quoted a remark, probably apocryphal, which he attributed to the then-President of the United States, Calvin Coolidge. The taciturn, literal-minded Coolidge, he wrote, when asked why he no longer rode a tall, lanky horse, had replied: "It gets my feet too far off the ground." In contrast, Barringer opined, "many of these scientific fellows evidently like to soar around in the empyrean, caring not a rap whether their theories will stand the acid test of common sense and observed facts."[40]

Barringer's concern here was heightened by the fact that by this time Colvocoresses was also worried about the possibility that the meteorite might have been vaporized and had exploded. On his return to Arizona, he had talked with Milton Updegraff, a former U.S. Navy astronomer and mathematician living in retirement at Prescott, who was preparing a paper for the American Association for the Advancement of Science "to show mathematically," in

Barringer's words, "that our meteorite was volatilized upon impact and *utterly disappeared* [Barringer's emphasis]."[41] Moreover, Colvocoresses had undertaken the task of rewriting the draft of the prospectus for the proposed exploration company and now felt, to Barringer's dismay, that the possibility should be at least mentioned in the new version. In this, ironically, he was simply following Barringer's own dictum that "We should prepare this statement with great care, so as to produce the best possible effect upon our friends whom we shall ask to go into the enterprise without misleading them in any way."[42]

Barringer quickly marshaled his forces. To Magie he wrote that Colvocoresses wanted, "evidently for the purpose of proving Professor Updegraff wrong, the results of such mathematical reasoning as you have devoted to the possibility of the mass having been volatilized, bearing in mind always the known facts which to my unmathematical mind are absolutely conclusive against such a possibility."[43] Magie was also to bear in mind their "old guess" that the meteorite had struck at a velocity of no more than two miles per second. "I do not think the mass was greatly retarded by the earth's atmosphere," he explained, but that there had been "a contest with gravity . . . indeed the mass may have almost encircled the earth before it came in." To Thomson, he expressed his alarm with full candor. The "poisonous seeds of doubt sown by Updegraff have taken root and have grown in his [Colvocoresses's] mind," he wrote.[44] "Now to have this contingency suggested in the Prospectus is not at all to my liking. . . . I do not wish to defeat its purpose by having included in it even a reference to these fool theories of volatilization or explosion. While I admit they are unworthy of discussion, yet the appeal they make is wonderful. . . . If such theories make an appeal to such a clever man as Colvocoresses, they will undoubtedly appeal to prospective investors, especially locally, and quite possibly scare them off. . . . It does seem to be an almighty job to get others to interpret the evidence as we interpret it."

To Colvocoresses himself, he dispatched two lengthy letters summarizing his arguments against volatilization and explosion on any grand scale, while conceding that "We have evidence that a very small portion of the outside members of the cluster was partially volatilized."[45] Stains on the Variety B of the metamorphosed sandstone, he noted, reacted for nickel, and thus "Here we have an undoubted effect of vaporization on a very minute scale and, what is important in this connection, this effect is conclusive as to the inevitable staining effect." How then, he asked, can we explain the snow-white silica? To argue that millions of tons of iron oxide "has utterly disappeared," he added, "is, in my opinion, to argue in favor of an impossibility." Moreover, the "finding of undoubted meteoric material underneath the southern wall makes any such theory of explosion untenable, because the southern wall would be blown away and the crater would not show the remarkable symmetry and other evidences of the meteoric mass having approached from a generally northerly direction."

Colvocoresses also inquired about Gilbert's early explosion theory. Barringer replied that Gilbert had published only one paper about the crater "in which he gives it as his opinion that the hole is due to a steam explosion, which opinion has been followed ever since by the United States Geological Survey, for he was a big bug among them," and then reviewed his objections to Gilbert's hypothesis at length.[46]

RED TAPE

The problem was resolved by a compromise. When the new version of the prospectus went to the printer in August, it contained the parenthetical statement that "A few scientists have suggested that a large portion of the impacting cluster was volatilized or thrown out by an explosion, but the evidence is conclusive against such theories."[47] It also contained other caveats that reflected Colvocoresses's reservations. Not only did it state unequivocally that an investment in the enterprise "will be purely a gamble," but it noted that it "is possible, but extremely unlikely, that the main body of the meteoric mass may be of different composition from the fragments found immediately surrounding the crater, or it may have been altered in composition by leaching and the values so disseminated as to make mining unprofitable."

Even with these concessions to the uncertainty of the venture, however, Barringer's lawyer, Herbert R. Smith, after reading the text, sounded a prophetic note of warning. "Mr. Smith," Barringer advised Colvocoresses, "seemed to object to the use of the word, 'Prospectus,' saying that conceivably it might in some way make trouble for us because of the 'Blue Sky' laws of certain states."[48] Barringer does not seem to have taken the warning too seriously, but some three months later he would be battling the Pennsylvania bureaucracy over that state's "Blue Sky" law regulating stock subscriptions and, furious over the delay in his plans, would be forced to go to court to settle this particular controversy.

Barringer, indeed, was now confident that the final obstacle to the new venture had been cleared away and even thought that work at the crater "may be started this fall."[49] But initially, at least, finding backers willing to risk substantial sums in the scheme again proved difficult. Once the lease was signed early in August,[50] and even before the prospectus was in hand, Colvocoresses had sounded out Barringer's old friend, Ricketts, and still another prominent

Arizona mining engineer, John C. Greenway, on the proposition. But the two men, he soon found, thought that Standard Iron's share of the profits—50 percent after the original investment in the exploration company had been repaid with interest—was too high. Colvocoresses forwarded their complaint to Barringer, who was unsympathetic. "I cannot agree to any alteration in the royalty terms of the lease," he replied, suggesting that Colvocoresses drop his negotiations. The "terms are not unfair and they are exactly the same as were agreeable to the U. S. Smelting, Refining and Mining Company who did not have the evidence of the last drill hole. . . ."[51]

Barringer also directly encountered the same complaint. Early in August he wrote to Barbour, the Harvard naturalist, enclosing his papers on the crater and explaining the new venture in great detail.[52] He quoted Magie and Thomson that the mass was "probably" not less than 400 feet in diameter and weighed "in the neighborhood" of 10 million tons. Once the mass was reached, he and Colvocoresses planned to mine not less than 500 tons per day, or 150,000 tons a year, for an annual profit of $3,750,000. "It is not improbable that the profit will be greater than I have stated, unless some unexpected discovery is made. . . ." Nor did Barringer neglect the scientific aspect of the enterprise, even borrowing a phrase from Gifford for the purpose. "I need not point out to you that the actual encountering of the meteoric mass with shaft and drifts would throw a flood of light on the science of cosmology and, in connection with what is observable on the lunar surface, would go far toward proving the accuracy of the planetesimal theory." Barbour, he urged, should not only join the venture himself, but should send him names of others in the Boston area who might be interested.

Barbour eventually did invest $5,000 for fifty shares of Meteor Crater Exploration and Mining Company stock;[53] but more immediately, he asked and received Barringer's hearty approval to send Barringer's letter and his papers to George Russell Agassiz of Boston, grandson of the noted naturalist Louis Agassiz.[54] George Russell, who styled himself simply as a "capitalist," quickly wrote Barringer to express his interest, to ask when they might meet to discuss the matter, and to note that he had visited the crater himself in 1909 with astronomer Percival Lowell, a fellow Boston Brahmin and founder of the Lowell Observatory at Flagstaff.[55] Agassiz, Barringer exulted to Barbour, "tells me they predicted that the meteoric mass would be found exactly where it has been found";[56] and in a second letter, he added that "It is no small satisfaction to me to realize that he and Lowell agreed years ago with Thomson, Magie and myself as to the probable location of the meteoric iron mass which we all agreed undoubtedly made the crater."[57] Barringer and Agassiz soon arranged to meet in Boston in October;[58] and the event was quite a gathering, with

Barbour and Thomson attending along with Colvocoresses and Barringer's geologist son, Reau, coming from Arizona.[59]

Agassiz, however, although deeply interested in the proposal, also objected to Standard Iron's eventual 50 percent share of the profits. "After conferring with some big capitalists in Boston," Barringer reported to Bennitt, "Mr. Agassiz has taken the same attitude that Ricketts and Greenway took, namely that the owners of the company are asking too much," adding that "they overlook the fact that more than $350,000 has already been expended on exploration."[60] But a few days later Agassiz had "second thoughts," as Barringer jubilantly informed Colvocoresses, and forwarded his agreement to subscribe $30,000 for 300 shares of stock.[61] Thomson now also subscribed $5,000[62] and the subscription campaign seemed well on its way. Only one minor problem must now be settled, Barringer felt. Late in October he sent Reau to the state capital at Harrisburg "to get permission from the so-called 'Blue Sky Commission' of the state for me to speak to a number of my friends here in Philadelphia."[63]

Clearly Barringer expected such permission to be granted as a routine formality, but he soon found he was mistaken. Now he warned Colvocoresses that "from the difficulty which I am having here I think it would be well for you to find out at once from the chairman of the Corporation Commission of Arizona just what you can or cannot do. Under Pennsylvania law, the view which you take of the situation is not a proper one. The physical delivery of the stock is not the thing with which our banking commission is concerned. It is the soliciting of subscriptions from any citizen of the state and we are not allowed to solicit these subscriptions without the approval of the commission. I find these rules are much more strict than I had supposed."[64]

Yet subscriptions kept coming in. Early in November, two friends of Barbour's, John F. Harris and H. R. Winthrop, partners in a New York brokerage firm, subscribed $15,000;[65] and a few days later Barringer received, perhaps with some trepidation, a $10,000 subscription from his Philadelphia friend and fellow Jefferson Medical School board member, Alba B. Johnson, former president of the Baldwin Locomotive Works.[66] "I had done little more than mention the subject to him," he explained to Barbour, "since I have been advised that I should not try to get subscriptions in Pennsylvania until I get permission to do so. . . ."[67] He remained confident, he assured Barbour a few days later. "I am still awaiting the decision of the Banking Commission at Harrisburg before tackling my friends here in Philadelphia. . . . I have no doubt that the necessary permission to solicit subscriptions will arrive shortly. Certainly there is no reason why it should not be granted."[68]

But now he was also complaining to Thomson that "being obliged to comply with all this red tape and to put up with the delay is very annoying."[69] His frustration increased late in November when the deputy handling his application, a bureaucrat named Einar Barford, sent a series of questions to U. S. Smelting officials about the crater[70] and directed Barringer to submit a financial statement for Standard Iron along with other documents, including a list of all assays made of the meteoritic material at the crater.[71] "These fellows in Harrisburg," he fumed to Colvocoresses, "are so ignorant that they seem to be rather appalled at the novelty of the undertaking. On the other hand, all kinds of trolley, banking and power schemes get through without any difficulty. Even if I am refused permission to sell stock in Pennsylvania I am by no means discouraged at the prospects of selling it elsewhere in considerable blocks." Subscriptions, he added, including his own for $10,000, now totaled $75,000.[72]

On December 2, a coldly furious Barringer learned that his application would be denied, and in terms which characterized the venture as "unfair, unjust and inequitable."[73] "I regret to tell you," he wrote Colvocoresses immediately, "that an ignorant deputy on the Banking Commission, to whom the matter was referred, is going to turn down my application . . . largely because the U. S. Smelting did not proceed to expend $250,000 to sink the shaft. On this theory whenever a mine is turned down or abandoned, it can never be taken up again. . . . You can easily see the animus of the deputy by the language he employs. . . . However this fact, just as I was greatly encouraged, puts me in a very difficult position for almost certainly he will notify the other states like New York and Massachusetts of his action." It was important, he added, to get permission in Arizona, but this should not be done as yet "as we are going to try to have the decision set aside. To say that I am 'hot under the collar' is to put it mildly."[74]

Two days later he notified Jennings, Thomson, and Magie of the decision, excoriating the "ignorant deputy, discussing possible courses of action, and urging them to be present at the formal hearing on the matter that was set for December 17 at Harrisburg.[75] "There may be another way," he suggested to Magie. "Colvocoresses will get registered in Arizona. Once this is done he can approach anybody anywhere in the United States, Pennsylvania included, and as I now think he can refer people to me for information. He, and not I, will be the solicitor." Still, he added, "I think I am better able to raise the money than Colvocoresses, and I wish to be allowed to do so." The decision was a "terrible setback" and he was "greatly enraged that an ignoramous of a Norwegian . . . should write as he has done." But: "Be all this as it may, we are up against it good and hard, so let us try to get him reversed if we can."[76] He was now also convinced that Barford was personally biased against the enterprise, writing Alexander that Barford's father "lost all his fortune in fraudulent mining or wildcat oil investments."[77] And in thanking Jennings for a letter strongly supporting his position, he wondered skeptically "whether such a zealot, who seems to be fanatical in his opposition to all mining companies on account of the loss of his father's fortune in them, will be converted by this letter when it is presented at the hearing. . . ."[78]

The hearing itself proved to be routine; Barford was not converted by Jennings's letter or by others Barringer submitted. Barringer, however, was advised that he could make changes in the prospectus and refile his application within thirty days. By early January, revisions in the financial plan, to which Barford had objected strenuously, were completed. A major change provided that stockholders would not only recover their investment with interest from 75 percent of the profits, but would receive an additional $2,500,000 before the profits were distributed on a 50-50 basis with Standard Iron.[79] The revised prospectus, along with his new application, was submitted to the Bureau of Securities late in January. If he was turned down again, Barringer now resolved to appeal to Pennsylvania's Dauphin County Court of Common Pleas; and when Colvocoresses questioned this course, he informed him bluntly that in such an event he would be forced to appeal "because of my reputation, whether you approve or not."[80] Barringer himself, he advised Colvocoresses a few days later, would pay the costs of the appeal, "but you must back me."[81]

Barringer, however, was cautiously optimistic that the new application would be accepted. "There can be no doubt," he wrote to his nephew Paul B. Barringer Jr., a New York City attorney, "that the ignorance and perverseness of the deputy who is practically in charge of the Bureau has tied my hands so that I am unable to make any effort toward obtaining subscriptions until he is made not only to understand the situation much more clearly than he does at present, but until his violent objections to the old financial plan are removed. It would seem in the [new] financial plan that there is nothing which even the most captious can justly criticize. . . . So we are hoping to get the Bureau to reverse its decision under the new set-up and, if not, I personally feel we shall win out on appeal to court." But he was also worried about the attitude that New York officials might take, and requested his nephew to contact "the appropriate assistant Attorney General at Albany . . . and get him to understand that this matter is sponsored by the very highest grade people in this country, both scientists and business men . . . very substantial people, intellectually and financially, men who could not for a moment think of becoming mixed up with anything which could be criticized. . . . Do not fail to tell him," he added, "that the offering is in no sense to be a public one and that only men of large means, who can afford to take the gamble, will be approached."[82]

Moreover, to Barringer's satisfaction, his subscribers continued to support him and were willing to abide with the enforced delay in the enterprise. Indeed, only a few days after he learned that his application would be denied, he had received Barbour's promised subscription, along with a $5,000 subscription from still another of Barbour's friends, Boston financier Philip Stockton.[83] Only Harris and Winthrop withdrew from the venture following the December 17 hearing. "Capital is proverbially timid," Barringer reminded Barbour in reporting their defection, "and as they know very little of the merits of the proposition, I can hardly blame them for refusing to go ahead when they suppose that the Bureau of Securities . . . had carefully gone into the matter." Nonetheless, he added pointedly, "I naturally do not wish Agassiz and your other friends to get cold feet where there is not the slightest justification for doing so. . . ."[84]

"DOODLE-BUGS"

Barringer largely suspended his money-raising activities during the winter of 1925–26 while the question of whether he would be allowed to sell Meteor Crater Exploration and Mining Company stock in Pennsylvania was pending. But a number of other matters relating to the crater now occupied his attention. In December, physicist David L. Webster of Stanford University asked to visit the crater; and Barringer, after receiving assurances from Russell, made the arrangements.[85] Webster spent New Year's Day there,[86] and that same night thanked Barringer for the opportunity "which is indeed almost the equivalent to a visit to the moon. . . . [T]he whiteness of the powdered silica," he wrote, "impresses me as it did you as strong evidence against the idea that much iron was vaporized. As to where the energy did go, however, I am inclined to question how much of it could be spent in pulverization without appearing as heat. . . ."[87]

Through the winter, too, the crater received a flurry of publicity and the promise of still more to come. In the December issue of the *Engineering and Mining Journal,* editor Josiah Edward Spurr took note of U. S. Smelting's abandonment of its crater operation and remarked that "repeated attempts have not located any buried meteorite; and it is improbable that further efforts will be made."[88] Barringer immediately fired off an answer which was published in January. U. S. Smelting's exploration, he replied, could not be dismissed as "unsuccessful" for its drill had "encountered what is beyond doubt the upper part of the buried cluster of iron meteorites" and had thus located the impacting mass "just where I predicted."[89] Spurr's editorial was picked up by the popular magazine *Literary Digest* in February, and Barringer promptly urged the *Digest's* editor to publish his reply as well.[90]

Barringer was far more intrigued by brief items that appeared at this time in both the *Journal*[91] and the *Arizona Gazette*[92] in Phoenix, reporting that a self-styled inventor named William A. Sharpe "claims he has located the great meteorite," in the *Journal's* words, at a depth of 1,400 feet under the south wall with a device he called a "radiocameraphone." "I can hardly believe that Mr. Sharpe has secured accurate 'photographs' with his machine," Barringer wrote immediately to Thomson, "but he certainly seems to have found something and that something can only be the buried meteoric mass."[93] Sharpe, he noted, claimed to be the great-grandson of the inventor of the Sharpe's rifle and to have sold a patent for an automobile starter to General Electric. Would Thomson check on both Sharpe and his device? "If he is not faking his results, it may be very worth while to have him return to the crater with Mr. Colvocoresses and spend a week there making 'Radio Photographs' around the entire circumference. . . ." He was "anxious to know whether his radio camera is based on sound principles and whether its results are even measurably accurate. If so, it will help tremendously in getting the money to sink the shaft."

Barringer also telegraphed a number of his associates in the West, asking them to make inquiries about Sharpe and his invention.[94] To Colvocoresses, he confided that he was "quite skeptical," pointing out that Sharpe could have known the location of the mass before he went to the crater, but adding that "Magie sounds very ready to believe that Sharpe had invented an instrument which will give the results claimed for it."[95] To fellow mining entrepreneur John Hays Hammond, once associated with Cecil Rhodes in South African gold mining ventures, he explained that although Sharpe's "doodle-bug" might be of some help in selling subscriptions, "Naturally . . . I do not want the project in any way to be identified with a fake device."[96]

The verdict came within days, and it was against Sharpe and his "doodle-bug."[97] On February 12, Barringer telegraphed Colvocoresses that he had been "Reliably informed that no dependence should be placed on him [Sharpe]." He was, he wrote geologist R. C. Coffin, one of his informants, "relieved to know from you that his results were probably faked."[98] The incident, however, stirred Barringer's interest in such technological devices, and he continued to be curious about such methods for another year until he concluded that no further proof was needed as to the location of the meteorite. "I have investigated doodle-bug machines to a finish," he wrote then. "We have located the mass, or at least a portion of it, and found it not only where we expected to find it but that it was, also as expected, in a thoroughly oxidized condition."[99]

No sooner was Sharpe disposed of, however, than the *Engineering and Mining Journal* published a letter from petroleum geologist Dorsey Hager of Seattle, Washington, who reported that he had examined the crater five years before and had then written an unpublished paper in which he had concluded that it was simply a sinkhole.[100] "The so-called Meteor Crater, so far as I can see, is a big hole in the heart of a perfect elliptical shaped dome in sedimentary rocks," he now wrote, and "can be explained as due to solution of the underlying gypsum, salt beds and limestone beds which occur in the Supai and Red Wall formations. . . . Numerous sink-holes occur in this area," he pointed out, adding that "so far as I can see there is no evidence pointing to either volcanic action or meteoric action. Meteoric fragments are quite common in that part of Arizona and are not confined to Meteor Crater."

Barringer, or rather his son Reau in his father's name,[101] replied to this with heavy sarcasm in a letter to the *Journal* in March.[102] "We all owe Mr. Hager a debt of gratitude," it began, "I, because he has shown me that all my twenty-three years' work, which I had thought proved the origin of the crater to be meteoric, has been wasted; and all other geologists as well, for he has proved that sink-holes can occur in sandstone, and can cause rocks to fall upward out of a hole as well as downward into it. . . . What a pity it was that Mr. Hager's paper . . . was not published, for it would have saved some of the most distinguished men in the country from the ignominy of forming opinions which were in direct conflict with the result Mr. Hager obtained in his investigation. And what would not the Crater Mining Co. have given for the knowledge which Mr. Hager so unkindly withheld from them, allowing them to go to the expense of drilling for the meteorite when a word from him would have shown them that no meteorite could be there. . . ?"

And as to meteorites being "quite common" in Arizona, he exclaimed: "To think that I have continuously encountered meteorites in Arizona without recognizing them, thinking in my ignorance that the meteorites around Meteor Crater (within five miles of which have been found 90 percent of the known iron meteorites of the world) and the aerolite which burst over Holbrook some years ago were the only true visitors from the sky to this region. It is obvious that we mining geologists have a great deal to learn from some of our petroleosophical colleagues." Hager published nothing more about the crater for twenty-three years.[103]

Barringer learned of still another article concerning the crater only through a Spurr editorial in the *Journal* early in April.[104] Spurr, after recalling Barringer's earlier statements, found it "startling . . . to read in the *Santa Fe Magazine* for March, under the caption of 'The Eighth Wonder of the World,' an article credited to the *American Weekly Magazine* that proclaims in most sensational language that a 'monster meteor of iron, liberally mixed with diamonds and precious metals of priceless value,' has been found beneath the crater's rim." The article, he noted, reported that the "meteor" was 500 feet in diameter, weighed 12 million tons, and had been "finally reached by driving an inclined tunnel for some distance outside the east rim, westward and downward at an angle of 45 degrees." Its author, Spurr added with a hint of incredulity, had calculated that the platinum in the meteor alone would be worth $1 billion! Despite the errors and gross exaggerations in this article and their repetition by Spurr, however, Barringer had no complaints and complimented Spurr on an editorial that was "very much to the point."[105] Clearly, such reports could only help his enterprise.

Early in February, Barringer also received requests for information on the crater from H. J. Minhinnick, editor of the *Verde Copper News* in Jerome, Arizona, and William D. Boutwell, who was writing an article on the crater for the *National Geographic Magazine*.[106] Boutwell's project seems to have originated in correspondence back in November with *National Geographic* editor Gilbert H. Grosvenor concerning the aerial photographs of the crater taken earlier by Army Captains Stevens and MacCready. Grosvenor then had shown interest in publishing such an article, but Barringer had been wary. Noting Gilbert's early investigation, the U.S. Geological Survey's tacit support of his conclusions, and thus his own "fear that the authors . . . may incline to the belief in the steam explosion theory," he had suggested that scientists outside the Survey, such as Fairchild, Thomson, and Russell, be consulted.[107] Now, however, eager for favorable publicity to bolster his case against the Pennsylvania "Blue Sky" commission and to aid in selling stock subscriptions, he responded enthusiastically to both Minhinnick and Boutwell, offering his full cooperation.[108]

Minhinnick's article, which appeared on April 20,[109] described in straightforward terms the crater, the history of its exploration, and its commercial possibilities as outlined by Barringer, and proved to be effective far beyond the confines of the small mining town of Jerome. Almost certainly, it inspired a similar report in the *Arizona Gazette* in Phoenix a week later.[110] Barringer had also urged him to send it to the *Literary Digest;* and Minhinnick, after revising it somewhat, now submitted it to both the *Digest* and *Progressive Arizona*.[111] The latter journal published it in July;[112] and in August the *Digest* printed lengthy excerpts from it along with brief passages from Thomson's

1912 *Scientific American* article, and from "a recent letter from Mr. Barringer" citing the scientific benefits to be derived from the further exploration of the crater.[113]

Boutwell's article, on the other hand, has a longer and more turbulent history. Barringer read his first draft early in April, making a few minor corrections and suggestions but finding it generally satisfactory.[114] "I do not know how much Dr. Grosvenor is going to 'edit' it," he confided to Russell at Princeton, "but if he will let it go as it is, it will be an interesting article and accurate enough for all practical purposes."[115] Barringer expected that it would appear quite soon, but the *Geographic*'s editor did, in fact, order some significant revisions. When it finally was published more than two years later, it would stir still another bitter controversy.

A LEGAL VICTORY

More immediately, however, his controversy with the Pennsylvania Banking Commission had come to a boil. In mid-March the Bureau of Securities again denied his application for permission to solicit stock subscriptions, and he had moved swiftly to appeal the decision to the Dauphin County Court of Common Pleas.[116] He now began to marshal evidence for the trial, writing first to his early ally, Fairchild, and then to Merrill and Campbell to request letters certifying the meteoritic impact origin of Meteor Crater.[117] Magie was to write such a letter himself and get similar letters from Russell, Webster at Stanford, and Princeton geologist William Berryman Scott, while Thomson would seek supporting statements from geologists William M. Davis of Harvard, H. E. Gregory of Yale, and Farrington, the Field Museum's expert on meteorites. Barringer was, he assured Magie, confident of the outcome of the case, as Common Pleas Judge Wickersham, who was to preside at the trial, had reversed Barford nine times in fourteen appeals during the previous year.[118]

Through April and early May, Barringer plunged vigorously into preparations for the trial, corresponding and conferring with his Philadelphia and Harrisburg attorneys, asking for additional letters of support, obtaining new assays of the meteoritic material at the crater, and assembling necessary documents. The prospectus also was to be revised again, for Barringer now agreed to liberalize its financial terms by cutting Standard Iron's share of the eventual profits from 50 to 35 percent, a change Agassiz, in particular, heartily approved.[119]

The letters Barringer sought from prominent scientists soon began to come in, and he was greatly encouraged by the response. He was particularly pleased with Russell's letter, not untypical, in which the Princeton astronomer declared the meteoritic origin of the crater to be "not only entirely reasonable, but to have strong observational support and to be the most satisfactory scientific explanation of the facts. There is no absurdity, and what is more, no improbability, in the belief that huge masses of meteoric matter have struck the earth from time to time. The attempt to find it by mining operations," he added more specifically to Barringer's point, "appears to me to be a thoroughly legitimate venture."[120]

Barringer and Magie, of course, would go to Harrisburg for the trial, but as Barringer explained to Thomson, "he and I are interested witnesses," and thus it was desirable to have as many other witnesses as possible.[121] Neither Thomson nor Russell were able to make the trip, however, nor could Jennings, on whom he was relying to explain U. S. Smelting's position in view of Barford's reference to its abandonment of its crater operation.[122] But Agassiz and Alba Johnson could go, and Barringer persuaded Jennings's colleague, Moore, to take Jennings's place on the witness stand.[123]

The trial was held May 25 and 26, 1926, with Arthur H. Hull of Harrisburg serving as Barringer's local attorney. Agassiz, Magie, Johnson, Alexander, and Moore all testified on Barringer's behalf, while Barford was the sole witness for the state. Although Judge Wickersham reserved decision pending arguments on the interpretation of the Pennsylvania Supreme Court's language in an earlier case involving the limits of the Banking Commission's discretion, he strongly indicated that he favored granting the appeal.

Barringer was jubilant. "The result," he wrote Thomson, "was that the court, at the conclusion of the trial yesterday, said that the state had utterly failed to make out a case and that it would render a decision immediately in favor of the appellant (ourselves) if the interpretation of a recent decision of the Supreme Court as to Barfod's [*] discretionary powers is such as Judge Wickersham . . . has placed on it. . . . The judge was evidently greatly interested in the testimony and arguments in a case so far out of his ordinary experience."[124] To Agassiz, he declared that the "testimony of Barfod was the most wonderful exhibition of venomous and almost insane prejudice against mining in general, and ourselves in particular, in my experience. . . . He tried to find fraud and concealment in nearly every paragraph of the prospectus. He even complimented Colvocoresses upon his ability to prepare a prospectus which was legally unattackable but was full of intentional misrepresentation and concealment of important facts." The judge, he added, "promptly and

---

* Barringer consistently referred to Barford in his correspondence as "Barfod."

very properly ruled out any testimony with regard to the Standard Iron Company as being irrelevant; otherwise I suppose I should have been in for all kinds of vituperation."[125] And to Moore, in thanking him for his testimony, he opined that Barford was "more or less insane" and chortled that Hull "tore him to pieces" on cross-examination.[126]

The legal question of interpreting the earlier court decision, however, tempered this triumph somewhat, for Barringer was annoyed at the further delay. Judge Wickersham, he explained to Agassiz a week later, was concerned whether the Banking Commission's approval of the "business expediency" of the enterprise was required by the higher court's interpretation of the state's "Blue Sky" law. "I fear that he has concluded that this was an excellent opportunity to have the Dauphin County Court once [and] for all interpret the meaning of the language of the Supreme Court and set a definite limitation to the power of this ignoramus and fanatically prejudiced man, Barfod. . . . [A]pparently he wishes it argued before the full bench next October, probably thinking such a decision would be more effective. . . . So it seems that we are to be made the goat because of the righteousness of our case." The "rather curious legal situation," he added, "to a great extent ties my hands until next October, but not those of Mr. Colvocoresses."[127] And to Hull, who was to argue the legal point, he wrote that "only one" interpretation was possible, i.e., "that the Act does not contemplate any investigation by the Commission into the actual security of the investment . . . but an investigation to determine whether securities are being offered to the public 'honestly and in good faith' without 'an intent to deceive or defraud.' Contrary to Mr. Barfod's sneering insinuations, there is not a single deliberate misstatement of fact of a character on which investors would be apt to rely. On the contrary, they have been told the simple truth and the whole truth, and moreover distinctly informed that their investment should be regarded as a gamble."[128]

In the event, his attorneys succeeded in having the arguments on the legal point moved up to June 25, when they also planned to urge that a decision "be handed down before the court adjourns for the summer" on the ground that Barringer and his associates were "suffering from the delay."[129]

THE ODESSA CRATER

The nuances of the law and judicial rulings, however, were not Barringer's only concern at this time, for he now learned that Meteor Crater was not unique. Early in June he was startled to read a brief letter to the editor in the *Engineering and Mining Journal* from Arthur B. Bibbins of Baltimore, Maryland, under the heading, "A Small Meteor Crater in Texas."[130] Noting the recent publicity that the Arizona crater had received, Bibbins recalled that in 1921, while he was prospecting for potash and oil near Odessa, Texas, a rancher had brought him a "sample of iron ore" he had found "at the edge of a so-called blow-out." A simple test showed that the sample was meteoric, Bibbins wrote; and when the rancher had taken him to the site, he found a second, smaller sample. The "blow-out" was 10 to 30 feet deep and "a few hundred feet in diameter"; and at one point where there was an outcrop of subsurface rock, the strata were "dipping rather steeply away from the center of the blow-out." Bibbins had sent the larger sample to Merrill in Washington, who analyzed it and subsequently pronounced it meteoritic in a report in the *American Journal of Science*.[131] Ironically, perhaps, Merrill found no further significance in Bibbins's find and thus missed his second opportunity to participate in the discovery of a terrestrial meteorite crater.[132]

Barringer and his son Reau, on the other hand, immediately recognized the possibilities, both scientific and commercial, presented by Bibbins's "blow-out." Barringer quickly contacted an associate in Tulsa, Oklahoma, who was planning a trip to El Paso, asking him if he would "stop at Odessa, find the hole mentioned, and if it bears any resemblance to our Meteor Crater get an option. If the description is correct," he added, "it might be very valuable to us, and of great scientific interest."[133] Reau, who was about to leave for Arizona to take a job inspecting mines there for Colvocoresses, wrote Bibbins directly for more information, and then rearranged the itinerary of his trip west to include Odessa.[134] Barringer's oldest son, Brandon, also sought verification from an Odessa bank that the "blow-out" actually existed.[135] The "blow-out," Barringer advised Thomson, might well be a "baby Meteor Crater."[136]

Yet Barringer himself remained somewhat skeptical. "I hastened to find out . . . something about the author of the article," he explained to Colvocoresses a few days later, "and found out that he is distinctly of the same type as our late friend Mr. Sharpe; that no reliance is to be placed on what he says. However, he got an iron meteorite from someplace, and sent it to Dr. Merrill in 1921. . . ."[137] Barringer, in fact, was overly hasty in this assessment; Bibbins's account soon proved to be accurate enough as far as it went.

Reau visited the "blow-out," some nine miles west and south of Odessa, on June 24; and the following day Barringer confided "a remarkable piece of news" to Thomson.[138] Reau, he wrote, "telegraphed last night from Odessa, Texas, in code, that the so-called 'blow-out' IS a meteor crater 500 feet in diameter, that he had found meteoric iron around it, and that he was trying to get an option on it." From Reau's data and his own knowledge of the Arizona crater, Barringer estimated that the buried mass "should be a solid sphere or spherical cluster 50 feet in diameter and should be reached by a shaft at a depth of considerably less than 200 feet. Such a shaft could be very cheaply sunk." Reau, he added, had not been able to determine the direction from which the meteorite had come, and thus it had probably struck "at a high angle, and therefore distributed the rock fragments fairly equally on all sides."

To Reau, Barringer sent his congratulations at length.[139] "Other people may have suggested that it was a meteor crater, as they did in the case of the Arizona crater," he pointed out, "but you are the first to distinctly know, and I hope to be able to prove, that it is an impact crater. . . . In other words, this Texas crater is in exactly the same position as our crater was when Grove Karl Gilbert and his Geological Survey party visited it." It was "desirable," he added, "that little or nothing be said about this new find until we get title to it." But Reau should write a paper for the Philadelphia Academy of Natural Sciences immediately—"it will serve the purpose of your right of discovery. That is to say, your right of precedence over anyone else, the Geological Survey for example." Reau quickly sent a straightforward account of his investigation and his conclusions to Barringer, who promptly forwarded it to the Academy with his request to withhold its publication until further notice.[140]

In his letter to Reau, Barringer had also suggested in passing that the meteorite that had formed the Odessa crater "may easily have been a travelling companion" of the Arizona meteorite. Thomson now voiced the same idea, wondering "could not after all the new discovery have been formed at the time of the Meteor Crater, in which case the direction would be from the north? It seems unlikely that *two* [Thomson's emphasis] large bodies should fall in that vicinity when there is so much other earth's surface to hit. . . . In reality the distance apart of the two falls is not so very great as compared with the size of the earth itself."[141] And in a postscript, he added that he had "often noticed that shooting stars often come in groups of 2, 3 or more a few minutes apart."

Barringer also expounded this idea of simultaneous impacts in a letter to Agassiz. The Arizona, Texas, and other meteorites found in the Southwest and in northern Mexico, he speculated, might be "all from the same swarm."[142] Moreover, to the northeast, he had heard of a "round lake" somewhere in the upper Midwest, and meteorites were found in numbers in Kansas and Nebraska. "Perhaps the Anighito and other meteorites in Greenland were part of the same swarm." All this, he hastened to add, was "pure theory" and there was "not much to support it." But clearly he liked the idea. More than two years later, when Reau's paper on the Odessa crater was finally published,[143] he advised Merrill that "Personally, I think it not unlikely that the Arizona meteor and this meteor in Texas fell practically at the same instant. May it not be so that the earth did actually pass through the head of a comet?"[144]

The long delay in the publication of Reau's paper, incidentally, resulted from Barringer's inability to obtain an option to buy the forty-acre tract containing the Odessa crater from its owners, the Texas and Pacific Land Trust. These "hard-boiled hombres," he found, were well aware of the oil, gas, and mineral potential of the property, whether or not a meteorite was buried there.[145] After two years of frustrating negotiations, he finally concluded that further efforts would be futile, explaining to Merrill that "the syndicate is averse to selling such a small piece of land at any price which we could afford to pay," and adding that "the chances are many to one that the mass is too small to possess commercial value."[146]

Surprisingly, in view of this delay, Reau's revelations regarding the origin of the Odessa feature were not anticipated by any other scientists. Yet at least one eminent geologist, E. H. Sellards of the University of Texas, also made an on-the-spot investigation of Bibbins's "blow-out" in 1926, only to pronounce it an enigma before the Geological Society of America in December.[147] Sellards offered no less than five possible explanations—a volcanic explosion, the collapse of a salt dome, slow upward pressure exerted by the hydration of minerals, the explosive escape of gas from within the earth, and the explosive rebound from the fall of a meteorite. But: "Between these several hypotheses no conclusive choice can be made at present." As to the meteoritic hypothesis, he added, "It is, of course, possible that the occurrence of the meteorite fragments at this locality . . . is entirely fortuitous and without significance, so far as this structural feature is concerned."

With Judge Wickersham and his Dauphin County Court colleagues mulling over the legal niceties of Pennsylvania's "Blue Sky" law and Reau's paper on the Odessa crater safely on file with the Philadelphia Academy, Barringer left for Arizona early in July for a long-planned vacation with his family at a cabin he had rented for the summer at Iron Springs.[148] His first order of business, however, was to visit Meteor Crater; and he was apparently there when, a few days later, he received word that the court had reversed Barford's decision in what one of his attorneys, Charles H. Scott, called "a complete victory for Mr. Barringer."[149] His secretary, Isabelle Heller, immediately notified his friends and associates;[150] and his son Brandon sent him a copy of Judge Wickersham's written opinion, commenting that "I am a little disappointed that he did not say what he thought of Barford, but it is certainly extraordinary that he found 'the meteor.'"[151]

Barringer himself welcomed the decision, but he recorded his immediate reaction only in a letter to his Philadelphia attorneys dealing mainly with other matters, noting tersely that "Judge Wickersham's opinion is wonderfully satisfactory. . . . I did not dare hope that he would go as far as he has done."[152] Indeed, Barringer seems to have been more interested in hosting visitors at the crater and in enjoying northern Arizona's summer climate than in writing letters, for barely a half-dozen are recorded in his letterbook over the next two months.[153]

Moreover, the legal battle had cost him much in time and money—the bill from his Harrisburg attorneys alone was $2,020[154]—and he was now weary of it. There soon would be more important things to do. On August 14 the Pennsylvania Banking Commission issued a license to the Meteor Crater Exploration and Mining Company to solicit stock subscriptions in the state,[155] and Barringer quickly assured Barbour that "Now that the legal shackles have been removed, I shall get to work next month and do what I can to make a number of men of means realize what an opportunity is offered them. . . ."[156]

TEN

THE FINAL SHAFT

> The conclusion . . . is that (a) the mass of the meteorite cannot have exceeded 3×10^6 tons (I think this conclusion is reliable); (b) the meteorite was a *compact swarm;* (c) the mass may be as low as 50,000 tons.
>
> FOREST RAY MOULTON, 1929

Barringer's summer sojourn in northern Arizona may well have assured his survival through the next few years of minor triumphs and major frustrations over Meteor Crater. He was now sixty-six years old; and his health, although still quite robust, was not what it had been. He had long suffered from a painful and irritating eczema—"I have been poisoning myself eating eggs," he once remarked to Bennitt[1]—and his hearing was beginning to fail. Moreover, the pressures of the past few months had been particularly heavy, and just the prospect of his Arizona vacation seems to have helped to sustain him through the arduous preparations for his court challenge to the administration of Pennsylvania's "Blue Sky" law.

His two months in the sparkling summer climate of Arizona's high plateau country left him "greatly refreshed," he reported to Thomson on his return.[2] But he had not been idle. He had, in fact, made a "critical review" of the evidence at the crater "which has only solidified my conclusions." The crater, indeed, remained a preoccupation even during his occasional sightseeing excursions in that wildly scenic area. "I have never seen anything so beautiful as Oak Creek Canyon—25 miles south of Flagstaff," he exclaimed to Thomson. But he was more interested in the canyon's geology than in its aesthetics, and specifically in the fact that its brilliantly colored walls displayed an unobstructed panoply of the full sequence of rock strata found at Meteor Crater. "I wished for you time and time again as I studied these wonderfully beautiful cliffs," he explained to Thomson in a second letter, "for they afforded an opportunity to see just what the meteoric mass did before it came to rest."[3]

His excursions included two extended visits to the crater itself, the first in July. As a result, he and his eldest son, Brandon, had each written brief summaries of the arguments for the impact origin of the crater and the presence of a large meteoritic mass under its southern rim to be used for the edification of prospective investors.[4] Both suggested that the probable weight of the mass was at least ten million tons; Brandon, in addition, pointed out that it could be "profitably mined," for it was "not a solid body of iron, but a cluster of many millions of small pieces, such as have been found on the surface."

Barringer could now also add the names of two more prominent astronomers—Vesto M. Slipher and Carl O. Lampland of the Lowell Observatory—to his lengthening list of scientists who, in general at least, favored his views about the crater. "The Flagstaff observatory staff were at the crater while we were there," he reported late in July, "and went away 100% convinced as to the correctness of our deductions. The astronomers all thought that the mass must have been more nearly 800 feet in diameter rather than 400 feet, which diameter calls for 10,000,000 tons."[5]

His second visit to the crater, late in August, was with Colvocoresses and Fred M. Searls Jr., chief geologist for Colonel William Boyce Thompson's Newmont Mining Company.[6] Barringer had been trying to interest Thompson in the enterprise to mine the buried meteorite since the previous November, but had found that the ailing mining magnate preferred to handle such matters through his associates.[7] Searls seems to have been favorably impressed by what he saw and heard during the visit and presumably reported his impressions to his principal. Over the next nine months, Barringer engaged in what at one point he characterized as "on-again, off-again, Mulligan" negotiations with Searls and Philip Kraft, another Newmont executive.[8] Thompson initially asked that all the remaining stock in the Meteor Crater Exploration and Mining Company be reserved for him until he could fully investigate the proposition and decide whether or not to invest in the scheme.[9] But by November Barringer broke off the negotiations because, as he confided to Barbour, "the big gun's advisers . . . have indicated they will not recommend his taking over control of the enterprise unless I am willing to transfer to him 75% of ownership in the Standard Iron Company. This I have refused. So I am now 'on my way' to get subscriptions. . . ."[10] The negotiations resumed again early in February, however, when Thompson, through Searls, indicated he was still interested; and they continued sporadically through the spring.[11]

240   THE FINAL SHAFT

The final break came late in May 1927, with Thompson still insisting on at least a 51 percent ownership interest in Standard Iron, a condition that Barringer, of course, could not accept. At this point, Barringer bluntly informed Searls that all further negotiations would be "de novo."[12] But there were no further negotiations; Barringer had decided that he could raise the funds needed to sink the shaft without Thompson's help.

Barringer had returned from Arizona in mid-September by way of California, where W. W. Campbell, now the president of the University of California, had invited him to lunch and later discussed the crater with him at length. Campbell, he reported to Thomson

> is as certain that it was made by a meteoric mass as we are, but remarkable to say, he asserted time and time again that while the mass made the penetration proved by the last drill hole, it afterwards was blown out. . . . Campbell had to manufacture some other explosive force than steam and he then said that the rocks at the end of the flight of the meteoric mass would be gasified and this gas could furnish sufficient explosive force to blow a ten million ton mass out of the hole! Ye gods & little fishes! I did not argue with him in the first person singular, but told him that such distinguished physicists as yourself and Magie had said without reservation that such an expulsion of the mass as he had imagined was literally impossible, etc. His answer was that "these men do not have such an appreciation of the cosmic forces as I have!"

Campbell, he fumed, "stayed by his guns and we left him just as *sot* [Barringer's emphasis] as when we began our talk. . . . I rather hope he will write a paper and go on record in favor of such a fool theory. However, I hope he will not publish the paper before we succeed in raising the money to sink the shaft!"[13]

AN EXPLOSIVE REVELATION

But remarkably enough, it was Barringer himself who soon modified his thinking on this point and accepted the idea that the meteorite's impact had produced an explosion of sorts. Back in April, Stanford physicist David L. Webster had planted the seed for this particular conversion in his letter supporting Barringer's appeal to the Dauphin County, Pennsylvania, court. Webster, Barringer had remarked to Russell then, "advances a new thought, namely that there was a certain amount of excavating effect due to the heating and expanding of the air between the sand grains. . . ."[14] Now Thomson pointed out also that "there must have been, if any moisture existed in the rocks, a production of steam at high pressure on the impact. . . . That does not make the [crater] the result of a steam explosion; it still leaves the whole crater as the result of the impact of a meteoric body, and what is an explosion anyway? . . . It is the sudden expulsion of material outward from a center or line in all directions, according to the resistance in these directions, and whether that sudden expulsion be made by one means or another makes no difference. It is an explosion. . . ."[15]

The idea thus stated seems to have struck Barringer with something of the force of a revelation. "I entirely understand now your theory of the steam explosion," he replied quickly. "I did not know that the rending of individual sand grains would at the same time generate sufficient heat to cause any moisture in the rocks at the time of impact to expand to steam."[16] The effect, he added, would be "enormously assisted . . . by what I have called, for lack of a better term, 'shock' or 'concussion waves,' and the 'waves of force' which immediately followed after them. . . . All these effects, however, as you so well point out, are in the nature of an explosion, but to claim, as Merrill claims, that they expelled the meteor out of the hole is ridiculous." And to Merrill that same day he wrote: "I am afraid we have been at cross purposes all along. There can be no doubt that one can say with absolute propriety that the effects of a projectile entering absolutely bone dry rock strata would be in the nature of an explosion; also that it would be even more proper to use this term if there were any moisture in the rocks. . . . But I cannot for a moment believe that sufficient steam was ever developed in front of it to throw out a ten million ton meteoric mass, the major (interior) portion of which was certainly as cold as the cold of outer space when it came to rest."[17]

Once back in his Philadelphia office, Barringer found that one of his first chores would be to deliver a lecture on the crater, prepared by his son Reau, to the Harvard Club in Boston in mid-October. Reau had arranged to give the lecture himself back in May,[18] but as he was now working in Arizona, his father agreed to substitute for him.[19] Barringer had finally obtained slides of the aerial photographs of the crater taken by Captain Stevens, and these he used to illustrate Reau's text.[20] His presentation seems to have been well-received. He drew prolonged applause and, over the next few weeks, a number of interested inquiries from scientists who had heard the lecture. Harvard astronomer H. T. Stetson, for example, wondered about the velocity at which the meteorite had struck; and Barringer replied that Thomson, Russell, Magie, and Agassiz had all "estimated it at 2–5 miles per second."[21] To another Harvard astronomer, W. J. Luyten, he explained that "the so-called thumb marks, which are so characteristic of Canyon Diablo meteorites, are not due, as has been suggested, to burning in the passage of the meteorite through the air, but are due to the oxidation of what has been termed 'shale ball' meteoric iron which undoubtedly filled these cavities and surrounded the entire ordinary Canyon Diablo unoxidizable meteorites."[22] And W. Spencer Hutchinson, a professor of mining engineering at Massachusetts Institute of Technology, wrote to say "how much I enjoyed your lecture on your Meteor Crater. . . . Now that you have the meteor driven into a corner when will you begin to pull it out?"[23] Hutchinson, in fact, would himself soon become involved in the effort.

Barringer now also concerned himself again briefly with the craters on the moon. On his return from Arizona, he had learned that the Carnegie Institution had appointed a prestigious committee to make an intensive study of the surface features of the moon, an action he considered to be a "direct result" of his 1924 *Scientific American* article on the lunar craters.[24] He hoped, of course, that the committee would quickly confirm his own view, but he soon found that one member at least, Fred E. Wright, a petrologist in the Carnegie Geophysical Laboratory, favored the volcanic analogy. "He seems to be a little skeptical about such round craters having been made by projectiles hitting the surface of the moon at an angle," he reported to Thomson in mid-October.[25] Barringer sent Wright a copy of his *Scientific American* articles, explaining that "I do not by any means exclude what might be termed 'volcanic agencies.' The point I try to make is that the *original* cause in every case was impact and the *sequentia* often included volcanic or igneous phenomena [Barringer's emphases]."[26]

The Carnegie committee did little to resolve the controversy between the volcanic and impact theories of the origin of the lunar craters. Despite many years of work, its members were unable, or unwilling, to arrive at any unequivocal answer to the question.[27]

In November, after breaking off his negotiations with Thompson and his associates, Barringer began in earnest his campaign to raise money for the new shaft, sending nearly a score of letters soliciting stock subscriptions to his friends and other likely prospects in a single week.[28] But almost immediately, the very basis of his planned enterprise came under attack. On November 20 the *Engineering and Mining Journal* published an article by geologist F. LeRoi Thurmond, which strongly reaffirmed Gilbert's original volcanic steam explosion hypothesis.[29] Thurmond had visited the crater in 1925 and, stimulated by a "lurid full-page feature article in a recent Sunday issue of a notorious saffron journal of San Francisco," had recently visited it again. "The meteorite has not been found," he reported, "and things are in the *status quo* of a year ago; no work was being done, nor had any been done in the interim."

Then, after briefly describing the crater and Gilbert's findings there, he cited the evidence against the impact theory of its origin. "First, no meteorite has been found in the crater," he declared, "although much exploration has sought to discover one. Secondly, there is no evidence on the rock debris of the metamorphic effects that would be caused by the impact of such a large super-heated metallic mass as the meteoric theory predicts." Moreover, he added, the assumption "by some observers" that the meteorite had been volatilized and blown out of the crater on impact "does not coincide with observed phenomena."

There was, on the other hand, a "plausible" theory; it was "perfectly conceivable that the crater was formed by an explosion of super-heated steam due to the invasion of the Coconino sandstone by an intrusive mass which does not come to the surface." Gilbert, he noted, had shown that superheated steam could provide the necessary explosive force; and "upon the sudden release of pressure through an explosion, the internal tension would rend the rock into very fine particles such as found in the great quantities of rock flour on the floor of the crater and on its flanks. . . . The concurrence of the meteoric iron with the crater would then be merely a coincidence. . . . I believe," he concluded flatly, "the evidence is competent to explain the crater on the hypothesis of a volcanic steam explosion, and that the proponents of the meteoric theory have failed to discover adequate support for their contention."

Barringer was furious, not so much because Thurmond's article "lamely attempts to revive the theory of a steam explosion" but because Thurmond's statements that no meteorite had been found might discourage prospective investors.[30] As a case in point, he informed Thomson that August Hecksher, "a tremendously wealthy man" with interests in steel, utilities, mining, banking, and Florida real estate, had seemed ready to invest in the venture but "evidently a doubt has been raised in his mind as to the correctness of our conclusions. The article," he added, "may have raised the same doubt in the minds of other rich men." Would Thomson write an immediate reply? Thomson, for once, would not, citing the pressure of other work and suggesting that "perhaps it is not worth while to discuss the matter with one who is so *inaccurate* [Thomson's emphasis]. He [Thurmond] argues from ignorance which never has proved anything."[31]

Nonetheless, Barringer was determined that "that ass Thurmond" should be answered[32] and the task fell to Magie.[33] Within weeks, indeed, he found that he was "not having the luck in raising money" that he had anticipated.[34] Early in January, he told Colvocoresses that "this damn fool Thurmond has really done us a lot of harm," explaining that he had submitted Reau's Harvard club lecture to the *Scientific American* and that one of its editors, Albert G.

Ingalls, while agreeing to consider it for publication, had cited Thurmond's article as "casting serious doubt" on the impact origin of the crater.[35]

Magie's reply, expanded at Barringer's urging from a letter to a full article[36] and then shortened to a letter again at Spurr's insistence,[37] appeared in the *Journal*'s "Discussion" column early in February.[38] The occurrence of meteoritic irons at the crater, Magie began, simply could not be coincidental, as the odds against such an event were "a billion billion to one," and this "enormous probability is in favor of the meteoric as against the steam explosion hypothesis." Nor could a steam explosion have produced the rock flour, he declared. "Superheated steam in the sandstone might have blown the rock to pieces and made sand, but it would not have shattered the individual grains to powder. The shock of a swiftly moving mass would give just such a shattering blow needed to produce this formation." As to the absence of "metamorphic effects" at the crater, Magie pointed to the metal-stained "pumice," or Variety B of metamorphosed sandstone, and noted that "Hot steam does not melt quartz or vaporize iron." Moreover, while Thurmond's "intrusive mass" had not been found, it was "no longer true" that the meteorite had not been found, and he cited the meteoritic material brought up from a depth of 1,370 feet beneath the south rim by U. S. Smelting's drill.

Reau's lecture, unlike Magie's rebuttal, survived the editorial process intact. For one thing, the doubts raised by Thurmond's article in the mind of *Scientific American*'s Ingalls were quickly quieted by Russell, a regular writer on astronomical subjects for the magazine.[39] Reau's argument against coincidence, Russell assured the editor, "appears to me to be irrefutable. How much meteoric iron there is, and whether it is in such a form as to be a paying mining proposition, cannot, of course, be a matter of certainty, but it is my serious judgment that the search for it is a thoroughly legitimate mining venture. I would not put money in it personally," he added, "but this is because I am not personally prone to invest in any venture, even a thoroughly legitimate one."[40] Ingalls at first also thought Reau's lecture was much too long and wanted to cut it from more than 4,000 to 2,000 words. But clearly he was impressed by its completeness, its readable style, and its lack of dogmatic pronouncements. When Barringer suggested that it be run serially, he quickly agreed.[41] It would be published, Barringer proudly announced to Thomson, "without changing a word."[42]

The lecture appeared as three articles under the running title of "The Most Fascinating Spot on Earth" in the July, August, and September 1927 issues[43] and provided, perhaps, the most comprehensive discussion of the crater, of Barringer's long explorations, and of the meteoritic impact theory of its origin available to the general public up to that time. The articles, moreover, were tastefully displayed and liberally illustrated with photographs, along with a map of the crater and a cross-section drawing, done by Reau, that showed the "Supposed Position of Meteoric Body" beneath the southern rim and the location of the proposed 1,500-foot shaft still farther to the south.

Surely to Barringer's delight and almost as surely at his suggestion,[44] Ingalls provided the first article with a boxed editor's note designed to justify publication of the series in view of its controversial subject. "Because certain people, reluctant to believe the unprecedented, regard as sensational the theory that Meteor Crater was formed by the impact of a giant meteor," he wrote, "we have obtained the following definite statements from two well-known scientists." He then quoted Magie as being "perfectly willing to make a strong affirmative statement in support of Mr. Barringer's article, but there is no need for it. There is no reasonable doubt that the Crater was formed by the fall of a meteor and that the meteor is buried in it." Thomson, too, had been "very willing" to be quoted. "There can be no question," he declared, "of the Crater being made by masses of meteoric iron, and that an enormous mass of such iron remains buried under the south wall of the Crater." Reau's own carefully phrased conclusions were hardly so unequivocal. Ingalls, incidentally, also quoted Russell in a photo caption in the final article to the effect that "I have examined the Crater on the ground, as well as the other evidence, and I am thoroughly convinced of its meteoric origin."

Reau himself remarked at the outset on the sensationalism of the public press regarding the crater. "It appears that most people have only a hazy knowledge of the subject at best. This is due, I think, to the fact that the technical publications about it have never had wide circulation; and to the fact that the more popular articles on the subject which have appeared from time to time have often been woefully distorted. It is a subject," he opined, "that seems to challenge the imagination of the average newspaper writer, and several of them have entirely outdone the actual facts in their sensational descriptions of it."

Reau's exposition contained little that was new; he simply spelled out in accurate detail and in nontechnical terms the facts about the crater and the case for its meteoritic impact origin. His conclusions, thus, largely emerged from his facts; and when they went beyond the facts, they were carefully qualified.

He did, however, outline his conception of the present condition of the buried mass: "At or near its center there will be nothing but rounded pieces of iron, more or less in an oxidized condition. The average size of the individuals . . . will be in the neighborhood of five or ten pounds. . . . Near the edges of the cluster, sand and crushed silica will begin to appear, mixed in between the individuals. This will be more abundant as the outside of the cluster is approached, until there remain only a few meteorites scattered here and there through a jumbled mass of rock-fragments and silica dust." Near the bottom of the mass, he added, he expected to find "quite a concentration of Variety B of the metamorphosed sandstone," and the rock around the mass would be "completely crushed and dislodged" except on the south, where "the rocks within a very short distance of the projectile should be solid. . . ."

And in conclusion, citing Barringer's articles on the lunar craters three years before, he noted a "very interesting by-product of the investigation . . . the resurrection of the theory that the moon's craters are not volcanic, but are the results of impacts similar to that which made the Arizona Crater." This latter theory, he concluded cautiously, "if not proved, is at least worthy of most serious consideration."

"A LITTLE BAND OF ADVENTURERS"

Reau's articles undoubtedly contributed substantially to the eventual success of Barringer's campaign to sell stock subscriptions. Since he had launched this campaign back in November, Barringer had worked feverishly to promote the enterprise, but he found that the very rich men whom he hoped would take large blocks of stock were either skeptical of his proposition or were simply indifferent. Hecksher, alarmed by Thurmond's article, had begged off at least temporarily, pleading his heavy losses in the collapse of the Florida land boom;[45] John Hays Hammond, although he took a small subscription, similarly cited losses from investments in Mexican mining properties.[46] In December Barringer again approached the Guggenheims, this time through Harry A. Guess, vice president under president Simon Guggenheim of the American Smelting and Refining Company.[47] But Guess was "satirical" and promptly turned him down on the basis of U. S. Smelting's costly experience at the crater.[48] He next urged Henry S. Pritchett, now in his twentieth year as president of the Carnegie Foundation for the Advancement of Teaching, to convince his fellow directors of the Atchison, Topeka and Santa Fe Railway to participate in the new exploration, but to no avail.[49] Through the winter, indeed, Barringer approached more than a half dozen mining companies and other firms, and more than a dozen individuals, some of whom had heard his pitch before, but with little success. At one point, hoping to lengthen his list of prospects, he asked Secretary of Commerce Herbert Hoover for "a few names of men of ample means," but Hoover apparently did not oblige him.[50]

By February, however, the outlook began to brighten. His son Brandon subscribed $2,000;[51] and a few days later, Barringer's friend George Bird Grinnell, a noted editor, author of nature books, and then president of the Boone and Crockett Club, invested $500.[52] In March still another friend, John Gribbel, a wealthy Philadelphia gas meter manufacturer, signed up for fifty shares at $5,000.[53]

A far more significant event, however, occurred at this time while Barringer was on a brief business trip to Boston. There, at the Somerset Club, he met an astute Boston capitalist and mining entrepreneur named Quincy A. Shaw. Shaw, president of North American Mines, Inc., had learned about the crater enterprise more than a year before from Agassiz and had sent in an unsolicited subscription for 200 shares.[54] Now, hearing about the venture directly from Barringer, his interest intensified. When Barringer returned to Philadelphia, he sent Shaw copies of all the pertinent literature.[55] Shaw responded enthusiastically, even suggesting that the proposed shaft be planned to accommodate 1,000 rather than only 500 tons of meteoritic ore per day.[56] Eventually, Shaw would expand his holdings to 450 shares and, as president and largest stockholder in the Meteor Crater Exploration and Mining Company, would preside over the great debate more than two years later that would mark the climax of the Coon Mountain controversies.

Barringer himself now also gave added impetus to the subscription campaign by finally agreeing to liberalize the terms of the lease of the crater property by cutting Standard Iron's share of any profits to a flat 25 percent. He first had broached the idea early in February in connection with Colvocoresses's attempts to persuade several mining executives in Arizona to join in the crater enterprise.[57] But he soon decided to incorporate the change into a new lease and a drastically shortened prospectus which, incidentally, is notable for the absence of even the suggestion that the venture might be a gamble and of all of Colvocoresses's other earlier caveats as well.[58]

This new lease would be effective June 1, 1927, for one year, and would be extended for an additional ninety-eight years if subscriptions totaling $200,000 were received by the exploration company before June 1, 1928, the figure being Barringer's estimate of the minimum amount needed to begin mining operations at the crater. As organizers, Barringer and Colvocoresses proposed to sell no more than 2,500 shares each of the $100 preferred and the $1 common stock, in units of one share of each. The lease, they declared in the new prospectus, "provides that the lessee company shall operate the property and shall distribute the net earnings when and if the same are earned, 75% to the lessee and 25% to the owner."

The original subscribers welcomed the new provision and not only renewed their subscriptions but in some cases increased them. Prospective investors also seem to have now looked upon the venture in a more favorable light. Through the summer Barringer continued his intensive solicitations, while Colvocoresses, in Arizona, prepared a comprehensive report describing the crater, the history of its exploration, and its commercial potential to supplement the abbreviated prospectus.[59] Colvocoresses also found time to write Slipher, the Lowell Observatory director, to inquire whether the observatory might be interested in such an investment,[60] and to Percival Lowell's widow in Boston to ask for names of more of Lowell's friends who might be willing to join the venture.[61] "Mr. Agassiz repeatedly told us that his interest in the enterprise originated from his association with Professor Lowell and his visits in company with Professor Lowell to the Crater many years ago," he pointed out, adding: "Our enterprise is a peculiar one, and we have found that it only appeals to persons who are interested in scientific research."

By mid-September the amount of subscriptions received had grown to $167,000.[62] New subscribers included Hecksher for $10,100, and Barringer's old friend and hunting companion, author Owen Wister, who took fifty shares for $5,050. Agassiz increased his subscription to $40,400, Shaw to $25,250, and Thomson to $10,100. Even astronomer Charles P. Olivier, the authority on meteors who was now at the University of Pennsylvania in Philadelphia, subscribed for $202. The amount was too small, Barringer felt, and he urged him to increase it, adding, however, that he was nonetheless "very glad . . . to know that you are willing to join the little band of adventurers who are going to dig into and, as we hope, dig up, the buried head of a comet."[63]

By the end of November, the total had climbed to $201,495, and the extension of the lease to a full ninety-nine years, as well as the sinking of the shaft, was assured.[64] A month later, Barringer advised Colvocoresses "to make no further effort" to solicit subscriptions "as only $20,000 more is needed" to sell out the 2,500-share stock issue.[65]

In January 1928 the Meteor Crater Exploration and Mining Company, with Shaw as president, Barringer as vice president, and Colvocoresses as general manager, began operations; the crater once again became the scene of bustling activity. There was much work to be done; the entire property was resurveyed, roads to the crater and around to its south rim were improved, new pumping machinery was obtained and installed, the dams in Canyon Diablo and the water line to the crater were repaired, and two new water tanks erected on the south rim. Some of the old buildings were refurbished and new fencing was installed.

In January crews began testing the strata to the south of the crater with a diamond drill, putting down a hole 800 feet south of U.S. Smelting's drill site at the edge of the south rim. The Coconino sandstone here was found to be softer and more granular than had been expected; and, after obtaining some new equipment, a second hole, at an angle from the vertical, was sunk to a depth of 426 feet, the bit encountering rock of the same general character. The first hole was then resumed to a depth of 721 feet and "found no evidence of the ground being fractured." Finally, a third hole was drilled to 500 feet "west and south 1200 feet from the supposed meteor" where conditions "better for shaft sinking" were found.[66]

By mid-April the exploration company's board of directors approved termination of the drilling program and directed that work begin on the shaft.[67] Barringer and Shaw visited the crater now to assess the results of the drilling and to survey the ground for the best possible site for the shaft. They concluded that it should be located near the third drill hole at a point 1,100 feet to the southwest of the U.S. Smelting hole.[68]

In May the *Engineering and Mining Journal* quoted Colvocoresses's estimate that it would require "a year to 15 months" to reach a depth of 1,600 feet.[69] Work began late in June, and by August the *Journal* could report that thirty-seven men were working in three shifts around the clock and that the shaft had reached a depth of 140 feet.[70] A month later Barringer informed Barbour that "the sinking of the shaft, as Shaw will tell you, is progressing rapidly," adding that its depth was "about or over 500 feet" and that Colvocoresses reports everything in fine shape and going well."[71]

There were other developments during this period relating to the crater that gave Barringer additional cause for optimism. In March *The American Mineralogist* printed an address by Austin F. Rogers, president of the American Mineralogical Society, in which Rogers strongly reinforced Barringer's 1914 claim that Variety B of the metamorphosed sandstone was a "knockdown argument" for the impact origin of the crater.[72] Rogers, a Stanford University mineralogist, had visited the crater in January 1927 and had written Barringer then that the silica glass, or lechatelierite, "gives the most conclusive evidence of its origin."[73] Now he described its occurrence at the crater as "unique" and pointed out that "it is almost inconceivable that a temperature in the neighborhood of 1600°C (about the melting point of quartz) should have been reached during any part of a steam explosion even if the steam were superheated. . . . Taking all the facts into consideration, it seems most reasonable to attribute the formation of Meteor Crater to the impact of a huge meteorite and the fusion of the sandstone to form lechatelierite to the heat generated by the impact."

250   THE FINAL SHAFT

In August mining engineer J. A. Reinvaldt of the Estonian Ministry of Commerce and Industries published the results of his investigation of five small, shallow crateriform features on the Estonian island of Ösel and concluded that they, too, were formed by meteoritic impact.* Reinvaldt had found no meteoritic material at the craters. Nevertheless, he argued his conclusion on the basis of comparisons with the Arizona crater as described by Barringer and Merrill, citing the distribution of the rock debris on and around their rims, the outward dip of the surrounding strata of dolomite, and the discovery, in excavations and borings in three of the craters, of finely pulverized rock flour.[74]

No less pleasing to Barringer at this time were the numerous references to Meteor Crater in sensationalized reports in the popular press of Russian geologist L. A. Kulik's expeditions in 1927 and 1928 to the remote Tunguska River region of central Siberia to make the belated first investigation of the great meteoritic fall there of June 30, 1908. These articles, often luridly illustrated, were based upon vague and fragmentary reports many times removed from their original source and perforce relied heavily on the imagination of the individual writers. But they nonetheless served to condition a fascinated if incredulous reading public to the idea that a large meteor could indeed penetrate the earth's atmosphere with devastating effects. Barringer, however, was disappointed at the dearth of actual data they contained concerning the Siberian event. Nor did Kulik's own reports provide much additional information, as Barringer discovered some months later after reading a translation of one that Olivier obtained. "It is simply rotten," he advised his son Reau, "and in no sense compares with Reinvaldt's excellent report of his craters. The man had an opportunity to be the first to describe even a greater wonder than our Arizona crater!" he marveled. But no matter; "the eastern papers are full of it, always containing a reference to our crater. . . ."[75]

THE *NATIONAL GEOGRAPHIC* AFFAIR

Against these positive factors, however, Barringer could set his extreme frustration over Boutwell's article on the crater which finally appeared in the June 1928 issue of the *National Geographic Magazine* under the title, "The Mysterious Tomb of a Giant Meteorite."[76] He had read a revised draft of this early in November 1926 and had then pronounced it "admirable."[77] To his great dismay, he found that in the interim some significant changes had been made in Boutwell's text. Most disastrously, it now contained the unembellished state-

---

*A sixth crater has since been identified.

ment that "the uniqueness of this world wonder escaped notice until Grove Karl Gilbert, the geologist, suggested the meteoric theory in 1895." Nowhere in the article was there any reference to Gilbert's rejection of this theory and his adoption of the steam explosion hypothesis. Moreover, there was no mention of Barringer at all. The entire tone of the article, Barringer realized to his consternation, created the impression, for the uninitiated at least, that Gilbert and his U.S. Geological Survey colleagues had discovered and demonstrated the crater's impact origin. This impression was further strengthened, he felt, by photographs used to illustrate it, which had been taken by Gilbert in 1891 and were now credited by the *Geographic* to the Survey.

Barringer turned immediately to Merrill, who only a week before had assured him that "You of all, deserve great credit for your persistent advocacy of the Crater's meteoric origin."[78] "I have just read the article on the Crater in the National Geographic Magazine, apparently published with the approval of the U.S.G.S., which does not even mention my name and inferentially gives the credit to Gilbert!" he fumed. "I believe you agree with me that the credit for the *discovery* [Barringer's emphasis] of the meteoric origin of the Meteor Crater of Arizona belongs primarily, and certainly to a greater extent than to anyone else, to me. That is to say, that such scientific distinction as is due to anyone for the discovery of the true origin of the crater is mine. . . ."[79] Merrill, he pointed out, had acknowledged as much in his "excellent" 1908 paper and "have bravely—*in Washington*—held to that opinion ever since [Barringer's emphasis]." It was his hope, he concluded, that "you, who know many of these facts to be true, will be glad to place the credit for the discovery of the true origin of the crater where it belongs and write me to that effect."

Merrill's reply was prompt, but noncommittal, and could have only deepened Barringer's dismay. Boutwell, he revealed, had shown him his manuscript and "I advised certain changes which he adopted, but I failed to note the omission of your name. He has since written me expressing regret at the omission." But, he added, "why worry? The article is of but ephemeral value. It will never be consulted as forming a part of the literature—historical or scientific. . . . My own vindication was enough for me. I waited a long time for it though."[80] Merrill's name, incidentally, is not on Barringer's list of more than a dozen eminent scientists who subsequently protested Boutwell's article to the *Geographic*'s editor, Gilbert H. Grosvenor, and to the National Geographic Society's board of trustees on Barringer's behalf.[81] "I have long known," he later confided to Reau, "that he is *very* jealous [Barringer's emphasis] of my having been the real discoverer of the origin . . . and [is] very loath to admit it."[82]

Barringer now wrote a long letter pleading the injustice of Boutwell's article to George Otis Smith, the director of the U.S. Geological survey as well as a member of the National Geographic Society's board.[83] "I do not know whether you have subscribed to the opinion of Dr. Grove Karl Gilbert and his followers in the Survey, notably Mr. Darton, or whether you have been inclined to believe in the impact theory of origin (rejected by them in favor of a steam explosion). . . . Whatever your personal opinion may have been or is, I am sure you know that neither Dr. Gilbert, nor any member of the Survey except Dr. George P. Merrill . . . is in any way entitled to the *scientific* credit [Barringer's emphasis] of having favored, much less having proved, the correctness of the impact or meteoric theory of origin. . . . We all know Dr. Gilbert carefully considered the meteoric or impact theory in the early 1890s . . . and is on record, after examining the facts, as having rejected it in favor of the steam explosion theory." He then requested Smith to "state in a short article to be published in the National Geographic that the survey has not to date advanced the meteoric or impact theory" and "that such scientific credit as attaches to the discovery . . . should not be given to the Survey or Dr. Gilbert, even by inference, but should be placed where it undoubtedly belongs. . . . [N]ot in the slightest degree," he added, "has this request any commercial bearing. . . . Sufficient money is now in hand to do further important work. No additional funds are being asked or accepted." This disclaimer, indeed, he appended to his letters to other Geographic Society board members.

Smith's reply a week later only added to his frustration. Referring to Gilbert's 1896 paper, the Survey director declared:[84]

> In the address above mentioned . . . Dr. Gilbert specifically assumes responsibility for the meteor impact theory of the origin of the features displayed at Coon Butte. To be sure, after discussing this and other theories fully . . . he expresses a preference for the explosion hypothesis as more satisfactorily accounting for all the facts at hand. He credits his associate, Willard Johnson, with the explosion theory, but his scientific disinterestedness is particularly well illustrated by the fact that, although he does not reach definite conclusions, he favors a hypothesis proposed by an associate in preference to one which he himself originated. There have, of course, been many papers written since Dr. Gilbert's address in 1895, but I do not understand that any of them have presented any hypothesis not considered by him. Like your own series, they have presented evidence—some of it, of course, not available to Gilbert—supporting one or another of the theories which he discussed in so searching and fruitful a way.

As to Barringer's request, Smith added that "I should consider it inappropriate for me to intervene in a matter in which you and the Editors of the National Geographic Society find themselves out of accord. . . . Moreover, the facts, as they appear in the printed record, render it impossible for anyone to deny that the meteoric impact idea originated with Mr. Gilbert, even though he did not adopt it. Furthermore," he concluded, "I find no mention of Gilbert's pioneer work in your 1909 paper, except the title in the bibliographic list. It may therefore not be inappropriate to suggest that the propriety of demanding for yourself a type of recognition which you have not deemed it worthwhile to extend to another may reasonably be considered to be open to question."

Barringer replied in cold anger.[85] "You claim," he wrote Smith, "That 'Mr. Gilbert specifically assumes responsibility for the meteor impact theory.' I understand you to mean that he suggested it and is, therefore, entitled to credit for it, even though he rejected it. The theory would suggest itself to any intelligent man visiting the locality." Barringer went on to say that

> You cannot deny that since Gilbert's paper was published the followers in the U.S. Geological Survey had publicly opposed and even privately ridiculed the impact theory, quoting him as their authority. Surely it is a new thought that the upholder of an incorrect theory is entitled to credit for the correct one, simply because he attempted to refute it. Yet this is the claim you make for Mr. Gilbert. . . . No one has ever questioned the fact that I was the first to announce the correctness of the meteor impact theory and that I am responsible for such acceptance as it has finally met with. . . . I regret that you do not state which hypothesis the Survey now believes to be the correct one.

Barringer also took "very decided exception to, and resent the innuendo contained in" Smith's remark about the propriety of his request to set the record straight. He had, he pointed out, referred to Gilbert three times in his 1906 paper and there was "no reason to refer to him again in my National Academy paper." Moreover, there was, he suggested, another matter of propriety—"that you, as its head, seem so willing to allow the Survey to be credited with an important scientific discovery when for thirty years it has rejected the theory on which this discovery was based."

Although Smith was not heard from again, there ensued a virtual flood of correspondence over the issue. Barringer himself wrote to Grosvenor and all the members of the Geographic Society's board seeking redress and asking that a correction be published in the magazine to counter the inference of Boutwell's article. And once more, he marshaled supporters among scientists of his acquaintance to write in the same vein to Grosvenor and to Society trustees whom they knew. The list was now quite a distinguished one and included not only those who were close to him, such as Thomson, Magie, Russell, and Alexander, but Olivier, Webster, Fairchild, the American Museum of Natural History's Henry Fairfield Osborn, and Howard McClenahan of Philadelphia's Franklin Institute. [86]

This outpouring of correspondence, however, brought few replies and no satisfaction. Barringer did get what he called "a disgusting letter" from "this contemptible fellow,"[87] John Oliver LaGorce, vice president of the Society and associate editor of the *Geographic,* who suggested that Barringer was trying to use the magazine to advertise the crater and sell stock. "You give the commercial side of the present exploration as the reason my name was omitted," he replied, "and state that I had agreed with Mr. Boutwell as to this. You even insinuate that 'pressure has been brought to bear on me' and that I now find it 'expedient to demand that a statement be published giving me scientific credit for the discovery.' In other words, you insinuate that my demand has some commercial motive. This is a gratuitous insult. I have made the claim on my own initiative and with no thought of commercial gain."[88]

Barringer also suspected a conspiratorial element in the situation. "The *prima facie* evidence is very strong," he confided to Barbour, "that the magazine has been made a catspaw by someone either in or friendly to the U.S. Geological Survey. . . . Incidentally, the article gives credit to Gilbert of the discovery of the fact that an impact by a body approaching obliquely . . . makes a round hole, though he never knew it and it was first suggested by me and I gave the idea to Boutwell."[89]

In October Olivier took up Barringer's cause in his regular column of "Meteor Notes" in *Popular Astronomy.* The *Geographic* article, he noted, "nowhere mentioned the fact that the meteoric origin was first proved by D. M. Barringer, to whose efforts indeed the scientific world owes most of the exploration carried out at this locality. An even more serious matter," he added, "is that this article inferentially gives the credit to another prominent scientist therein named, who in fact used his influence in trying to prove quite a different theory, having rejected the meteoric or impact theory."[90]

The controversy simmered through the fall and winter of 1928–1929 to Barringer's growing discouragement. "What's the use!" he exclaimed to Merrill in December. "Nothing could be more unjust than to give the credit for the discovery to the very man who repudiated it."[91] His discouragement deepened a few weeks later when the Carnegie Institution's Fred E. Wright, as Merrill had done before him, advised Barringer to ignore the *Geographic* article as of little consequence. "I am convinced," Wright wrote, "that your theory of the meteor-impact origin of Meteor Crater is correct, also that you deserve great credit for having the courage of your convictions and gone ahead to test them out in a practical mining way." But, he added, "I do not believe that statements in popular articles of the National Geographic Magazine carry much weight. . . . The fact that Mr. Boutwell made a slip . . . should not be taken too seriously."[92]

Redress of sorts, however, was in the offing. In January 1929, when it became evident that no satisfaction would be forthcoming from the *Geographic*, Barbour suggested that it was time to send a full statement of the facts to the journal *Science*.[93] Barringer at first was hesitant but soon agreed; undoubtedly, a letter from Grosvenor that Howard McClenahan forwarded to him helped him make up his mind. "Grosvenor," he advised McClenahan, "seems to endeavor to mislead you by his persistent habit of quoting only that portion of a publication or letter which is favorable to his contention. . . . This is exactly the trick which was apparently employed by him in his statement to Dr. [John Barton] Payne and the Board of Trustees. . . . Why did he not show my whole letter . . . instead of just the concluding paragraph, and why does he again say that the manuscript . . . was submitted . . . to me, evidently trying to make you believe . . . that the manuscript . . . *as published* [Barringer's emphasis] was submitted to me?" Grosvenor's letter, he added, "is a clever but strained attempt to get you to interpret Gilbert's paper . . . differently from the way in which the entire U.S.G.S. and men of science have interpreted it for 30 odd years, and incidentally to place me in a false position. This attempt to write a new meaning into Gilbert's paper . . . is ridiculous as well as contemptible."[94]

Fairchild, he was glad to learn, was willing to write such a statement; he had been particularly impressed with Fairchild's letter of protest to the Society's board chairman, John J. Edson, in which the geologist emphasized his long association and friendship with Gilbert.[95] His plan, he advised Magie early in March, was that "I first write a dignified letter and then Fairchild follows with an article." Fairchild, he added, "has hit the nail on the head when he states that the U.S. Geological Survey is largely responsible for Boutwell's article. I have always believed this was true."[96]

As it turned out, *Science*'s editor, James McKeen Cattell, refused to publish Barringer's letter, a fact which Barringer found "a little surprising" in view of "the present world wide interest in the phenomena,"[97] but he would consider Fairchild's article. Barringer read Fairchild's lengthy draft early in April and returned it to the Rochester geologist with more than three pages of corrections.[98] It appeared in *Science* on May 10, 1929.[99]

Fairchild, while censuring the *Geographic*, directed his sharpest attacks at the Geological Survey. Boutwell's article, he declared at the outset, "fails to give proper credit for discovering and publishing the evidence of [Meteor Crater's] impact origin," and on the contrary, "appears to give credit for the discovery to G. K. Gilbert and the U.S. Geological Survey."

> Naturally any account of the crater exploration and the discovered evidence would have required the name of Daniel Moreau Barringer, which was entirely suppressed. Naturally and properly the implication in the article has been resented by Barringer who at great cost in time, effort and money explored and probed the crater, marshalled and first published the facts, and so proved beyond any reasonable doubt the meteor-impact genesis. A very spirited correspondence, involving a number of friends of Barringer, has not yet obtained the *amende honorable*. There is here a question of scientific and journalistic ethics. . . .

Fairchild then detailed the history of the scientific investigation of the crater, including Gilbert's visit there in 1891 and his subsequent conclusion of "a volcanic genesis of some sort of a steam explosion," Barringer's and Tilghman's explorations and conclusions, his own and Merrill's investigations in 1906 and 1907, Barringer's later papers on the crater, and even Reau's 1927 *Scientific American* articles. From all this, he declared flatly, "The impact origin of the crater must be accepted as fact, no longer as theory. . . ."

Yet despite this, "From 1893 to the present time, the U.S. Geological Survey has by its silence tacitly held to the volcanic or steam explosion hypothesis, by entirely ignoring all the work of geologists since Mr. Gilbert's report, while the negative attitude of some members of the Survey toward the impact origin is well known. This attitude of the Survey deserves criticism," he added. "The scientific evidence is before the court of scientific men. The writer as mutual friend of all the parties, and especially as a close friend of Dr. Gilbert, will now assume the unsolicited and delicate task of summarizing the case and of pronouncing verdict."

While some at the Survey may consider Barringer a *persona non grata* and not wish to advertise his personal, commercial venture, still another reason, "perhaps the chief one," for the Survey's silence, Fairchild added, "is that the history and the facts show that a mistake was made by an eminent and beloved member of the Survey. Dr. Gilbert certainly did form an erroneous opinion." Yet:

> Such a reason for the attitude of the Survey implies either that the workers on the Survey are considered infallible, or if fallible, that the Survey never admits an error. The writer had intimate personal and scientific relations with Dr. Gilbert and yields place to no one in regard and admiration for him as a man and geologist. It is difficult to understand how he came to favor volcanism as the cause of Meteor Crater. Most certainly he later knew his mistake. . . . [F]ollowing publications by Barringer and myself he never questioned the impact theory, as he surely would have done had there been any doubt in his mind. . . . No luster is added to Gilbert's name by neglect-

ing to admit the evident truth. Nor would admission of error hurt his reputation. It is human to err. . . .

"Confession by the Survey," Fairchild concluded, "will of course be painful. The public as father-confessor is waiting. And really, a little evidence of humility and admission of fallibility by a great bureau of the government would be something new. . . . [T]he Survey has no ethical or legal right to suppress or withhold any geological truth, for personal or any other reason. . . ."

Barringer was only briefly elated. Fairchild, he chortled after the article appeared, "has certainly put the editors of the National Geographic Magazine and the Great and Only U.S.G.S. in a hole, from which I think they will have a good deal of trouble excavating themselves."[100] But the *Geographic*'s editors and Survey officials simply ignored the attack, and a few months later Barringer ruefully conceded to Olivier that "no *amende honorable* from the National Geographic or statement from the U.S.G.S. has appeared. That Washington crowd seems to be a pretty poor lot."[101]

A WATER PROBLEM

Throughout much of this running dispute with the *Geographic* and the Survey, Barringer was concerned with a far more serious matter, one so grave that it would soon prove fatal to the crater enterprise. Early in November 1928, to Barringer's consternation, the crews working at the crater had encountered water in the shaft below the 600-foot level. He had long believed that, while the crater was holding water much like a sealed basin, the surrounding strata were dry, basing this conviction largely on the inability of the Santa Fe Railway to find water in drill holes along its mainline tracks to the north. Moreover, drill supervisor C. W. Plumb had reported no water problems in the course of sinking U.S. Smelting's deep hole on the south rim. But now water was pouring into the shaft, and Barringer concluded that it must be coming from the crater.

"You have seen from Colvo's reports what trouble we are having with the water," he wrote Reau in mid-November.[102] "How serious the situation is we do not know. . . . I am simply appalled at the idea that we shall be obliged to drain the crater before we can work the meteoric mass, and yet this looms up as a very dangerous possibility. . . . This will, of course, mean the raising of a great deal more money and unless Mr. Shaw's and Mr. Agassiz's interest has been so greatly aroused that they will undertake to help raise it, I do not know where it will come from."

Shaw's and Agassiz's attitude, however, proved to be "all that we could desire," he wrote Reau a few weeks later.[103] The board was "determined to keep up the fight against the water" and had "authorized Colvo to get additional power and pumping equipment." Moreover, with Shaw and Agassiz in the van, the directors "personally underwrote, or rather guaranteed, approximately $45,000" in new funds for the enterprise. "Colvo feels very strongly that we can handle the water," he added, but "Quien sabe? I would give much to know whether we are draining the crater or not, and how much of the territory to the east, south and west outside the crater. . . . Personally, I greatly fear that we are drawing water from the crater and I am slowly becoming convinced that in some utterly non-understandable way Plumb made a mistake when he declared his hole was dry clean to the bottom and remained so for some time. . . . Why the water does not run off somewhere through the Coconino sandstone to earth cracks or to the Little Colorado Canyon it is impossible to know, and why is there so little water to be found along the line of the railraod? . . ." Everything was going wrong, he added, and "There seems to be no way of diverting the tide of luck which seems to have set steadily against me."

Only the long-delayed publication of Reau's paper on the Odessa crater at this time seems to have brightened his spirits briefly, for apparently he still held out hope that he could obtain an option on the property and more fully explore the crater. "The actual locating of the mass in this little crater," he pointed out to Reau, "may have a very encouraging effect on the stockholders when we ask them for more money, as it would seem we must do if the shaft is to continue."[104] He had now concluded, he added, "that Plumb did not, at any time, definitely determine that the churn drill hole was dry. He clearly had the impression that the hole was dry, and gave that impression to everyone who read his reports. . . . Both Colvo and I think we shall have to fight water all the way down and the misery of it is that we think it will come to us in increasing quantity. . . . I am told that it is very difficult to sink a shaft when the flow is as much as 600 gallons a minute." And to Magie, he wrote that Plumb's insistence on a dry hole was "at present an unsolved mystery, but nothing could be more certain than that we have already pumped some three or four million gallons of water at a rate of 150 gallons a minute, and it is still coming in faster than we can handle it."[105]

Barringer's concern over the water problem was intensified, too, by his realization that he could no longer fully control the operations at the crater, as he pointed out in still another long letter to Reau. "It is getting a little beyond me," he confessed, "and I am very fearful that we shall make a wrong decision; that is, that Shaw and Agassiz will make a wrong decision, because it is very clear that what they wish will be done and should be done, because we cannot afford to run counter to their wishes."[106] Colvocoresses had, in fact, advised shutting down operations for several months until more powerful machinery could be installed, and while Barringer agreed, the other directors were "unanimous in wishing to continue our efforts to sink the shaft."[107]

In mid-January, a friend and veteran mining superintendent, Charles K. Barnes, from whom Barringer had sought advice,[108] suggested that a method known as the François Cementation Process, by which the shaft walls were sealed with cement as excavation progressed, be used at the crater. Shaw, Agassiz, and even Colvocoresses were first skeptical;[109] but by February the directors accepted a proposal by F. R. Dravo of the Dravo Contracting Company in Pittsburgh to apply the François technique. Dravo's charges "seem high," Shaw conceded to Barringer, who had been worried about expenses at the crater,[110] but "all fees are payable in stock on the same terms as to subscribers . . . a very sporting proposition on Mr. Dravo's part."[111] Application of the François process was begun late in March and initially worked quite well.[112] By mid-May, Barringer could report to Barbour that "things are finally going pretty well at the crater."[113]

During the spring of 1929, too, Barringer was heartened to learn that two more prominent astronomers now also agreed with his theories not only of the impact origin of Meteor Crater, but of the craters on the moon. In April he attended an American Philosophical Society dinner at which Harlow Shapley, director of the Harvard Observatory, spoke on meteors and meteorites. The next day he sent Shapley a copy of his 1924 *Scientific American* paper on the lunar craters, explaining apologetically that "because of partial deafness, I could not hear your address."[114] Shapley replied promptly. "Although I consider the matter not conclusively proved, I have for a long time favored the meteoric theory of the lunar craters. . . . The whole problem of meteors strikes me as one of the most important unsolved questions in astronomy. . . ."[115]

Late in May, Forest Ray Moulton, recently retired from the University of Chicago, wrote that he planned to refer to Barringer's work at the crater in a new edition of his textbook, *Introduction to Astronomy;* and Barringer, "not a little pleased," quickly sent him all his papers along with Reau's 1927 *Scientific American* articles and his paper on the Odessa crater.[116] Moulton briefly acknowledged them, noting that "When I wrote the earlier edition of my work, I was not familiar with the evidence that the Coon Butte was produced by a meteor, though I was of the opinion that such was the case. Neither did I know, at that time, that the impacts made by oblique projectiles are essentially circular, a fact I found out when I was in the Army during the war. . . ."[117]

After reading Barringer's papers, Moulton declared in a second letter that

"It is needless to say that the evidence you bring forward is conclusive and that no reasonable person can fail to accept your conclusions. . . . I shall refer to you as the one who established the meteoric origin of this crater. . . . Since reading your papers, I feel a deep resentment against the National Geographic Society for the paper they published a year ago without referring extensively to the enormous amount of work you have done." As to the lunar craters, he added, "I have always personally been inclined to the view that the craters on the moon are due to impacts of meteors, but in writing a text book for general use, I have been under obligations to reflect to a considerable extent, the views of other scientists. Certainly, until recently, most astronomers have not looked with favor on the meteoritic origin of the lunar craters, and some years ago I was criticized for speaking of this theory with so much respect. Moreover, my good friend, Professor Chamberlin, always thought they were of volcanic origin. This will, I think, be unexpected news to you. In the new edition of my book I shall express myself on the subject more frankly, particularly as it is now apparent that the lack of an appreciable number of elliptical craters is not so serious an objection to the impact theory as I formerly thought it was."[118]

But any satisfaction Barringer may have received from Moulton's frank confession was dissipated almost immediately by the deteriorating situation at the crater itself. Water was now flowing into the shaft in much greater volume, and it became clear that the François process could no longer stem the flow, that indeed, as Shaw later reported, it was "an entire failure."[119] More pumps were now installed and the work continued, but progress was painfully slow. On June 17 Colvocoresses reported the shaft at 699 feet, on July 3 at 703 feet, and on July 23 at 713 feet, a gain of only fourteen feet in five weeks.[120] At this point, a pumping breakdown again flooded the shaft, which would, in fact, never get any deeper.

REASSESSMENT AND NEW ANALYSIS

The financial condition of the enterprise had now become equally grave. The $250,000 raised through initial stock subscriptions had been spent; and the directors, at their May meeting, had made good on their earlier pledge by advancing $43,228 to keep the work going at the crater while new sources of funding were explored.[121] In June, Shaw proposed a plan whereby the existing stockholders would subscribe another $200,000, their subscriptions becoming binding only if and when an additional $200,000 worth of subscriptions was obtained from outside investors.[122] Under such a plan, Shaw advised Barringer, he was "inclined to take my full allotment if it will assure the $200,000 which it is necessary for 'M.C.' [Meteor Crater Exploration and Mining Company] stockholders to raise to carry out the plan." Shaw himself, however, would soon conclude that $400,000 might not be enough.[123]

In June, Shaw also decided that a full reassessment of the situation at the crater, as well as the overall feasibility of the enterprise, was in order. This was to be undertaken by the Boston firm of W. Spencer Hutchinson and Robert Livermore, engineering consultants for Shaw's North American Mines, who would report their findings to North American's board. Barringer had no objection to this. Hutchinson, in fact, was the Massachusetts Institute of Technology professor who had praised his presentation on the crater before the Harvard Club more than two years before, and Livermore was himself an "M.C." stockholder who would soon take the aging Agassiz's seat on the exploration company's board.[124]

In mid-July Shaw, Hutchinson, and Livermore spent five days at the crater observing the mining operation, surveying the crater itself, and talking with Colvocoresses. Shaw, at least, remained cautiously optimistic. "Think we must expect slow progress and high costs," Shaw telegraphed Barringer from the crater, "but hope to be able to suggest a plan for adequate financing."[125] And a few days later, he wrote that "I do think in general essentials Hutchinson and Livermore have accepted the situation as we believe it, but at the same time there is a certain amount of work, which I think they want to go over, before they write their report. . . . In reality," he added, "they represent new talent applied to the subject, both from scientific and mining points of view, and I think they will be of great help."[126]

Early in August, Shaw proposed a new financial plan designed to generate more than $500,000 through the sale of 5,000 additional shares of the exploration company's stock.[127] Under this plan, the company itself would obtain subscriptions for at least 2,000 shares, or $200,000, at which point North American Mines would underwrite the balance of the offering for up to $300,000. Hopefully, he noted, the company would raise more than $200,000 so that North American would have to underwrite only a proportionately lower amount. Shaw then launched into a lengthy discussion of the situation which, he admitted, "sounds pessimistic but is not so intended because I have not lost my faith in the undertaking. . . . Hutchinson and Livermore," he also assured Barringer, "as engineers for North American Mines, Inc., are greatly interested in the Meteor Crater as a possible opportunity for that company's funds, which are set aside for exploration and development. . . . On the way back, *they* [Shaw's emphasis] proposed the general scheme of N.A.M. taking an underwriting. . . ."

Implementation of this new plan would be contingent upon the approval of the directors of both companies; moreover, the approval of North American's board, it was quite clear, would depend on the nature of Hutchinson and Livermore's report. Shaw now emphasized the importance of their investigation in a mid-August draft of a report to the exploration company's stockholders, noting that they "have made a careful study of the whole situation" and "have consulted many authorities of high standing on the problem involved."[128]

On August 26 the exploration company's directors approved the new plan; and Barringer, as vice president, that same day drafted a formal agreement for submission to Shaw and North American's board.[129] This embodied the basic provisions of Shaw's plan. In addition, it stipulated that the exploration company would amend its certificate of incorporation to authorize the issuance of 15,000 shares of both preferred and common stock, at $100 and $1 par value per share, respectively, to be sold in units of one preferred share and one and one-fifth common shares at a unit price of $101.20. Certain shares would be assigned to the Standard Iron Company for modifications of its lease, while North American would receive a bonus of 2,000 shares of common stock and $15,000 in cash for underwriting up to 3,000 shares of the new offering. North American would also agree to furnish the exploration company "a full engineering advisory service through its own engineers, Messers Livermore and Hutchinson."

Barringer's hopes once again soared. As both Shaw and Livermore were also members of North American's board, the chances seemed excellent that once Hutchinson and Livermore's report was received, it would take favorable action on the proposed agreement. What little Shaw had told him of their investigation had been encouraging; and from the mood of his fellow directors on the exploration company's board, he anticipated few difficulties in obtaining the $200,000 in stock subscriptions required to put the agreement into effect. Colvocoresses was already getting estimates of the cost of the most powerful water pumps available anywhere in the country which, he was convinced, would drain the shaft and keep it dry.[130]

Barely a week later, however, Shaw sent him some wholly unexpected news that plunged him once more into despair.[131] That Shaw understood the impact his words would have on Barringer seems evident in the fact that he chose them carefully and tried, at least, to break his news as gently as possible.

"When I wrote you some days ago about leaving to go over some reports, etc., with Hutchinson and Livermore," he began, "I little thought that this letter would be written. Their results of checking assays, etc., had in the main been favorable to former results, in fact above the old figures in some cases, though the results of drill hole material are as yet not so satisfactory as conclusively showing the true *location* [Shaw's emphasis] of the meteoric mass.

"As I wrote in my draft of my proposed report to the 'M.C.' stockholders," he continued, "both Hutchinson and Livermore 'consulted eminent authorities' on the various problems involved." In line with this, he now revealed, his son, Quincy A. Shaw, Jr., who held fifteen shares of "M.C." stock, had gone to see Shapley, the Harvard Observatory's director, and then, with Hutchinson and Livermore, talked with Shapley's colleague, Willard F. Fisher. Both astronomers had suggested that they consult Moulton. Consequently, they had contacted Moulton in Chicago and asked him to determine through mathematical analysis the most probable dimensions of the Meteor Crater meteorite. Moulton had agreed.

"I thought this phase of the examination well worth while," Shaw explained, "since recent years' work in astronomy and celestial mechanics offers much new data and opinion. Professor Moulton's independent interpretation of the problem would be one of high authority and based on the most recent knowledge." However:

> The results given in his twenty-five page analysis are to say the least very disturbing. He has attacked the problem from various angles and comes to the same conclusion as others, viz., that probably the meteor was a "cluster" or, as he put it, a "swarm." His work is not yet completed, but his report has dampened Hutchinson and Livermore's enthusiasm, and I feel it necessary before we go any further that we dig up every piece of reasoning and all calculations that have come into your hands as to the probable maximum and minimum values for the velocity, mass and penetration of the meteor. . . . The generally accepted figures that we have been using, two to five miles per second (perhaps 2½ as the probable speed) and six to more than ten million tons, with a diameter of 400 to 500 feet, are so far out of line with Moulton's analysis that I want to get the actual reports and calculations of the various astronomers and physicists that led to the general conclusions just given.

The implications of Moulton's analysis would follow Barringer to the end of his life-long venture.

ELEVEN

THE GREAT DEBATE

> . . . if more than 500,000 tons of meteorite are found, take a long and hearty laugh at my expense and call on me to put up a dinner at the crater for all interested. If my reasoning and my results are so much in error, I will come across with the dinner.
>
> FOREST RAY MOULTON, 1929

Rarely if ever has a major debate among prominent scientists been at once so brief, so intense, and so obscure as that which followed Moulton's first report on his investigation into the dimensions and dynamics of the Meteor Crater meteorite.[1] This little-known controversy, tinged with bitterness and climaxed by tragedy, lasted barely three months. Late in November, Moulton's second report,[2] a massive analysis of the problem delivered with the confident challenge quoted above, followed a few days later by Barringer's death, put an effective end to it. A far more restrained discussion was carried on among a few scientists for another two months and until Moulton disposed of the lingering issues in still a third report.[3]

MOULTON'S CONCLUSIONS

The debate was conducted in private and almost entirely by correspondence, with Shaw acting as both moderator and commentator. Tens of thousands of words, many of them repetitious and some of them quite heated, were exchanged among the participants, several of whom produced more or less lengthy "reports" of their own. Yet this vast outpouring is represented in the literature only by an abstract of a brief paper which Moulton read before the forty-fourth meeting of the American Astronomical Society nearly a year later, and a few press summaries of what he said there. In this, he discussed mainly the theoretical aspects of the effects of the atmosphere on meteors and meteorites. He also announced his conclusion that a close swarm of meteors with a mass of 300,000 tons could account for all the phenomena observed at Meteor Crater.[4] That figure, interestingly enough, agrees well with recent (1983) estimates of the Arizona meteorite's mass.[5]

Moulton's first report, submitted to Hutchinson and Livermore August 24, was, as he freely admitted, incomplete, inaccurate in some of its premises, and too hastily prepared, and thus "has not all the clarity which might be desired."[6] But its import was clear enough—the meteorite's mass could not possibly have been anything like the ten million tons that Barringer had claimed for it for many years.

At the outset Moulton noted the dearth of data for the problem, pointing out that "the only methods of approach to a determination of the mass . . . are the dimensions and character of the crater and the work done by the meteorite upon impact." The "crucial quantity" required, he declared, was the meteorite's minimum velocity at impact; indeed, "no results that are at all reliable can be obtained without determining first a lower limit to the velocity of impact."

Much of Moulton's preliminary discussion was designed to support his theoretical approach. There were, he noted, three possible hypotheses respecting the meteorite's motion prior to its impact: That it had circled the earth at a distance much less than that of the moon and had gradually spiraled in to fall to the surface; that it had circled the earth in an elongated orbit with its perigee gradually coming closer and closer to the earth under the perturbative effects of the moon; and finally, that it had come directly from interplanetary space, striking the earth on its first very near approach. As the first two hypotheses must have involved a series of periodic grazing collisions with the earth's upper atmosphere, he dismissed them on the basis of Barringer's estimate that the crater was less than 3,000 years old, reasoning that such an object would have been so conspicuously bright during these grazing encounters that it could not have escaped the notice of astronomers in antiquity. Livermore quickly pointed out that Barringer's age estimate might be invalid, and that the crater might well be much older,[7] but in any event, Moulton's dismissal of the two hypotheses was not later called into question.

Only the final hypothesis, Moulton contended, was tenable and offered a basis for determining a lower limit for the meteorite's impact velocity and then only through "much more powerful methods than those heretofore used. . . . Under this hypothesis, the sun played a dominant role until the meteorite was far inside the orbit of the moon. This throws us at once into the three-body problem, and the impact velocity of the meteorite cannot be determined without facing the difficulties of the three-body problem."

These difficulties need not be faced here, however. Moulton's analytical derivation of a lower limit for the velocity, using the concept in celestial mechanics known as the partition of rotating space, was not challenged by any of the scientists who participated in the subsequent controversy. However, while his method was valid, most celestial mechanicians would consider it an unnecessarily complicated approach to the problem and somewhat akin, perhaps, to shooting squirrels with an elephant gun. Readers interested in the details of Moulton's method are referred to his *Celestial Mechanics,* to which he frequently referred in his report.[8] For present purposes, it is sufficient to note simply that he found a lower limit for the velocity of impact of 6.933 miles per second, a figure which, despite his many assumptions and the "numerical uncertainties" of the problem, he felt "can be accepted with confidence that it is in full accord with what happened." It is, of course, also approximately equal to the velocity from infinity under the earth's gravitational influence alone, but this fact, he declared, "was purely accidental and has no significance."

A meteor traveling at this velocity, he pointed out, would have kinetic energy equivalent to 42×10^9 foot-pounds per ton and, combining this with liberal estimates for the amount of rock crushed at the crater and for the amount of energy required to crush it, he computed an upper limit for its mass. His value for the amount of rock crushed was 1200×10^6 tons, three to six times higher than previous estimates. For his unit of energy, he used a figure that Hutchinson had given him and which came to be called the "stamp mill factor," i.e., that 23,760,000 foot-pounds of energy were required to crush one ton of ore in a stamping mill. This he multiplied roughly by four, using a figure of 100×10^6 foot-pounds per ton of Arizona rock to allow for any other kinds of work, such as heating, that the meteorite might have done.* His calculations yielded an upper limit for the meteorite's mass of three million tons, a figure which he also thought could be accepted "with confidence."

Conversely, to find the lower limit of the mass, it was necessary to determine an upper limit for the velocity. For this he assumed that the impacting

---

*Moulton consistently used the foot-pound as his unit of energy in his papers on Meteor Crater, and his usage will be followed here. For those who prefer to think in terms of ergs or megatons, it may be noted that one foot-pound equals 1.356×10^7 ergs; and that 4.19×10^{22} ergs, or 3.1×10^{15} foot-pounds, equals one megaton TNT equivalent.

body was a planetoid revolving around the sun in the same direction as the planets. "We shall have an upper limit to the velocity of impact if the perihelion distance of the meteorite was equal to the distance of the earth from the sun," he explained. "This will, in fact, give us an upper limit to the velocity of impact for all bodies moving in orbits of a given eccentricity which can strike the earth." Thus, he constructed an orbit, arbitrarily assigning a liberal value for its eccentricity, and then, with recourse to the three-body problem and with an additional liberal assumption for the inclination of the orbit to that of the earth, he calculated an upper limit for the velocity with which a body traveling in such an orbit would strike the earth. This, he found, was 13.977 miles per second (MPS), from which he computed a lower limit for the mass of 170,000 tons. But, as he now had used only the "stamp mill factor" itself and thus had restricted the work done by the meteorite to crushing rock alone, he conceded that this result was "probably too low," and that "Probably 250,000 tons would be closer to the truth."

If the meteorite had been a periodic comet from Jupiter's well-known family of comets, he added, "the discussion would lead to comparable results." However, if it had been a comet traveling in a retrograde direction around the sun, as some were known to do, or if it had been part of the November Leonid meteor swarm which meets the earth head-on with a velocity of the order of 40 mps, "the mass could not well exceed 30,000 tons."

In the course of his investigation, Moulton had also considered the nature of the meteorite. Assuming that its diameter "was not much less than that of the present crater," he argued that its density must have been low, perhaps only one-fiftieth that of water, and thus that the crater must have been formed "by a compact swarm of meteors and not by a single solid body." This conclusion he supported by what he called the "inverse method of attack." In this, he postulated a solid sphere 3,000 feet in diameter, with an average density of 3, a mass of 1325×10^6 tons, and a kinetic energy sufficient to pulverize $55,650 \times 10^7$ tons of rock. Then, by assuming a density of 2.75 for the rock at the crater, he showed mathematically that such a body would penetrate to a depth of 173 miles. "Even though the estimates on which this computation is made are many hundred per cent in error," he declared, "the results indicate that the meteoric mass could not have been a solid sphere, for there is no indication that the meteorite penetrated to such a great depth."

Moulton's mathematics were not an issue in the ensuing debate except in a general way—Thomson, for instance, expressing his "mistrust . . . of mathematical deductions where the premises are not fully known."[9] But a number of his assumptions, or "inferences" as he preferred to call them, were vigorously challenged, most notably his dismissal of atmospheric retardation as a factor in computing the meteorite's minimum velocity at impact. On this subject, Moulton had pointed out early in his report that "Since the whole mass of the air penetrated is equivalent to only 45 feet of water . . . while the meteorite penetrated something like 1000 feet of rock, the resistance of the air may be neglected as being unimportant," adding that this was "permissible particularly because the air is elastic."

Both Hutchinson and Livermore immediately grasped the implications of Moulton's report for further financing of the shaft-sinking operations at the crater. Moulton's "new ideas," Hutchinson now advised the astronomer,[10] "alter the picture in a number of ways; the most important appear to be the probable depth of penetration and the estimated magnitude of the meteoric mass. If it should appear that the meteor is deeper than 7,000 feet, it would equal the depth of the world's deepest mines, and the mass of it would need to be large indeed to warrant its exploration." To this point, too, he informed Moulton that laboratory work on material from the final hundred feet of U.S. Smelting's south rim drill hole "yields results which are not convincing that the meteoric mass is nearly as close-by" as Barringer and others had assumed. Additional studies could be made, he noted, such as shooting rifle bullets into the Arizona rock, but it was "futile" to carry out such inquiries "unless the uncertainty and doubt can be lifted which, as the situation now stands, render financing wholly impractical."

Hutchinson also corrected Moulton's assumption that the meteorite's diameter was "not much less than that of the crater," pointing out that it could not have been much more than 1,000 feet, even if it had been a swarm, "without infringing on rock masses which our evidence indicates are not crushed or much displaced. . . . We arrive at a minimum size based on economic considerations, for it would not be warranted to explore for the meteoric mass even at a depth of 1400 feet unless it should have a probable magnitude of 6,000,000 tons. We would choose 5,000 feet for the maximum limit of depth within which exploration might economically be undertaken. We should like very much to have your idea of the probable depth of penetration starting with the above economic limitations. . . ."

Livermore also noted that the mass "would seem to be nearer 1,000 feet in diameter than the possible 3,000 feet named in your report," and thus that "the meteor may have been of greater density and, by the same reasoning, would have penetrated deeper."[11] Yet, he wondered, "would not there be much more evidence of the deeper formations penetrated being thrown to the surface? Of this there is no sign; pieces of all the formations up to 1400 feet are in evidence but none of the distinctly different formations below that point."

Moulton quickly replied to these and other queries in a series of letters. First he reported a conversation with his son, Gail F. Moulton, a petroleum geologist with the Illinois Geological Survey, who "strongly recommended" using geophysical methods and particularly the magnetic or the torsion balance "as he thinks the meteoric mass should be easily detected by both means, provided it is as large as seems probable and does not lie at a depth of greater than 5,000 feet."[12] In a second, longer letter the same day, he conceded that he had perhaps taken "too seriously the conclusion that the boring of 1922 had actually penetrated the main mass of the meteorite," and that his finding that a hypothetical meteorite would penetrate to a depth of 173 miles "was influenced by this conclusion . . . ."[13] Hutchinson also wondered whether the friction caused by the meteorite in passing through the rock might not consume more of its energy than the crushing of rock in its path. But Moulton did "not think it is likely that the energy used in this way can be a considerable fraction of the whole. . . . Even if this should be true for a hole in rock one foot in diameter, he added, "it would not likely be true for a hole in rock 4000 feet in diameter."

Further progress on the problem of the mass and the location of the meteoric material, he suggested, "is most likely to come from a study of the effects of very high speed projectiles on rock similar to that penetrated by the meteorite. If you can determine the relation between the energy consumed in pulverizing the rock and that consumed in heating the meteoric material and the surrounding rock material, it would give us a basis for making more reliable estimates respecting the depth of penetration. . . . From the beginning I have been impressed by the fact that there is no clear evidence of extensive fusion of rock material. . . ." Still another line of approach would be to determine "the distance to which earth waves were transmitted in considerable rebound of the elastic rocks immediately after the impact, the magnitude of which, of course, depends upon the momentum of the meteorite. If we had data relative to the distances from the crater at which rock waves made observable displacements of rock, we might get some idea of the momentum of impact which . . . would lead to rather definite conclusions."

In a third letter a week later, Moulton now was "inclined to think that I overestimated, somewhat, the amount of work done by the meteorite," particularly in regard to heating.[14] "The heating would be the result of only the work done against the forces of the non-elastic type," he explained, for if "a force does work on an elastic mass, such as crystal, without shattering it, and the elastic mass retakes its initial state, no energy is changed into heat." Thus any heat generated would be the consequence of the shattering of the rock and in compressing it, and since 23,760,000 foot-pounds of energy required to crush a ton of rock in a stamping mill "would probably heat a ton mass of rock less than 100°F, it now seems unreasonable to expect to find any considerable evidences of fusion. This conclusion is drawn neglecting the compression of material," he added.

At this point Moulton became mildly alarmed at the effect his report and his subsequent correspondence had produced. Hutchinson had indicated that his "new ideas" might preclude further financing of the crater venture, and Livermore advised him that "Naturally, we cannot consider continuance of our proposed program of financing, in view of the authoritative results obtained by you; however, we wish to afford those who have spent much time and money in exploration, and particularly Mr. Barringer, every opportunity to offer counter-argument. . . ."[15]

"I was not aware," Moulton immediately responded, "that my conclusions were regarded with such weight in connection with your determination of what your next step would be. Let me say, on this point, that I realized from the beginning that the matter before us had important commercial aspects, as well as scientific, and that someone had to reach definite conclusions in order to form a basis for deciding what might be advisable in the future. With this point in view, I made direct statements of my conclusions. I expect, however, that you will check them in every way possible before basing any actions upon them, and if anyone should show that I am in any respect wrong, I shall, of course, be glad to acknowledge the fact."[16]

Shaw, of course, was fully aware of the contents of Moulton's report and of Hutchinson's and Livermore's reactions to it, when he wrote his letter to Barringer.[17] Barringer, surprised by its unexpectedly negative tone and confused by its vagueness, nonetheless immediately grasped its significance for the Meteor Crater enterprise, and just as quickly seized upon the point around which the subsequent controversy over Moulton's conclusions would center. Shaw's letter, he at once wrote Thomson,[18] "is indeed astonishing. Of course Messrs. Hutchinson and Livermore are justified in consulting Prof. Shapley, Dr. Fisher and Prof. Moulton with regard to the major problems of the size and present location of the meteoric mass. Although I have not seen Prof. Moulton's 25-page analysis, it is clear that he, like Campbell, has adopted the theory that the mass was of small size and hit the earth at something like astronomical speed. . . . Neither Shapley nor Moulton nor Dr. Fisher (I do not know who he is) seem to have stopped to consider the tremendous retardation of the earth's atmosphere, through 150 miles of which, at least, the mass traveled. . . . The situation is a serious one," he stressed. "All of our financial plans will come to naught unless the report of Hutchinson and Livermore . . . is favorable. There was no doubt whatever that it was in the main distinctly favorable until doubt has been aroused in their minds by the interviews which they have had with Messrs. Shapley, Fisher and Moulton. The $500,000 which we so badly need was practically in sight until then."

BARRINGER'S DEFENSE

Barringer did not cite these fears in his lengthy reply to Shaw the following day; instead, he reviewed many of his now familiar arguments for his estimates of the meteorite's mass and diameter, but focused primarily on the role of atmospheric retardation in meteoric dynamics.[19] He had, he began, already written Thomson, and while both Magie and Russell were in Europe, he would cable them, and also try to obtain opinions on the dimensions and dynamics of the meteorite from other authorities. Then he proceeded to establish what would be his first line of defense against Moulton's conclusions, declaring that "Probably the problem may call for an even slower speed than we have been calculating upon—say one mile a second."

Noting that Moulton "has assumed a planetary speed and calculated that a far smaller projectile would furnish the energy required," he argued that "the facts at the crater clearly show, however, that the mass did not strike at any such planetary speed." For if it had, "we could not have the penetration we find," nor could it "have produced *the kind* of crater we have [Barringer's emphasis]. Instead we would have a wide shallow crater with evidences of great fusion of both the individual meteorites and the arenaceous limestone, which would make a perfect slag. . . . The utter absence of fusion and volatilization of the impacting mass and the rock target (with the single exception of very small amounts of Variety B of the metamorphosed sandstone . . .) furnishes incontrovertible evidence that the mass did not strike with anything like cometary speed."

Had the meteor struck the earth head-on, Barringer agreed, its velocity might have been as high as 40 mps, but "if it was a following blow it might be a very few miles a second," and a "following blow would seem to be far more likely as nearly all the planetary bodies have the same clockwise revolution."

Moreover, meteorites weighing hundreds of pounds reached the ground

> with a velocity no greater than that of any similar body falling from a great height, the planetary speed having been entirely lost in the passage through not less than 60 miles of atmosphere. The great Bacubirito meteorite in Mexico, weighing 37 tons, which I have personally seen lying where it fell, penetrated only a few feet. Our meteorites approached at an angle, as we know, and probably passed through a thousand or more miles of atmos-

phere (the 1910 fireballs were visible from Manitoba to Bermuda). Our cluster may have partly circled the earth before the gravitative pull of the earth overcame its cometary speed. How much of a cushion of compressed air would be formed is well shown in the Siberian fall of 1908 where the released air knocked down and scorched a forest for thirty miles in every direction.... I do not know whether it is possible to theoretically allow for the amount of retardation in our case, but it must have been enormous.

As to Shaw's remark that Hutchinson and Livermore doubted that the actual meteoritic mass had been located by U.S. Smelting's drill, he declared that it "is evident that no meteoric material could have gotten 1300 feet under the southern rim unless the whole mass made a way for it. True, the larger portion of the mass may be ahead of or on either side of it, and not improbably is." His arguments, he added, represented the "consensus of opinion" of Thomson, Magie, Russell, Webster and others, "all of whom visited the crater and personally studied the evidence. . . . I wish Dr. Shapley, Dr. Fisher and Professor Moulton could go to the crater and see the facts for themselves." Hutchinson and Livermore, in fact, soon invited Moulton to inspect both Meteor Crater and the Odessa, Texas, crater; but the astronomer politely declined, explaining that "I cannot qualify as an expert in geology and related matters, and under the circumstances doubt if the value to you would be worth the expense."[20]

Barringer, it must be noted here, had misunderstood Shaw's letter in regard to Shapley's and Fisher's roles in the matter and had conveyed his misunderstanding to Thomson. The latter, in a long, rambling reply devoted primarily to generalities about atmospheric retardation, had also found it "indeed astonishing . . . that men like Professor Shapley, whom I know quite well, and Professor Moulton, whom I know only through his reputation, should come to the conclusions they do."[21] After Barringer sent Shaw a copy of Thomson's letter, however, Shaw quickly set both men straight on the point. "Neither Professor Shapley nor Dr. Willard F. Fisher . . . have made any reports to Messrs. Hutchinson and Livermore," or "have expressed any opinions as to the speed, mass, volume, etc. of this particular meteor," he advised Barringer. "It would be unfair to either . . . to have my letter to you of Sept. 4th [sic] so interpreted."[22] Thomson, acknowledging a copy of this letter, explained to Shaw that "Barringer in writing me was evidently laboring under a misconception of the import of Moulton's statements; and did not await details. I shall prefer to understand what Moulton takes as the basis of his calculations before having anything more to say on the subject."[23]

Barringer, incidentally, would later ask Thomson to try to persuade Shapley to their way of thinking on the atmospheric retardation issue, but Thomson rejected the suggestion with some vehemence. "I may say that I could not undertake to do that even if I had the time," he declared. "I cannot even *doubt* [Thomson's emphasis] that Shapley is aware of the effects of air resistance, nor assume that he needs any tutelage of that kind; and certainly persuasion as against his idea of the facts would not fit the case. . . ."[24]

If Barringer had been shaken by Shaw's first letter, his agitation now deepened as the result of a special meeting of the Meteor Crater Exploration and Mining Company directors, held in Philadelphia on September 11, at which Shaw spent nearly three hours summarizing Moulton's report and his subsequent correspondence with Hutchinson and Livermore. "I find it difficult, such is my state of mind, to write you," Barringer reported to Thomson. "Owing to my deafness, I may not have heard all that Mr. Shaw said. . . ."[25] But he had, in fact, heard Shaw's major points well enough, and the points he missed, or misconstrued, were minor ones. As Shaw and his engineers now felt that the mass was "distributed under the whole southern rim" and thus that "mining could be most expensive," Hutchinson and Livermore "apparently are unwilling to recommend the continuance of the exploration. . . . The net result is likely to be disastrous unless we can show Moulton to be fundamentally wrong. . . . The evidence is all against such hyfalutin [sic] theories as he has advanced. . . . Great is the power of ignorance. . . ." The situation was serious, he again warned Thomson. "For ten days ago, as I wrote you, the papers had all been prepared which would have secured a subscription of $300,000 from the North American Mines, Inc., and Mr. Shaw was going to underwrite $90,000 additional of the $200,000 reserved for the Meteor Company and get several of his friends to underwrite for smaller amounts so that, with what the old subscribers would take of the new issue, it was practically certain that this $200,000 would have been subscribed in a short time. All of this is now queered by what I, personally, have no hesitancy in terming fool theories. . . ."

That the meeting had been traumatic for all concerned, perhaps, is indicated by the fact that while the directors agreed to suspend immediately all operations at the crater, no one was sure who was supposed to notify Colvocoresses to this effect. Barringer thought Shaw was going to do it, but Shaw thought Brandon Barringer had been assigned the task. Not for several days did Barringer himself end the confusion by telegraphing Colvocoresses "to cease further efforts to sink or pump and reduce expenses to absolute minimum owing [to] financial situation. . . ."[26] Colvocoresses wired his compliance the next day.[27] For the record, a total of $275,916 had been expended in the abortive attempt to sink the shaft.[28]

THE DEBATE ENSUES

Even as the directors met, Hutchinson had moved to formalize the debate over Moulton's report, asking the astronomer if he "would be satisfied to have your report and letters go into discussion with Barringer and Thomson and perhaps also with Prof. Olivier . . . and Norris Russell, and will you join in the discussion?" Moulton immediately replied that "you are at liberty to use my report in any way that you wish. There are, of course, certain parts of it which are not numerically accurate as a consequence of my having some misconceptions respecting the certainty of previous conclusions."[29] He was also willing to participate in "a joint discussion with Mr. Barringer, Professor Thomson and others. . . . If there should be a disposition to continue developments that ignore or contradict the implications of my conclusions as to the minimum velocity of impact, I think it is desirable that I should be present in order to defend these conclusions."

As to future actions, Moulton suggested that Hutchinson and Livermore "get in touch with the U.S. Artillery in order to learn whether or not they have any information respecting the penetration of projectiles in rock." Finally, he noted, "it would be very hazardous, as a financial venture, to put more money into the project until the location of the meteoric mass has been determined with a higher degree of probability. I see no more promising way of securing results which would make a safe basis for operations than the geophysical methods respecting which you are fully informed."

On September 19 Shaw notified Barringer that he was having copies made of Moulton's report. "Professors Hutchinson and Moulton's letters contain a good deal of importance which must be read together with the report, and it should be remembered that Dr. Moulton was rather hurried, because of our anxiety to get together all the information we could for the proposed financing."[30]

Barringer had already begun to rally his allies. He had met with Olivier in Philadelphia and "the world's greatest authority on meteorites," as Barringer called him, had "volunteered to give his own judgment" after seeing Moulton's report.[31] He had also telegraphed Webster, the Stanford physicist, asking for his estimate of the meteorite's size and mass; and Webster had quickly replied that the problem "involves so much uncertain extrapolation," and the result "depends on so many unknowns that I doubt the reliability of any estimate." He did, however, "emphatically agree with your opinion that [the] iron did not volatilize much and is practically all still buried under the south rim."[32] A few weeks later, Barringer would ask Webster to prepare a formal report on his conclusions about the Arizona Meteorite "merely to meet Moulton with his own weapons and prove him, if possible, by mathematical and physical reasoning, to be wrong."[33] The physicist did, in fact, eventually submit a report,[34] but Barringer never read it, and probably would not have approved of it if he had.

While Moulton's report was being copied, incidentally, Shaw himself became concerned over a spate of patently erroneous articles in the press regarding the crater operation. He now wrote Barringer that both the *Engineering and Mining Journal* and the *Northern Miner* had published articles which, in the words of the *Journal's* headline, declared that "Meteor Crater Shaft Hits Iron at 1400 feet."[35] He had, he indicated, written to both publications pointing out that the articles were "wholly inaccurate," and asking for their source and for corrections. A few weeks later, the *Boston Evening Transcript* published a similar article, and Shaw protested again, advising the paper's editor that "I would be glad to hear something from you of the history of why this was published."[36] Shaw apparently never received a clear answer to his questions.[37] But almost certainly, the articles stemmed from a front-page story a few weeks earlier in *The Arizona Republican* in Phoenix.[38] This, datelined from Winslow and illustrated by a five-column aerial photograph of the crater, quoted "men at work on the project" that not only had "metal" been "located at 1400 feet," but that "the shaft has been extended to 1600 feet and is now deep enough to allow the miners to cut underneath the body to determine its extent."

Barringer received copies of Moulton's reports and the related correspondence on September 28 and quickly distributed them to Thomson, Olivier, and his other confidants and correspondents, including Magie and Russell in Italy, urging them to criticize Moulton's methods and reasoning and to refute his assumptions and conclusions in the strongest possible terms. Magie and Russell would not receive their copies for some weeks, and it would be several weeks after that before Barringer could expect their replies. But this would not delay the debate, which now began in earnest.

OLIVIER AND THOMSON RESPOND

Olivier was the first to respond, submitting his own "report" only three days later in the form of a long letter to Moulton.[39] His thesis was atmospheric retardation; his method, analogy, and more specifically, an extensive review of known bolides and meteoritic falls designed to prove that "meteoric objects carry with them an envelope of heated gases, and particularly a cushion of air before them;" that "all bodies are greatly retarded, even at great heights;" and that "meteorites have low velocities on striking the ground." He agreed that the mass had been a swarm and probably "the head of a small comet," and that it had "hit the outer atmosphere at a velocity of not less than 7 m/s, and probably greater." However, Moulton's neglect of atmospheric retardation in determining the velocity at impact was "the fundamental assumption about which I am forced to differ. Briefly, and without having any actual figures to cover the case, I believe that even a velocity of 10± m/s would be largely lost or neutralized in penetrating power, before the ground was reached." To this point, he cited more than a dozen examples of historic fireballs and meteorite falls where observed velocities in the atmosphere had been low, or where penetration on impact had been slight. These included the Tunguska event in Siberia in 1908, and, from recent reports, "particularly the immense mass found by [the] French" in the western Sahara desert which was "Said to be 100 metres long, by 40 × 40," and was "Only partly buried in sand."

On Moulton's assumptions, Olivier argued, penetration should have been far greater in these examples. Using Moulton's values for the work done by such an object impacting on rock, he calculated that a 500-pound meteor with a "striking cross section" of one square foot would penetrate rock with a volume of 12 cubic feet per ton to a depth of 1,260 feet, or "At least several hundred feet." From his cited examples, he declared, "this is from 100 to 1000 times too great a penetration. It is further fair to assume that these, on the average, had as high a cosmical velocity as the Arizona swarm. Atmospheric resistance in every known case retarded them immensely."

Olivier pointed out that a "cylinder or cushion" of air must have been formed in front of the whole cross section of the Meteor Crater swarm. "That such a cylinder or cushion was formed is certain, and its amount can be vaguely visualized by reference to what the comparatively small Siberian swarm carried with it. By all analogy with bolides, the main mass was constantly and rapidly losing velocity as it came lower. The final act may be visualized in two ways. . . . Either the mass of air, at great density, actually disrupted the rock ahead of the solid parts. . . or it rebounded in whole or part forcing the solid particles to pass through it in opposition direction and exercising a great retarding influence for the last fraction of a second. This last seems to me the more probable action. . . ."

Barringer, of course, found Olivier's report much to his liking. "It is certainly up to Professor Moulton to produce evidence contradicting your evidence, and that he cannot possibly do," he wrote the astronomer, enclosing a $75 check "as your very reasonable fee."[40] Thomson, too, thought Olivier's report was "remarkably fine," particularly as he had given "due value to the wad of air which must precede any fast moving body."[41] M. B. Huston, a San

THE GREAT DEBATE   277

Francisco mining engineer with whom Barringer was corresponding about the crater, declared that "they must consider Olivier a disinterested party and I do not see how they can get past his letter. He has proven with all the data the world has that meteors 'light' with low velocity and has indicated that great penetration can only be had by great *depth* of mass [Huston's emphasis], no matter what the facial area may be."[42]

Barringer's son Brandon even argued to Shaw that Moulton's report, given Olivier's data on air resistance, actually confirmed the earlier assumptions as to the meteorite's mass.[43] He agreed that Moulton's computations were "mathematically accurate" and that the mass had entered the atmosphere at 7 mps. But he contended that Moulton's inference that atmospheric retardation could be neglected "is unsupported by evidence. . . . If we accept Olivier's conclusions, we get a mass of the character and dimensions we have all along figured; i.e. if the speed at striking the ground was ½ Dr. Moulton's minimum, or 3½ miles a second, we get a mass four times his maximum or 12 million tons. The discussion would then have merely corroborated our original assumptions which now stand the test of mathematical calculations never before applied in such detail." Shaw would later point out that Brandon here had accepted Moulton's value for the total energy, i.e., $3,000,000 \times (7)^2$, rather than Olivier's assumption of $10,000,000 \times (2½)^2$, and that on the latter's assumption, when the velocity was increased to 3½ mps, "the mass becomes 5,000,000 for the same energy. . . . [Brandon] has simply made a set of figures to represent *Moulton's maximum energy*, which is 2½ times larger than Olivier's figure for energy [Shaw's emphases]."[44]

Shaw, indeed, had other questions about Olivier's "report" which he conveyed through Barringer. "The conception of a cylindrical body of compressed air in front of a meteor seems to me to deny the apparent evidence of photographs of high-speed projectiles," he pointed out.[45] "Such a conception apparently cancels all the phenomena of stream lining which we know about. It seems to infer that the effects of such speed as 10 miles per second are so great that they, to a very great extent, deny the elasticity of the air." He was interested, then, in Olivier's interpretation of Moulton's remark that neglect of atmospheric retardation was "permissible because the air is elastic." Shaw also questioned "the destructive effects of this cushion of air, as exemplified by the Siberian Meteor," and "especially since it is recorded in the report of that Meteor that some trees were left standing. Are not the charred effects on the trees well within the limits of the heat supposed to have been generated by the Arizona meteor?"

Olivier replied to these queries promptly.[46] "As to stream lines for projectiles, please remember that no projectile moves faster than about 1 km/sec [kilometer per second]; also that they are usually shaped to give the best stream line effect. These experiments are wholly inadequate to explain what takes place for velocities from 15 to 75 km/sec." In the Arizona case, he added, it was conceded that the diameter of the swarm was between 400 and 2000 feet, and "to apply stream line effects to such a moving mass" was "illogical." Olivier admitted that he did not "fully understand why Dr. Moulton applied the word elasticity in his argument," but noted that his colleagues in the University of Pennsylvania physics department "agree with me that whatever elasticity the air has, would tend to slow up the cometary mass; a result that entirely tends to strengthen my argument." Finally, he declared, his point about the Siberian meteorite "apparently was misunderstood. I tried to show that if such a sparsely scattered lot of meteoric masses caused as *much* [Olivier's emphasis] destruction, i.e. had as great a mass of compressed air in front as they did, then at least a thousand-fold effect would be expected in the Arizona case, where the front cross section must have acted almost like the face of a continuous solid."

Moulton, like Shaw, was also skeptical, but he was sufficiently intrigued with Olivier's arguments to explore the question of atmospheric retardation himself in greater detail. He wrote to him:

> Inasmuch as we agree on the velocity with which the meteoric mass arrived at the surface of the atmosphere, there apparently remains nothing but to consider the resistance of the air. I will not undertake to do it in this letter, but I should like to point out to you that you are taking the position that most of the energy of the meteoric mass was taken up in passing through an atmosphere which was the equivalent of something of the order of 30 or 50 feet of water, or 10 or 15 feet of rock, and that after the great reduction in energy by this relatively small mass it penetrated sandstone and limestone rock to a depth of at least 1000 feet. I think if we had no preconceptions on the problem induced by experience with small bodies such a position would on the face of it seem somewhat hazardous.[47]

On this latter point, Olivier conceded that he had relied on "analogy from small bodies," but he again cited the Saharan meteorite, noting that its mass had been estimated at 100,000 tons, "a mass strictly of the order in question. If our *present* [Olivier's emphasis] information is confirmed,* we cannot ignore the fact that it is not deeply buried. . . ."[48]

---

*But Olivier's information was not confirmed, as Shaw suspected at the time; see Q. A. Shaw to D. M. Barringer, October 12, 1929. A French official, one M. Ripert, had reported finding such a mass in 1916 thirty miles southwest of the Chinguetti oasis in Mauritania

Two days after replying to Olivier, Moulton advised Hutchinson and Livermore that "In view of the amount of discussion aroused by my report and its apparent weight in bearing on your future activities, I have decided to take up a discussion of the possible effects of the atmosphere in retarding the motion of meteoric materials through it. . . . In spite of the distrust that several writers experience respecting the results . . . of mathematical processes, I propose to look into the question somewhat thoroughly, if it meets the approval of you and Mr. Shaw, and although I cannot hope to determine the mass of the meteorite very precisely or locate it very precisely, I have much greater confidence in the methods that I use than I do in the generalities contained in the correspondence."[49]

Barringer and those few of his associates who had so far commented on Moulton's conclusions had indeed dealt entirely in generalities; even Olivier had conceded that he was unable to quantify the effects of atmospheric retardation. But Thomson, certainly, was the most grievous offender. The aging, much-honored inventor, after "looking through" Moulton's report, advised Barringer that Moulton's viewpoint "is so different from my own that I think it is not necessary for me to analyse it or controvert his statements."[50] Nonetheless, he then produced five single-spaced typewritten pages of gratuitous comments, unsubstantiated pronouncements, rhetorical questions, and somewhat vague recollections of meteoric phenomena he had observed himself during a long life, dating back to his childhood and the "famed" Leonid meteor shower of November 1867. For Thomson, it was enough to state that the "resistance of the air itself to a rapidly moving body is, in my experience of observations of meteors, quite sufficient to turn a high velocity body into a comparatively slow one." Again, Moulton's assumptions based on the amount of rock crushed at the crater were useless because no one could say "what energy was required in pulverizing, heating, transporting and, incidentally, communicating the tremendous earth wave," and thus "the stamp mill analogy might well have been left out of a case of this kind." As to Moulton's minimum velocity of impact of 6.933 mps, he noted only that it was "curiously, about the velocity that would be required to be imparted to a shot thrown out from the earth to carry it beyond the gravitational limit. . . ."

Much of Moulton's theorizing, Thomson suggested, was impertinent, and

---

and sent a nine-pound block found near it to Paris, where it was later determined to be a nickel-iron meteorite. The mass, however, was never relocated. Strangely enough, the Aouelloul Crater, discovered in 1920 and many years later found to be an impact crater, is only about four miles away from the approximate position of the Chinguetti "meteorite" as given by Ripert. See L. LaPaz and J. LaPaz, "The Adrar (Chinguetti), Mauritania, French West Africa, Meteorite," *Meteoritics,* 1(1954): 187–96.

particularly his postulate of a solid mass with an "absurd" density of 3 to show that the meteorite had been a swarm. "We have never doubted the swarm idea.... I think, therefore, it is not at all necessary to consider the case of a solid body as Moulton does. It seems to me he makes a number of assumptions, and then proceeds to smash them. What is the use of that? I cannot see it." Moulton's mathematical approach he simply dismissed as being based on false premises. "I have not had the time or the inclination, I may add, to go into these matters so critically, because I do not see that they lead us anywhere if the original assumptions upon which the argument is based are incomplete, defective, or the reverse of what we seem to have discovered."

Barringer clearly felt that such pronouncements, given Thomson's prestige, would be persuasive, but neither Shaw nor Moulton seem to have paid much attention to them. A few days later, it may be noted, and after he had seen Olivier's "report," Thomson confessed to Barringer that "I did not do the thing as thoroughly as I might have done."[51]

BARRINGER AND SHAW VOLLEY

Barringer's own first reactions to Moulton's report were general because he had perforce to rely entirely on his understanding of what Shaw had told him of its contents. But even after reading the report, he eschewed specificity in favor of what he considered "common sense" arguments, broaching one of his favorites to Shaw.[52] Tilghman, he recalled, had cited artillery experiments "made 80 to 90 years ago" with shot fired into masonry and rock at a velocity of "about 1800 feet per second," which produced "a comparatively shallow crater of conical form about five times the diameter of the projectile, within which, the projectile, or its wreck, was deposited." These experiments also showed, he declared, that the depth at which the shot had broken up "was a fraction under two diameters of the projectile used." Thus, and in line with his own long experience with high-powered rifles, he argued that "there is a limit of penetration in any yielding target at which a given speeding projectile attains its greatest penetration. Increase the speed beyond the critical limit and the energy begins to be exerted in the deformation and destruction of the projectile." Consequently, if "the projectile (our projectile) was moving too fast *it could not* have penetrated as far and done the work which it has done [Barringer's emphasis].... It does not take many such facts to overcome a ton of hypothetical reasoning which cannot be backed up by factual evidence," he added.

Moreover, if the Meteor Crater meteorite had struck at "cosmical or cometary speed we would have had a great amount of fusion and perhaps volatilization both of the projectile and the rock," and the crater "would necessarily have been a shallow one. . . . We argue that it did not strike at anything like the speed at which it entered the earth's atmosphere. Why? Because it was retarded by the thick blanket of atmosphere. The proof of this retardation which has been furnished seems to me to be conclusive. . . . From the work done and the depth of penetration," he concluded, "it is inferred that it was in the general order of size we have all along estimated. It could not by any possibility have been a beggarly 50,000 tons. . . ."

Shaw forwarded excerpts from this letter to Moulton, commenting dryly that "I am astonished that guns and powders were made 80 to 90 years ago which gave a speed of 1800' per second."[53] Barringer, he added, "gives you in his own words his postulate that there is a 'critical limit' to the velocity of a projectile beyond which an increase in velocity does not accomplish an increase in penetration. What factors he uses for 'yielding' etc. and how it is applied to this problem, I do not know." Barringer had also stated "more clearly than heretofore another reason offered for impact velocity being as low as 2 to 2½ miles per second, viz, that at 'cosmical or cometary speed' there would have been a great amount of fusion and perhaps volatilization." Finally, he noted, "Barringer and others, having established that $V = 2½$, infer that the energy of 'the work done and the depth of penetration' is equal to $10,000,000 \times (2½)^2$ [i.e., mv^2]. To date there has been no comment on the fact that this figure is only about ⅖ of your maximum energy of $3,000,000 \times (7)^2$."

Shaw himself, however, now offered data on Moulton's energy allowance for the work done by the meteorite, and specifically on the "stamp mill factor" which not only Thomson, but Brandon Barringer and Colvocoresses had contended in general terms was an inadequate measure of the energy expended at the crater.[54] Shaw had quickly taken up Moulton's suggestion that tests be made with high-velocity projectiles fired into rock similar to that at the crater and had obtained a 300-pound block of the crater limestone for the purpose. He had fired .30-caliber, 150-grain rifle bullets into the block and found that each bullet, striking at a velocity of 2,700 feet per second, "crushed 10 times its own weight and made its own crater."[55] Therefore, one pound of bullets, or forty-seven bullets, should crush 10 pounds of rock; and 200 pounds, or 9,400 bullets, should crush a ton of rock. And, as each bullet had a kinetic energy of 2,400 foot-pounds, 9,400 bullets would have a kinetic energy of 23,560,000 foot-pounds, or essentially Hutchinson's "stamp mill factor" of 23,760,000 foot-pounds. Shaw punctuated his statement of this result with an exclamation point. Barringer's son Brandon, at least, quickly conceded that the agreement was "quite remarkable and would indicate that, so far as it can be experimentally verified, the figure for the energy used is not very far out. . . ."[56]

Shaw pointed out that the rock crushed in his tests "represents only part of the energy," and that he did not know what percentage of the total energy was expended in "other work," i.e., breaking up the bullet, generating heat, moving the limestone block, and blowing away rock. However, he argued, Moulton's assumed tonnage of rock crushed at the crater had been 1,200,000,000 tons, four times Thomson's estimate of 300,000,000 tons, and his value for the energy required to crush this rock was 100,000,000 foot-pounds per ton, more than four times the "stamp mill factor." Thus, Moulton had allowed seventeen times more energy for "other work" than was needed to crush 300,000,000 tons of rock in a stamping mill, and the factor would be even higher if the estimated tonnage of rock crushed was lower. Later, he would note that even assuming that 400,000,000 tons of rock was crushed at the crater, Moulton had allowed 92 percent of the meteorite's total kinetic energy for "other work."[57] Moreover, "the *so far allowed* tons (by Thomson and others) of rock crushed per ton by *this* meteor are from 20 to 40 tons. These results are greatly in excess of what I think is inferred as the results *per ton* of known meteors in the examples given by Dr. Olivier." But on the other hand, they fell "short of the work of a bullet," for his tests indicated that a ton of bullets traveling at 2 mps would crush 160 tons of rock. This figure, indeed, was "really larger because the *factor used for bullet* was determined *for limestone only*; the friable white sandstone must increase the number of tons crushed [Shaw's emphases]."

Shaw also commented on the issue of air resistance, noting that if the velocity of a mass moving at 7 mps at entry into the atmosphere were reduced to 2 mps at impact, then, all other things being equal, "it follows that about 90% of this energy has been lost in passing through the air." But, he pointed out, even with liberal assumptions of the amount of rock crushed and the energy required to crush it, an increase of the impact velocity to even 3 mps "immediately places the mass of the Meteor at a critical tonnage which I consider the minimum that offers a fair gamble under the present known difficulties of exploration plus the probable difficulties of mining and the character of the material. . . . Granted that the retardation of the air is from 7 to 2 miles, I do not suppose it is contended that if the original velocity was 10 miles that the retardation would again reduce the velocity to this 2 miles per second. . . . Both Drs. Moulton and Olivier seem to indicate that the original velocity was in excess of 7 miles. The terminal velocities are the critical ones."

This letter from Shaw, with its accompanying notes, seems to have convinced Barringer that a broader base of authority was needed on his side of the debate, and to this end he quickly launched a brief but intensive letter-writing campaign seeking support from outside his immediate circle of confidants. At this point he urged Webster to undertake an analytical critique of Moulton's report. More significantly, in terms of the debate, he persuaded physicist Edwin Plimpton Adams of Princeton's Palmer Physical Laboratory to conduct a similar investigation, with an emphasis on the issue of air resistance. Shaw, he advised Adams, "has been greatly disturbed by the Moulton report. . . . I ask you to carefully examine his figures and point out wherein he is wrong, if it is possible to do so."[58]

In addition, Barringer dispatched copies of Moulton's and Olivier's reports to a number of well-known scientists and mathematicians, along with letters summarizing his now familiar arguments and appealing for support, particularly on the issue of atmospheric retardation. To the Carnegie Institution's Arthur L. Day, for example, he wrote that "If you believe as we do, that the mass which made the Arizona crater met with enormous and constantly increasing atmospheric resistance and was therefore greatly slowed up before it reached the earth, will you be good enough to write me briefly to that effect?" adding that "a noted astronomer takes the opposite view."[59] Similar appeals he sent to astronomers John A. Miller of Swarthmore's Sproul Observatory and Carl O. Lampland and V. M. Slipher of the Lowell Observatory, meteorite experts Alfred La Croix of the University of Paris and Oliver C. Farrington of Chicago's Field Museum, mathematicians William E. Echols of the University of Virginia and Ernest W. Brown of Yale University, physicist Augustus Trowbridge of Princeton, and Professor Alexander Klemin of New York University's Daniel Guggenheim School of Aviation. All but La Croix and Farrington seem to have replied sooner or later, but their responses were brief, couched in general terms, and not always what Barringer wanted.

Miller's reply was perhaps better than most from Barringer's standpoint. "I think the atmosphere of the earth offers great resistance to the motion of meteors through it," he declared. "Theoretically, it is impossible that it should be otherwise." Observations of "cannon balls and rifle balls," as well as meteors, "show beyond question that the retardation of the atmosphere is very great." And to his point, he quoted brief excerpts from an 1876 paper by Lord Rayleigh (J. W. Strutt) and an article on meteors in the 11th edition of the *Encyclopedia Britannica* by Lazarus Fletcher, the British Museum's expert.[60] Barringer forwarded Miller's letter to Shaw, and subsequently would cite Lord

Rayleigh in support of his position on atmospheric retardation.[61] Adams, too, would find the late eminent British scientist's half-century-old statement useful in preparing his "report" for the debate.

Not all the replies found their way to Shaw, however, or into the debate. Echols, for instance, wrote Barringer that "Professor Moulton's elaborate analysis is very scientific and interesting and so far as it goes, under the assumptions made, is all right." He was, he added, "inclined to think that Olivier is entirely right about the light weight meteors arriving at the surface with relatively low velocities due to air resistance," but he also felt that "Professor Moulton appears to be right about the relatively heavy compact metallic mass."[62]

Shaw's tests with rifle bullets, incidentally, stimulated Barringer to seek ballistics data from the Western Cartridge Company to support his contention, based on his big-game hunting experiences, that there was a critical limit to velocity vis-à-vis penetration. The company's R. F. Riggs replied that the "killing power in a rifle is largely a matter of bullet weight. . . . In going over a list of cartridges that are known to be excellent killers do we not find that practically without exception they have bullets weighing over 200 grains and velocity under 2700 feet per second? . . . We are always very much interested in such matters," he added, "and are indeed glad to assure you that our experience checks thoroughly with your own. . . ."[63] Riggs's letter does not seem to have impressed Shaw, but it may have assuaged somewhat Barringer's disappointment over Thomson's view of the issue. "I cannot use arguments based on ordinary ballistics," Thomson had advised him a few days earlier, "for I don't think they apply in this case. We are *not* [Thomson's emphasis] dealing with a case of a properly formed bullet or an assemblage of them of uniform size and round, like a load of shot. Our target too is a moist (wet further down) mass of porous rock as I have tried to show—which complicates the problem with steam pressures and flows. . . ."[64]

Beyond offering generalities and providing moral support, in fact, Thomson proved to be of little help to Barringer in the debate, even on the key issue of atmospheric retardation. After reading Olivier's report and the ensuing correspondence, he had attempted some simple calculations of the effects of air resistance but, perhaps because of his mistrust of mathematical analyses based on uncertain premises, he had no confidence in his results. "Anyhow," he explained apologetically to Barringer, "this series of guesses is as good as any other. . . . I am still pressed for time and have written only an outline of the case based on the idea that air resistance increases as the square of the velocity of a body moving in it. . . . I come back to the problem as it stood— naturally a gamble—and I for one would not bet that the water problem can be taken care of without prohibitive cost. All the rest seems easy compared to that. . . ."[65]

At the end of October, Shaw reviewed the correspondence he had received to date, pointing out contradictions, misconceptions and omissions in the reasoning of the various writers.[66] But his primary thrust again was to defend Moulton's use of the stamp mill factor, arguing that Moulton had made ample allowance for "other work" besides crushing rock done by the meteorite.

Barringer, however, was optimistic again. He had, in fact, been advised that Adams's report was almost ready and that it would strongly support his stand on atmospheric retardation. He confided to Colvocoresses in a long letter on November 1 that

> It can only be inferred from Professor Adams's argument that the weight of the mass is probably even greater than we have been estimating. . . . I write you this, more or less confidentially, with the idea of suggesting that too much discussion . . . may lead us deeper into the quagmire or maybe muddy the water more than it is muddied already. Let us first prove Moulton wrong and Olivier right as to air resistance. . . . Then let us take up the matter of penetration. . . . I suggest that you and I reserve our arguments to the very last, that is to say, let us convince our friends in Boston by degrees. If we can prove Moulton wrong with regard to his air resistance, we can certainly prove him wrong about his multiple of crushing energy, etc., all of which is Greek to me. I am trying to focus my mind on two things: Air resistance and penetration, and bring to bear my own experience with regard to such matters.[67]

Adams's report, which arrived a few days later,[68] was indeed cause for rejoicing on Barringer's part. After twelve pages of theorizing and equations relating to hydrodynamics and the motion of bodies in a compressible medium, the physicist concluded that "the maximum velocity of a sphere 200 feet in radius, specific gravity 7, on striking the earth will be about 1.2 miles per second, and this velocity does not depend upon the velocity . . . on entering the earth's atmosphere provided that this is as large as or larger than 10 miles per second." Then, assuming that 500,000,000 tons of rock had been crushed at the crater, and using the stamp mill factor as his unit of energy, "It follows that the minimum mass of the meteor is 9.5×10^6 tons, or between 9 and 10 million tons."

The details of his analysis are not necessary here. Suffice to say that he began with Gustav Kirchhoff's nineteenth-century formulation of the retarding force exerted on a plane in motion at right angles to itself in a fluid, deriving an approximate expression for retardation from which he computed the velocities at which spheres and disc-shaped bodies of various diameters and thicknesses entering the atmosphere at 10 miles per second would strike the earth. This yielded "very large retardations" for relatively small bodies, but small retardations for large ones. However, he added, this theory "ignores one very important observation with respect to meteors"—that they "must be raised to high temperature" to account for their luminosity. Ordinary friction, he declared, was "altogether inadequate" and a much more probable cause of this heat was "the rapid compression of the air in front of the meteor," in effect, Olivier's "cushion" or Thomson's "wad." And here, as Miller had done, he cited Lord Rayleigh's 1876 statement that the "resistance to a meteor moving at speeds comparable with 20 miles per second must be enormous, as also the rise of temperature due to compression of the air. . . ."[69] Adams now developed an expression for the retardation produced by the adiabatic compression of air in a meteor's path which, he declared, would show that this was "quite sufficient to retard even very large meteors to velocities of the order of 1 or 2 miles per second on reaching the earth."

Adams's theories and calculations would soon be demolished by Moulton, Russell, and others, and he himself would modify them quite drastically in an effort to salvage his conclusions. But for the moment, an enthusiastic Barringer set to work distributing Adams's report to his newly enlarged list of correspondents, urging them again to write him if they agreed with its results. Most of these replies too were brief and general in nature. New York University's aerodynamicist Alexander Klemin, for instance, declared simply that while "Professor Adams' treatment can only be regarded as approximate, it is extremely able and probably as near the truth as it is possible to get without extremely difficult experimentations."[70] Yale mathematician Ernest W. Brown had "no doubt" that Adams's result "is correctly worked out from the initial assumptions. The difficulty is to know whether those assumptions correctly represent the physical conditions. . . ."[71] Lowell Observatory astronomers Carl O. Lampland and V. M. Slipher gave a more satisfactory response in view of Barringer's own ideas on the subject. Adams's results, they telegraphed him, "appear to fit Meteor Crater facts as we know them from first hand examinations. Generally speaking, calculated low terminal velocities also fully confirmed by great mass of observation. . . . Finding great Greenland, Mexican, African meteoric irons on surface obviously indicates very great atmospheric retardation for such bodies [punctuation added]."[72]

Barringer's elation over Adams's report was short-lived, however, for he soon found that those of his associates on whom he had relied so heavily for so long would now fail him.

DETERIORATION OF SUPPORT

At first it was Thomson. While Barringer was away from his office and suffering from a cold, Thomson complained to Barringer's secretary, Isabel Heller, that "it takes time to analyse any reports such as Adams'. . . . The mathematics will probably be all right, even in Moulton's case, but what you do with mathematics depends on what you put into the formula as premises. . . . Kindly tell Mr. Barringer that he must not expect me to analyze all the mathematics that come through these reports." [73] Colvocoresses, too, was reluctant to come to grips with the issues in the debate. Although he had argued Barringer's cause in general terms in a number of letters to Shaw, he had been concerned primarily with the water problem and its solution.[74] Now, after receiving Adams's report, he confessed to Barringer that "I think it will never be possible to *prove Moulton* wrong any more than it will be possible for him *to prove* that he is right, until we actually develop the meteoric mass [Colvocoresses's emphases]. Perhaps you may be successful in satisfying everybody that Moulton made a mistake in practically neglecting the atmospheric resistance and therefore the initial speed of the meteorite would of necessity have been much reduced before it struck the earth, but I fear Moulton will come back and argue that the initial speed was in all probability so much greater than the minimum which he estimated that even though the atmospheric resistance had a substantial effect, the speed of impact was still very great and the size of the body correspondingly small."[75]

Moreover, no sooner had Barringer dispatched copies of Adams's report to Magie and Russell in Italy than Magie's "report" finally arrived.[76] This, to Barringer's consternation, was wholly unsatisfactory. "Am afraid [to] show Shaw your report," he immediately cabled his friend of fifty-odd years, "only because of one thing, namely, its practical agreement with Moulton on air resistance and velocity on impact regarding which now necessary far above everything else to prove Moulton wrong. We cannot mathematically answer energy calculations. Suggest you revise your criticism stating you understand Adams is dealing with air resistance and velocity problems and omit all reference to these except to express confidence in Adams. . . . Don't mention this cablegram but send your revised report as if original. Regret this seems necessary [punctuation added]."[77] Magie apparently found nothing unusual in this request and promptly revised his report and forwarded it to Barringer along with an apologetic letter. "I am very sorry my first report did not suit," he wrote. "I have tried to do what I can without lying in this new draft."[78]

Russell's long-awaited report, sent two weeks earlier from Florence, arrived only days after Magie's and produced an even more devastating effect. The astronomer had concluded in a three-page "Note on the Energy Relations Involved in the Meteor Crater" that "the mass of the meteoric swarm may have been as great as thirty million tons or as small as half a million, though either of these extreme values appear to be improbable—especially the higher value." Because of the uncertainties in the problem, he added, this result "should be treated with caution." But he had also accepted many of Moulton's conclusions, including his minimum velocity of 7 miles per second. In addition, he had made some calculations of the effects of atmospheric retardation, using liberal assumptions but finding that "even this extreme allowance for atmospheric resistance will do no more than double Moulton's estimates, except for the higher velocities." On this basis, he noted, the mass would be 5 to 5.5 million tons.[79]

It was Russell's accompanying letter, however, that threw Barringer into a panic. "I have read Moulton's report and the accompanying letters with care," Russell had written. "I am unable to 'come out strongly' against Moulton's view, because I have concluded that he is nearly (and perhaps entirely) right. . . . My estimate of the largest possible mass is higher than Moulton's because I have made a larger allowance for the energy expended in heating the rocks. But it would be *altogether unjustifiable* and, in my opinion, unethical, to proceed with the expenditure of the stockholders' money on the assumption that my estimate of the total energy (that is, *my guess at the greatest possible amount*) is bigger than Moulton's [Russell's emphases]." Moulton's most important recommendation, he added, was that a geophysical survey be made and "I consider it the plain duty of the Company to have such a survey made before spending more money in the mining work. I am very sorry it did not occur to me to advise you to do this long ago. . . . I am afraid this letter will disappoint you, but I can only state what I regard as the conclusions from the evidence. Why didn't you tell me, by the way, what Shapley thought as you told Elihu Thomson? I would take his judgment very seriously."[80]

Russell specifically requested that Barringer send copies of this letter and his "Note" to Shaw, and it surely is a measure of Barringer's distress that he did not immediately do this. Instead, he cabled the astronomer, now in Rome, that he was "holding" his letter until Russell had had an opportunity to review Adams's report. He also cabled Magie, urging him to "communicate" with Russell. "Russell unfortunately agrees with Moulton especially regarding energy. Feel sure his physics wrong. Suggest you both wait until you receive Adams's excellent report [punctuation added]."[81] And in a second cable, he informed Magie that Russell "fully agrees with Moulton and intimates unethical to proceed further because considers body unworkably small. Entirely due to Moulton's fool report. Wishes me to show letter to Shaw and Directors but cabled him that I am holding until he discusses with you Adams's report. . . . He [Adams], Olivier and Thomson certain of great air resistance but Thomson unwell, irritated at ignorant opposition and cannot give much time. . . . Russell must not be allowed through ignorance of facts to wreck enterprise as his letter would do [punctuation added]."[82]

Barringer followed up these cablegrams with lengthy letters to both Russell and Magie. To the former he explained that "It is in the interest of the stockholders . . . that I am not sending your letter to Mr. Shaw or calling it to the attention of the Board of Directors because, in the first place, practically all of the money so far subscribed has been expended, and that part of your letter which relates to the expenditure of the stockholders' money would not be pertinent." Again, "your letter supporting Moulton's views (you are the only one so far who has done so) would make it very difficult to raise new money to continue the exploration." And finally, he declared, all the physicists and geologists that he had consulted "are agreed that no small object, as Professor Moulton imagined, could possibly have produced the effects so familiar to many of us at the crater." He then launched into an extended review of his now well-worn arguments, noting in conclusion that "you seem to be under a misapprehension regarding any statement as to what Shapley thought. He has never expressed an opinion so far as I know. . . . When consulted by Mr. Shaw and the Boston engineers he refused to express an opinion and advised them to consult Prof. Moulton. Prof. Thomson has written me, however, that he feels that Dr. Shapley would agree with him [Thomson] on the subject of air resistance."[83]

To Magie, he expressed both consternation and indignation. "Why Russell, who has no financial interest whatever in our matter, should not only ask me to send a copy of the letter to Mr. Shaw, but indicate to me that it would be unethical to spend any more of the stockholders' money, is utterly beyond me to understand," he fumed. "I suppose it's due to his utter lack of knowledge of business and business methods. . . . From my letter to him, he must have known that such a letter as he chose to write, and asked me to send to Mr. Shaw, who is on the other side of the argument, would be fatal to our plans, at least as far as getting the additional $500,000 with which to continue the exploration of the crater. . . . I feel very strongly about this matter, not only for personal reasons, but because the interests of a great many stockholders whom I induced to put in their money are seriously jeopardized by the probable action of Mr. Shaw in being guided by a man like Moulton and in seemingly refusing to listen to the other side of the question." He was, he added in a postscript, no longer worried about the water at the crater. "The great Fairbanks-Morse Co. will *guarantee* [Barringer's emphasis] to sink 30" holes and pump from them sandy water to the extent of 1000-2000 or even more gals. per minute."[84]

By this time, Moulton had completed his second report and, as he advised Shaw, was checking over his figures and his equations for accuracy.[85] Barringer now suggested to Shaw that "Professor Adams has gone so exhaustively into the mathematics of the problem that I think it would be wise for us, for the time being, to confine our discussion to the accuracy or inaccuracy of his calculations and conclusions as to air resistance."[86] And to this end, he prepared a lengthy statement of his arguments, emphasizing atmospheric retardation, which eventually found its way to various participants in the debate.[87]

Adams, too, had gone over his own report and decided that he could strengthen his argument. In a two-page supplemental report, he noted that his value of 1.2 miles per second for the meteorite's velocity of impact had been the maximum figure and then proceeded to compute the value at which the acceleration of gravity became greater than the retardation and would increase the velocity of fall. This, he found, would be 0.6 miles per second.[88] Moulton had two immediate comments on Adams's supplement which he offered in a terse letter to Shaw. According to Adams, he declared, enough of the Arizona meteorite's kinetic energy "would be given up in the course of 3 or 4 seconds of its flight . . . to heat it and all of the air through which a sound wave could have penetrated in such a short time to at least 6000°F. Unless it is assumed that radiation was extraordinarily effective, it will be difficult, I think, to explain what became of this heat energy." Moreover, Adams's result that retardation "balances gravity at a speed of 0.6 mile per second" defied both experiment and experience. On Adams's theory, "since retardation is proportional to surface divided by mass, the atmospheric retardation of an iron sphere one foot in diameter (weight about 245 pounds), moving at 0.6 mile per second, would be 400 times gravity. This result is not remotely related to the facts and the formula is even more astronomical for higher velocities."[89]

Barringer, meanwhile, had had a three-hour talk with Thomson, and his confidence was somewhat restored. "He is with Adams, horse, foot and artillery," he confided to Barbour in a long letter on November 22, "and more or less confidentially, has no patience whatever with the theories which have been put forth by Moulton." He did not know, he added, "what effect, if any, Adams's masterly treatment of the subject of air resistance has had upon Quincy Shaw. Surely he cannot fail to be impressed by the fact that Moulton is standing alone and that great physicists are opposed to his assumption that the resistance of the air . . . can be regarded as 'negligible'. . . . I feel sure that Moulton's second paper will be even weaker than his first, if possible. In any event it will probably be only more mathematics along the same general line . . . mathematics based on wrong assumptions."[90]

MOULTON'S SECOND REPORT

Moulton's second report arrived the following day.[91] The astronomer had produced 127 pages of closely reasoned computations to support the conclusions in his first report, and to show that his neglect of atmospheric retardation was not only justified, but that even when liberal provisions were made for its effects, the results in the case of the Arizona meteorite were not significantly different. Moreover, he replied to complaints about his allowance for the amount of work done at the crater by reexamining the question and drastically reducing his value to 9×10^{15} foot-pounds (12.2×10^{22} ergs, or 2.91 megatons TNT equivalent). In all, he formulated eighty-two equations and, assuming various velocities, masses, densities, angles of impact, and coefficients of retardation, carried out 1,300 individual computations, displaying the results in sixty-two tables. In a fourteen-page foreword, written after his report was completed, he painstakingly demonstrated the deficiencies in Adams's formulas for air resistance and their disagreement with such experimental evidence as was then available, concluding from Adams's expression for atmospheric retardation that "we are driven to wholly absurd conclusions, and I have no hesitation in rejecting it on these grounds alone as having no validity in the present problem."

In his foreword he also explained that "I have taken great pains at every step to state precisely the assumptions under which the computations were being made. In fact, I have repeated such things until they may be tiresome. This care in stating assumptions has been taken partly because a number of readers of the first report seemed frequently to lose contact with my line of reasoning. . . . I wish to state that I stand by the results (barring some possible important numerical slip) given later in this paper. . . . Unless some large error in the theory or computations is discovered, it would be as hazardous to take a 'bull' attitude respecting this meteorite as it has been to take this attitude respecting the values of many stocks."* He then urged that his theory and calculations be checked carefully, that geophysical methods be used to prove "the existance and location of the meteorite," and, if the results were favorable, to "bore for it" and "put down shafts. . . ." But he was nonetheless confident that no mass greater than half a million tons would ever be

---

*The reference, of course, is to the Wall Street stock market "crash" of October 1929.

found, and thus he issued his provocative invitation to a dinner on the crater's rim.

Barringer, bewildered, immediately turned to Thomson for answers to Moulton's massive arguments. "Moulton's tremendous report has just been received from Mr. Shaw," he wrote that same day. "It is 127 pages of typewriting and incredibly intricate mathematical computations."[92] He then quoted the two final paragraphs of the astronomer's report. In the first of these, Moulton concluded flatly "that the mass of the swarm cannot have exceeded 3,000,000 tons and that it was probably considerably less than 1,000,000 tons. It may well have been of the order of 100,000 tons, depending on the density of the swarm and its velocity of impact." In his final paragraph, Moulton summed up his view of what had happened at Meteor Crater:

> My interpretation of the probable event is roughly as follows: The dense part of the swarm was something like 2000 feet in diameter and its mass was from 100,000 to 500,000 tons. It crashed into the rock to a depth of something like 800 to 1000 feet, carrying with it a large mass of greatly condensed (and consequently heated) air, which was further condensed on penetrating the rock. The ejected rock was thrown out by the condensed air and the steam generated, the finer fragments coming from the pulverized materials in the central track and the larger masses from the broken surrounding walls. And, as Mr. Shaw Jr. has suggested, essentially all the material which entered may have been blown out by these explosive gases. . . .

"He admits," Barringer pointed out to Thomson, "the highly compressed and heated wad of air, but states that it went into the rocky strata . . . and then blew out the meteorite and rock fragments. He seems to utterly forget the fact that a similar wad of compressed air did not enter the Siberian swamp when the cluster struck there but was forced aside by the meteoric swarm and went out over the surface of the earth with the force of a hurricane in every direction. . . .[*] This would seem a sufficient answer to the conjectures of young Shaw. Again, I contend . . . that any great explosive force such as he has imagined would have emptied the crater of the great many millions of tons of fine silica. . . . One thing seems certain," he added, "that Moulton is not possessed of 'that sense uncommon called common sense. . . .'"

At this point, Magie's "original" report arrived from Italy.[93] The physicist, after receiving Barringer's cablegram, had scrawled seven pages of generalities about the crater in longhand, along with a four-page "Supplementary State-

---

[*] It is now generally believed that the Tunguska object exploded in the atmosphere several miles above the earth.

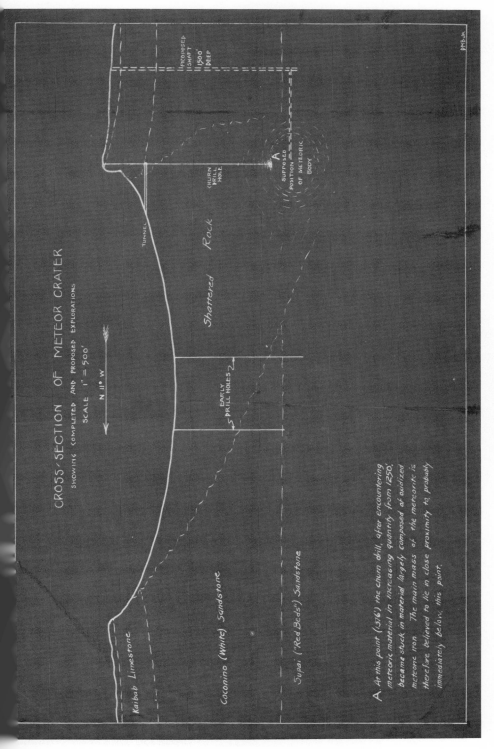

35 D. M. Barringer's sketch cross-section of Meteor Crater.

36 H. L. Fairchild

37 F. R. Moulton

38 Elihu Thompson

39 E. P. Adams

40 Henry Norris Russell

ment" in which he attempted some calculations based on what he called "the principle of dynamic similarity" and derived results which even Barringer must have considered wholly unrealistic. Moreover, he had not fully carried out Barringer's instructions. He had not, for example, waited for Adams's report nor, indeed, did he mention Adams at all, and what he had to say about atmospheric retardation hardly was supportive of his colleague's efforts to quantify its effects. He conceded he was "not able" to criticize Moulton's derivation of the minimum velocity of the meteorite, but "the astronomical velocity would obviously be cut down by the air resistance, and if a wad of compressed air were caught between the meteor and the earth just before impact it might make a considerable reduction in the velocity. This, however, could hardly be calculated because the conditions are so uncertain."

Magie's "main criticism" of Moulton's first report was that it "seriously underestimates the work done, and so the energy delivered." Even though Moulton had made liberal allowances for the work done, Magie declared, he had failed to take into account the energy used up in "setting up waves of shock," and the energy "spent in this way was immensely greater than he estimates." An "extreme case," he pointed out, was "the energy delivered to a bell when it is struck by a hammer" where the energy of the blow "has gone into waves of shock almost altogether." Unfortunately, Magie conceded, "we cannot estimate the ratio between the energy that did visible work and that which went off in waves. We can, however, decide that any calculations based only, with a small factor of 'safety,' on the visible work will bring out a result which is much too small for the mass of the meteor."

In applying his "principle of dynamic similarity" in his "Supplementary Statement," Magie conceded that the only experiments on which he could base his computations were those he and Barringer had made with bullets during their visit to the crater in 1909, "and it is stretching things a little to compare the holes thus made with the crater." Nonetheless, he proceeded to extrapolate from these, deriving a value for the energy expended at the crater of 30×10^{25} ergs.* From this, he calculated that at a velocity of 60 kilometers (37 miles) per second, the meteorite's mass would have been 17 million metric tons, and at 8 kilometers (5 miles) per second, it would have been 900 million tons. He admitted that the masses for lower velocities "come out excessively and impossibly large," but he argued that even if his estimate for a velocity of 5 miles per second was one hundred times too large, "it would still leave 9 million tonnes for the mass. I am strongly of the opinion that this represents the minimum mass of the meteor."

*An incredible figure, equal to 7,160 megatons TNT equivalent.

BARRINGER'S LAST LETTER

Events now crowded in on Barringer. On Wednesday, November 27, the day before Thanksgiving, the morning mail brought him two letters that clearly disturbed him deeply. In one, Thomson advised him petulantly that he did not intend to analyze the computations in Moulton's second report for errors and suggested that in any case, it was futile to attempt to convince the astronomer that he was wrong. "The truth is I have no time to waste on these matters or on the calculations of men who do not have the same view point as I have," Thomson wrote. "I think I have a clear conception of what took place at the crater, and I have had it for a long time. . . . Perhaps this is why the new *entrants* [Thomson's emphasis] do not interest me much, certainly not enough to waste valuable time needed sorely for other uses. . . . I think Moulton might come around at last but it would be a long pull—I certainly wouldn't be ready to try to persuade him."[94]

For Barringer, however, the second letter, written by Russell from Rome on November 14, was surely far more distressing. "I find two letters from you containing further correspondence upon the meteor problem," the astronomer began, "and I have today received a long cablegram, which causes me serious disquietude."[95]

> You say you are "holding" my letter, pending my receipt of further arguments. . . . If this means that you have not communicated my letter to Mr. Shaw, the president of the company, I can only express my great surprise at what I must regard as a breach of the express conditions upon which I gave you my technical opinion. I must now make it absolutely clear that I can continue to answer your requests for expert opinion only upon the definite understanding that my findings shall be transmitted to all who are concerned, and that this shall be done with equal promptness, whether my conclusions agree or disagree with your own. . . .

Letters such as Miller's, and encyclopedia quotations, Russell noted pointedly, "have no bearing on the problem of air resistance to so large a mass as is here in question. . . . I am very glad that Moulton is to attempt an independent solution of the problem of air resistance." He had, he noted, met Magie in Florence and "judged from his remarks that he was in considerable agreement with my conclusions. But you have doubtless heard from him directly as he spoke of hearing from you repeatedly." Russell would, he added, give Adams's report "as much time as I can reasonably afford. . . . But I must repeat that your 'holding' my letter, because you think that this report may change my opinion, was quite unwarranted. If I find reason to change my conclusions I will say so . . . but the responsibility for both my earlier and later views is my own. And I cannot consent to have other views suppressed or pigenholed because of its [sic] consequences." And in a postscript, he advised Barringer: "Please do not cable me again. . . . The cable tolls from London, which I would have to pay, are heavy."

Barringer was stunned, but his reaction was immediate. He called in his secretary, Miss Heller, and dictated lengthy letters to Russell and to Shaw, and a shorter one to Thomson that would be the last letter he would ever write.

To Russell, he confessed "no little surprise that you should express yourself as you have done. You do not seem to realize that I told you that I was withholding your letter for your own protection," and here he cited "such eminent men" as Magie, Thomson, Adams, Olivier, Slipher and Lampland as opposed to the views that Russell had expressed.

"I also withheld your letter, again for your own protection, because you had not yet had an opportunity to read Prof. Adam's report or to talk over the matter with Prof. Magie, who is far more familiar with the facts at the crater than you are. Finally, I withheld your letter because for some unknown reason you, to my surprise, seem to have completely changed your opinion as to the character of the projectile, an opinion which you have expressed to me in many letters which I have from you, some of which were used with your permission in raising the money for the exploration among a few selected men. . . ." He then launched into an extended review of his arguments against Moulton's conclusions, adding that "I am, of course, sending a copy of this letter, together with copies of both your letters, to Mr. Shaw so that he may be fully advised."[96]

To Shaw, he referred to Russell's October 30 letter "written after he had received Prof. Moulton's first report but before Prof. Adams's report reached him, and also before he had an opportunity to consult with Prof. Magie. I think you will be surprised, and perhaps amused, at the fact that Prof. Russell sees fit to indicate to me where my duty lies. . . ."[97] Then, he pointed out that Russell "as I remember, has made but one short visit to the crater of at the most a few hours' duration, but I had supposed from his many letters to me on the subject, written many years ago, and even as late as when we were raising the money for the Meteor Crater exploration, he was in entire accord with the views held by his associate and friend, Prof. Magie, Prof. Thomson and myself. Prof. Moulton's mathematical theories must be very captivating to have intrigued so many intelligent men, but I say again with all the emphasis in me that they do not fit the facts, and therefore must be wrong."

And finally, in his letter to Thomson, he confessed that "The controversy seems to have reached the acute stage. . . . It is utterly beyond me to understand how Professor Moulton has been able to influence Mr. Shaw, his son, and the engineers Hutchinson and Livermore to such an extent."[98]

Barringer apparently did not feel well when he left his Philadelphia office that afternoon while Miss Heller was still typing these letters. He spent Thanksgiving at his suburban home in Haverford with his family. Three days later, on November 30, 1929, he died suddenly of a massive coronary thrombosis.[99]

The New York Times, in reporting his death the next day, noted that among his other accomplishments he had "discovered the origin of the famous Meteor Crater of Arizona and proved that it is due to the impact of a meteoritic mass."[100]

TWELVE

THE AFTERMATH

> I cannot recommend the subscription to our stockholders for the purpose of further exploration and so invite their continued confidence in a venture which I think presents such remote chances of financial profit. The Meteor Crater presents an extraordinarily interesting problem and would seem still to promise great scientific reward from exploration. . . .
>
> QUINCY A. SHAW, 1930

Barringer's death effectively muted the commercial aspects of the Coon Mountain controversies; henceforth no one would seriously contend that profits amounting to hundreds of millions of dollars could be realized by mining the nickel, platinum, and iridium contained in a meteoritic mass supposedly buried beneath the crater. But Barringer's death had no effect whatever on the scientific issues in the debate which, once raised, had acquired a momentum of their own. They continued to fascinate some of the debaters and have intrigued many other scientists up to the present time.

The debate, however, now became more subdued, and the urgency that had marked its early stages largely disappeared as the participants grappled with the theoretical aspects of meteoritic impact in general, and the corollary problems of energy expenditure and atmospheric retardation in particular. The issues that Moulton raised in his reports were largely unprecedented and transcended the specific phenomena at Meteor Crater. Indeed, they are still the subject of investigation and discussion largely because, as Moulton found in the case of the Arizona meteorite, the available data are rarely, if ever, sufficient to resolve them in more than a general way.

Those scientists who were most closely associated with Barringer and who, incidentally, held at least a small financial stake in the outcome, no longer

298   THE AFTERMATH

actively took part in the continuing discussion; neither Thomson nor Magie apparently felt they could shed further light on the issues, and Olivier seems to have dropped out of the debate altogether.[1] Now, the principals were Russell, Adams, Webster, Moulton, and, finally, the distinguished Harvard physicist Percy Williams Bridgman who, a few years hence, would win a Nobel Prize for his pioneer work in the field of high pressure physics.[2]

One matter that had caused Barringer much distress, however, was disposed of almost immediately. Shortly after Barringer's funeral, his son Reau found the letters that Barringer had dictated just before his death and promptly mailed them, sending Shaw copies of Russell's earlier correspondence as the elder Barringer had intended. "As you have no doubt heard," Reau wrote to Russell, "my father died very suddenly. . . . When Father cabled you that he was holding your letter until you should have the benefit of the later reports, he naturally assumed that if you did not wish the letter withheld from Mr. Shaw or anyone else, you would so wire him. In that case the total delay in Mr. Shaw's receiving the letter would have been hardly more than a day. I feel I should say this," he added, "because you have intimated that Father attempted to withhold the contents of your letter from the President of the Company. Had he so wished, he certainly would not have cabled you. While I think that Father's letter is a complete answer to yours, I will nevertheless welcome any further expression of your views or any further assistance you may give us in throwing light on this problem."[3]

Shaw, after receiving the Barringer-Russell correspondence, immediately reassured the Barringer sons, urging Brandon to "Please dismiss from your mind everything of the discussion between Prof. Russell and your father as to what may have been the interpretation of the duty in sending correspondence to me. I think I understand the situation. . . ."[4] And Russell himself would later advise Shaw that "Mr. Barringer's confidence in Professor Adams's conclusions really justified his action in holding up my letter until I could see Adams's report."[5]

As it turned out, however, Russell was not at all impressed with Adams's report when he did see it. The day before Barringer's death, Russell had written Adams that "I am forced to conclude that there is a slip in the reasoning of the latter portion, dealing with motion in a compressible fluid. . . . There is evidently something seriously wrong. . . . Such huge effects as you calculate could only occur if the effective diameter of the meteorite was increased in some way to be hundreds of times the actual diameter. Now there may be an air cap, in front of the mass, but I don't see how the gas in such a cap . . . can possibly be compressed to this enormous degree, nor how the momentum of the central mass can be transferred to it. . . . The question of air resistance," he added, "does not look as important to me (or to Magie) as it does to him [Barringer], for all our estimates of the energy required to produce the crater are mere guesses (Moulton's being a very low one) and, if one factor is so uncertain, no accuracy is required in the other."[6]

The following day, Russell had written in much the same vein to both Barringer and Shaw. Adams, he wrote Barringer, had made a "serious slip," explaining that "the whole theory is illustrated by a very simple piece of engineering. Air at 10,000 pounds' pressure is a very powerful agent, but a small quantity (say a cubic inch) of it can't do much work." He had discussed the matter with Magie, he noted, "and he agrees with me that Adams has greatly overestimated his resistance. Moulton regards it as negligible. For larger masses (ten million tons or more) I agree with him. . . . The whole energy delivered by the meteorite may therefore have been many times greater than Moulton's value. But it is impossible to *prove* this conclusively." Because of this uncertainty, he added, "I dare not say . . . that the search for the mass will *certainly* repay the cost, unless the engineers decided that a million tons or less would pay [Russell's emphases]."[7]

To Shaw, Russell pointed out that because of the uncertainty of the energy factor in the problem, "no reliable estimate of the mass . . . can be arrived at. It appears to me, however, that it is decidedly improbable that it was less than one million tons, or more than fifty million." As to a future course of action, he suggested that "If the limit of profitable operation is not far above a million tons, success would be probable enough to make future work a legitimate and attractive mining venture."[8]

WEBSTER'S ANALYSIS

Russell did not describe in detail the "slip" he had detected in Adams's calculations. But now Webster, in a lengthy analysis of the problems of the meteorite's mass and of atmospheric retardation, completed the day before Barringer's death, devoted fully half of his report to a mathematical critique of the physicist Adams's reasoning.[9] The Stanford scientist questioned two specific assumptions that Adams had made in deriving his equation for the retarding force per unit area of the cross-section of the meteorite, or meteoritic swarm. One was that of a "steady" motion, i.e., that "there is no change with time in the density or velocity of the gas passing any given point in space." This, he noted, "calls for the accumulation of a 'wad' (as Olivier calls it) of compressed air sufficient to raise the pressure at the center of the front side of the meteor by a factor of many millions. Does the meteor run into enough air to accumulate this amount in the time it has available?" he wondered. "If the meteor is small, and the wad of compressed air is correspondingly thick, it probably does; but if the meteor is several hundred feet thick, as Adams assumes it to be, and if the wad is even a few percent as thick, all the air the meteor has before it in its whole path may not be enough to supply material for a wad at such pressures. In this case, and also if the rotation of the irregularly shaped meteor causes the wad to spill off every once in a while, then this is not a case of 'steady' motion, and the equation involves an error of unknown magnitude."

Adams's second questionable assumption, Webster declared, was that the average pressure exerted on the frontal surface of the meteor was one-half that at a central point where the streamlines divide. "At low velocities," he opined, "I have no doubt that this is a very reasonable value, but I doubt seriously if it still holds under these conditions of extreme compression. . . . [I]t seems not inconceivable that the air expanding sidewise from this high-pressure spot may acquire such a speed that the centrifugal effects tend to keep it away from the rest of the front surface, and thus prevent it from transmitting to that surface the pressure of the air lying before it. . . ." From an observational standpoint, he added, the tendency for large meteors to burst in midair "may well be an indication of the extreme concentration of pressure and heat in a small spot, just such as I have suggested here."

For a "first approximation of the retardation," then, Webster assumed that "the wad at any instant must practically consist only of such air as has been reached at that instant," and that this "reduces the problem to a simple case of dead impact" to which the law of conservation of momentum could be applied. For a "second approximation," he assumed that the compressed air in the wad "must be continually spilling off," and that "all momentum transferred to air picked up by the meteor is soon lost, because that air is lost." Then, using a velocity of entry into the atmosphere of 10 miles per second and the conservation of momentum law, he computed velocities of impact for various sized meteors which "differ radically from those of Adams." Where Adams found an impact velocity of 1.2 mps for a sphere 200 feet in diameter, for example, Webster's computations gave impact velocities of 9.35 and 9.32 mps for his first and second approximations, respectively. His second approximation, he concluded, "summarizes the findings with sufficient accuracy for all present purposes, at least as applied to individual meteors, travelling alone."

For swarms, he conceded, the matter was more complex. But in general, "the front members of the swarm will be retarded much like lone individuals, especially if well separated, while the rear ones will move in air, much of which has acquired a considerable velocity in the direction of the motion of the meteors. The rear ones must therefore overtake the retarded leaders, pass them and take their own turns at leading." There would thus be "a contraction of the thickness of the swarm" and this "must give the swarm, initially spherical, a sort of lens shape. . . . Still further modification of this sort must eventually give a swarm a shape like that of a jelly fish, with the thin edge trailing far behind the thick center." The meteors at the edges would strike many seconds after those in the center of the swarm, he added, and this "would account for the presence of many stragglers at the very surface of the debris around Meteor Crater." But, Webster noted, the velocity of the center of the swarm could not be calculated definitely, and he concluded merely that "the swarm will be retarded less than an average individual if it were travelling alone, and will therefore land with a higher velocity; and exceptionally large individuals, if present, will land with almost their full initial velocity."

The possibilities of his expression for retardation were not exhausted when the meteors reached the ground, he noted,

> for it may contribute something to the much-discussed question of the expenditure of energy there. If the speed of the meteors was such as to crush the rock to fine powder, then the fine powder must have been brought up to speed, much as the air was, and it may even happen that the forces needed to overcome the inertia of this rock flour were greatly in excess of those needed to overcome solid stresses and produce the fine powder. . . . In any case, however, we may be sure Newton's second law of motion still holds, and that the momentum given to the rock flour to get it out of the way must be abstracted from the meteors. They must, therefore, lose velocity at least as fast as this transfer of momentum demands, and an equation of this type therefore sets a lower limit to the rate of retardation.

The "difficult question" here was at what velocity the solid stresses became more important than the inertia effects, and he now calculated the mean thickness of a swarm penetrating rock, assuming that "it cannot get far beyond the point where this change occurs, and then say it went 2000 ft. in being retarded to this speed." He found that "any reasonable assumption" for the initial velocity yielded values for the mean thickness "in the neighborhood of 150 to 300 feet." Then, assuming it to have been 200 feet thick, and the diameter of the central part of the swarm to have been 400 feet, the volume would be 25 million cubic feet, and the mass six million tons. Such a swarm, said Webster, would be very dense, and "there is room at Meteor Crater for a diameter at least 5 times the 400 ft. assumed above. . . . Thus, assuming say 1000 ft., we find a mass of 40 million tons. *But we must not ignore the very large probable errors, all through this calculation. All we can say is that the mass was probably in the millions of tons, maybe 5, 10 or 20, but not surely even 5* [Webster's emphasis]."

Webster made two additional points regarding the energy expended at Meteor Crater that Moulton later would explore. As to "earth shock, or seismic waves, I think it may be possible to get the order of magnitude of this item, but no accurate value, by assuming that the rock was crushed beyond its elastic limit throughout a hemisphere of, say, one mile diameter, and that the stresses at the boundary of this hemisphere were maintained at somewhere near breaking stresses for a second or two." But not having data at hand on such factors as crushing stress, "I leave it to someone else to make this calculation if it is needed." A more important expenditure of energy, suggested by his earlier discussion, was that at impact, "when the rock flour was pushed out of the way. . . . it must have been shot up into the air as an enormous cloud of white dust, possessing enormous kinetic energy. It could not have stopped very short, but must have carried this kinetic energy far up into the air and must have scattered both dust and energy to the four winds" as shown by the "effect of the smaller impact on the Siberian forest. Under such conditions, it is no wonder that the material in which to locate this expenditure of energy is no longer present at Meteor Crater."

Copies of Webster's report and Magie's "original" report reached Moulton during the second week of December, and he found them of interest. Both, of course, had cited "earth" or "seismic" waves as a possible means of accounting for the energy expended by the meteorite, Magie contending that the energy consumed in this way would be "immensely greater" than that used up in crushing and excavating the rock. Now Moulton advised Shaw that if Magie was right, "the surface rock would be thoroughly fractured all around the crater for many thousands of feet in every direction. I do not understand that this is the case."[10] And in a second letter the following day, he reported that preliminary calculations supported this conclusion. "Under these circumstances I think we should expect to find the rocks completely broken up for a distance of two or three miles from the point of impact. . . ."[11] In mid-December Shaw provided the directors of the Meteor Crater Exploration and Mining Company with a "rough analysis" of the data and opinions he had so far received from no less than fifteen "authorities" on the problems at Meteor Crater.[12] "To date," he wrote, "I think no one has given any better method of attack for a reasonable solution of our problem than Moulton, viz: to determine the velocity of the Meteor and the Energy needed to form the crater." He also felt that no one had offered a better measure for this energy than Hutchinson's stamp mill factor.

His analysis was "preliminary," he declared, and "I do not care at this time to give any specific figures" in interpreting the various opinions. But he nonetheless had come to three general conclusions pertinent to the issues in the debate. First, "I am convinced that we are dealing with far higher velocities than anything like 2 miles a second;" second, that "certain phases of the discussion and some data offered now lead me to think that much less *and not more* energy was needed to make the crater than Moulton's maximum Energy, which was represented by 3,000,000 tons @ 7 miles per second;" and third, that "consequently the mass is more probably some *small* fraction of the 10,000,000 tons which has heretofore been generally accepted [Shaw's emphases]."

BULLETS IN FLIGHT

Adams, by this time, had found some high-speed photographs of bullets in flight that he considered of great significance, in a book on ballistics by one C. Cranz.[13] These he brought to the attention of Shaw in another of his brief "supplementary" reports.[14] Two weeks later, after receiving Moulton's second report with its sharp criticisms of his earlier conclusions, Adams now produced a second full report of his own on air resistance.[15]

In this curious document, Adams charged that Moulton, in computing the resistance encountered by meteors, had ignored "the fundamental principles of physics," and thus had seriously underestimated the effects of retardation. He then discarded his earlier hypothesis on retardation while defending somewhat ambivalently his earlier reasoning, advanced a new argument for air resistance based on the photographs of bullets in Cranz's book, and ended up by declaring that "the whole question of atmospheric retardation is a problem incapable, at present, of anything approaching exact solution."

The only physical principle that Moulton had used in calculating air resistance, he contended, was that of linear momentum; and, although he had appealed "in a very elementary way" to the kinetic theory of gases, he had ignored the "principle of energy" as well as those of thermodynamics. "Work must be done in bringing the molecules of air close together against their mutual repulsions, since air is compressible. Part of this mechanical work is converted into heat energy. Anyone who has experimented with a bicycle pump knows that a large amount of work has to be done to get any appreciable heating of the air. All of the work done in compressing and heating the air must come from the kinetic energy of the moving body, and this necessarily causes a diminution of its velocity. All of this Dr. Moulton ignores."

Rather, Adams contended, the principles of hydrodynamics were more germane. "Instead of considering the air as made up of discrete molecules, we must consider it as a continuous, compressible medium. . . . The air immediately in front of the meteor is set into motion with the velocity of the meteor. Since this velocity is greater than the velocity of compressional, or sound waves in air, the air at a distance in front of the meteor can have no knowledge that the meteor is coming along. It, therefore, does not make way for the meteor but stays where it was, and the approach of the meteor compresses it." Since the conduction of heat in the air is a slow process and the motion of the meteor is very rapid, "there will not be enough time for all the heat produced by the compression to be dissipated. . . . I assumed no dissipation and so made use of the law of adiabatic compression. It is readily seen that if all the heat of compression were dissipated by conduction and radiation, with a resulting isothermal compression, the retardation of the meteor would have been even greater than in the adiabatic case."

The validity of this argument, he thought, would be "generally accepted on purely theoretical grounds," but when it was applied to high-velocity, rotating projectiles, he conceded, "it leads to wholly wrong results." The explanation for this, however, was provided by the photographs of bullets in flight. They showed that "a bullet travelling at a velocity much greater than that of sound is accompanied by what appears to be a cone of steady compression, with its apex just in front of the projectile, and another similar cone, nearly parallel to the former, which would have its apex a short distance to the rear of the bullet if it were projected forward. Trailing back from the rear of the bullet is a region of turbulent, vortex motion, which prevents the rear cone from being complete." From this he concluded that "suction" or "drag" contributed to the projectile's retardation.

At the time he had submitted his first report, Adams noted, he had been convinced that the rapid rotation of a symmetrical projectile was a necessary condition for the formation of such cones of compression, or "discontinuities," and had decided, "perhaps wrongly, that they would not exist in the case of irregularly shaped bodies like meteors." But now he felt that at least some of the resistance in the case of meteors "must be due to the drag of the turbulent air behind it." Lord Rayleigh's formula for calculating pressure at the nose of a projectile moving four times the velocity of sound, he pointed out, when applied to a projectile's full cross-section, "leads to wholly wrong results, owing partly to the neglect of the partial vacuum at the base." Indeed, according to the English physicist Sir Joseph Larmor, "the whole of the retardation is due to the drag at the base," he declared, citing two not entirely pertinent quotations from Larmor's works to his point. Thus there was no satisfactory theory to account for the motion of even symmetrical bodies in the air, and the problem was "enormously more difficult" for irregular masses such as meteors. "All we can say with certainty is that the resistance to the motion of such a body will be very much greater than the resistance to the motion of projectiles. Absence of any stream-line form will greatly increase the turbulence at the rear and thus the drag."

The point must be significant in relation to a meteoric swarm as well, he argued. In a swarm, the larger and thus faster bodies would move to the front of the swarm, and the "suction" behind them would tend to increase the speed of the slower meteorites, and decrease the speed of the faster ones. "We should expect, therefore, that the speed of the whole swarm would be something between the speeds of the fastest and slowest meteorites if these were travelling alone. The whole argument comes down to this; that whatever the law of resistance to a meteorite is, that law should be applied to each individual meteorite in a swarm and not to a surface enclosing the swarm. . . ."

Despite all this, however, Adams insisted that he was "by no means convinced that the adiabatic compression of the air in front of the meteor does not give a fairly good account of the retardation, and that the calculations I gave in my first report are not substantially right. However . . . for the present I shall assume that even with meteorites we have to deal with these discontinuous waves of compression. It appears at first sight, then, that we are where we started from, and know nothing at all about the atmospheric retardation to the motion of meteors."

By late December Shaw had distributed copies of all these reports and letters among the various participants. Clearly, Adams was now standing alone on the issue of atmospheric retardation. Webster, for example, advised Shaw that Adams's second report "does not appear at all convincing, and I am more than ever certain that Moulton, Russell, and I are about right on this phase of the question."[16] Thomson too accepted Webster's concept of the dynamics of a meteoric swarm and his conclusion that it would strike the earth at a higher speed than its individual components. "The views of Russell appeal to me," he wrote Shaw, "but more so does the treatment by Webster interest me as it comes very near to an expression of my own views. The dispute as to the velocity of striking the ground is really of less consequence than it appeared to be; for the reason that very large masses, especially if deep in the line of traverse, will not be so greatly retarded by the air resistance as the smaller bodies and these, if they fall into the track of the larger ones, will retain their velocities and follow closely upon the larger masses. The greatly elongated jelly fish is a good comparison. . . ."[17]

Russell, whose comments came in later from Cairo, Egypt, briefly informed Shaw that the letters he had sent just before Barringer's death reporting a "slip" in Adams's reasoning "contain almost all that I have to say about the present Meteor Crater problem." But he repeated his belief that Moulton had "seriously underestimated the total energy," and thus that his conclusions represented minimum rather than maximum values for the mass. The most important new material he had received, he added, was Moulton's foreword to his second report. In particular, his "practical suggestions" therein for a geophysical survey were "obviously good."[18]

MOULTON'S THIRD REPORT

Moulton, in contrast, had much more to say on both atmospheric retardation and energy expenditure, and this he presented in a third report, sent to Shaw early in January 1930.[19] In this, he sharply refuted criticisms that his estimates in his second report for air resistance and for the energy expended were too low. That he savored the exercise is evident; his method, in effect, was that of *reductio ad absurdum*.

At the outset, Moulton noted that in addition to the reports and letters that had been forwarded to him, he had had "the advantage and the pleasure of a long conference with Mr. Colvocoresses" who had "corrected certain somewhat erroneous views I previously held respecting the local conditions at the crater." These, with the "misconceptions of others respecting my ideas . . . are the occasion for these supplementary remarks." In particular, those who argued that he had not accounted sufficiently for atmospheric retardation had "misunderstood" his position, he declared. "The retardation which I derived and used in arriving at my results is enormous, as I will illustrate."

Then, "in order to come as close as possible . . . to what is generally regarded as the probable nature of the meteoric swarm," he assumed a swarm with a density of 1/80 that of iron, weighing 6,400,000 tons, and entering the atmosphere at an angle of 45° with a velocity of 14 miles per second and a kinetic energy of 11×10^{17} foot-pounds. From his tables taking atmospheric resistance into account in his second report, such a mass would strike the earth at not less than 8.88 mps with a kinetic energy of 442×10^{15} foot-pounds. Therefore, he had allowed for a loss of 658×10^{15} foot-pounds of energy to the atmosphere. Dividing this number by the number of pounds in his assumed swarm, "we find that, on the average, the atmosphere took from *each pound* [Moulton's emphasis] of meteorite over 51,000,000 foot-pounds of kinetic energy. This energy would raise the pound to a height of 51,000,000 feet, or nearly 10,000 miles. About half of this resistance was encountered in the last 20,000 feet of the atmosphere, which the meteorite traversed in about 0.4 of a second. In other terms, the resistance I used during the last 0.4 second was so great that if an equal force had been applied to the whole 6,400,000 tons at rest, it would have given it a velocity that would have shot it up in a vacuum to a height of nearly 5,000 miles." Moreover, the 51,000,000 foot-pounds of work done on each pound "is equivalent to about 232,000 horse power working on each pound during the flight of the meteorite through the last half of the atmosphere."

Moulton turned next to Adams's second report, briefly discussing the "apparent waves" in the photographs of bullets in flight that the physicist had cited. These, he noted, were "not waves, strictly speaking. The tip of the projectile is a point at which there is in effect a discontinuity in the density of the air . . . from which spherical waves radiate. . . . The particular apparent condensation photographed is due to refraction of light as it passes through the conical envelope of rapidly varying density." It was well known, he pointed out, from both experiments and experience in ballistics, that these waves appeared only for velocities above that of sound, and that they traveled at the velocity of sound. But, he added, it was "unsafe to push the analogy of projectiles with meteorites very far," for projectiles were elongated, rapidly rotating bodies describing curved trajectories whose flight depended upon several other dynamical factors.

"But I quite agree, however, with Professor Adams that 'the air at a distance in front of the meteor can have no knowledge that the meteor is coming along,'" Moulton declared.

> I add that, when the meteorite moves with a velocity above that of sound, the air in any other direction from the meteorite can have no knowledge that the meteorite is in the neighborhood and can have no effect upon the meteorite. There is no more "suction" behind a meteorite than there is in a pump, which fails to act when there is no outside pressure. When an object moves through the atmosphere with a velocity above that of sound, the only atmosphere that affects it is that which is immediate to it. In the case of a projectile, it is the head resistance and the drag along the sides—the turbulence in the rear is an evidence of what the projectile did to the air (not what the air did to the projectile), but has no more effect upon the flight of the projectile than does the smoke trailing behind a train upon the performance of the locomotive. Such ideas, when applied to high-speed bodies, as that quoted from Larmor, it seems to me serve only to fog our minds. All of the following statements in the [second] report of Professor Adams which refer to "suction" in the rear seem to me to be involved in the same sort of ideas, which are directly opposed to his statement which was quoted above.

Adams, Moulton continued

says I appear to think the problem of atmospheric resistance is one of simple mechanics. As a matter of fact, I did not attempt to determine the resistance, but simply an upper bound to it. Then he says I used only linear momentum and made no allowance for the energy used up. In the first place, I remark that if the linear momentum mv of a mass is reduced, its kinetic energy $\frac{1}{2}mv^2$ is reduced. In the second place, I did account for an enormous loss of energy of the meteorite. It seems to me that the difficulty with all of Professor Adams's criticisms of my results is that in them he falls back into what would be true for a slow-moving object, but the meteorite came into our atmosphere with a velocity somewhere from 35 to 200 times that of sound. I repeat that since the atmosphere at a distance could not know what the meteorite was doing, it had no effect on the meteorite. Only the atmosphere essentially in contact with the forward half of a spherical meteorite could affect it.

Moulton pointed out that when a meteor strikes a molecule of air it can only move it against its inertia, and that other molecules could not act directly against the first one except in proportion to their inertia "and so on to all the molecules in the path of the meteorite. Consequently, the inertia of the whole column of air in the path of the meteorite, under the assumption that the coefficient of rebound is unity, is the measure of their whole capacity for resistance. . . ." Moreover, he contended, at velocities of 35 to 200 times that of sound, "There would not be time for appreciable interactions between molecules in the column and those outside, and no way has been suggested by which the air outside the track of the meteorite could subtract appreciably from its momentum."

"In conclusion," the astronomer declared, "I do not admit that I ignored energy or that I used 'hypotheses in contradiction to the fundamental principles of physics.' I shall be confident that I determined an upper bound to the resistance until something more specific than generalities and appeals to the principle of 'suction' are advanced."

Moulton, however, was not finished with Adams yet. In Adams's second report, he noted, the physicist had concluded that a swarm, whatever its size, would reach the surface of the earth with the average velocity of its individual members, suction or drag slowing up the faster ones and accelerating the slower ones. "This conclusion is of so great practical importance in interpreting the phenomena . . . and is so remarkable that it calls for examination."

Here again Moulton postulated a 6,400,000-ton swarm 1,600 feet in diameter falling through the atmosphere.

The front meteors produce waves, but the work is done at the points where they impinge on the air. Professor Adams assumes the meteorites lose energy; therefore they lose momentum. Consequently the air encountered must acquire an equal amount of momentum. Since the swarm moves its entire diameter of 1600 feet in a small fraction of a second, the momentum contributed by the first meteorite to the air cannot be scattered outside the limits of the swarm before those immediately behind this first layer arrive in contact with the same air. . . . Adams assumes that, when the second layer encounters the air to which the first layer has already contributed much momentum, the effect on the second layer is the same as it would be if the air were at rest. This simply is not true, for the reactions between meteorites and air depend on their relative velocities.

Any number of things of common observation proved the falsity of Adams's idea, he noted. "If Professor Adams were correct, the wind pressure behind a building or a dense grove of trees would be as great as it is out in the open, but this is not true for these moderate velocities, and much less is it true for the high velocities with which we are dealing. The best I can say of this argument is that I think it is wholly unsound and that the results are entirely erroneous." Webster, considering this same point, "disagrees completely" with Adams, he added, "and I accept all of the essential views of Professor Webster, and differ only in the minor detail that I think it likely that the meteoric swarm carried so much atmosphere along with it at its own speed that the smaller fragments were not in general lost. The idea is illustrated by the downrush of air which accompanies a sudden heavy local shower."

THE ENERGY PROBLEM

But all this was only preliminary to the main thrust of his report, which was to show that multi-million-ton masses such as those suggested by Barringer and his associates produced amounts of energy far exceeding what could be accounted for by the phenomena at Meteor Crater. His approach here was to determine the energy expended by a large meteorite or swarm in penetrating rock to a depth where its velocity would be reduced to two miles per second. The figure, he explained, was the probable upper limit of the velocity of "seismic," or "compressional" waves in "porous and faulted Arizona sandstone." Above this limit, the meteorite would keep up with its compressional wave, and thus no energy would be dissipated in this manner. He considered an iron meteorite 400 feet in diameter weighing 8,000,000 tons striking the earth at 13.72 mps with a kinetic energy of 13×10^{17} foot-pounds. Using a formula equivalent to the one Webster had used in considering the penetration, Moulton found that such a mass would penetrate 700 feet before being slowed to two miles per second.

The energy absorbed in shearing the rock, he noted, was "probably wholly unimportant" in comparison with that required to crush the 16,000,000 tons of rock in the 400-by-700 foot cylinder of rock penetrated by the meteorite. This energy, on the basis of a slightly modified stamp mill factor, he found, was 32×10^{12} foot-pounds. But the 16,000,000 tons of rock itself, under his hypothesis, now also had a velocity of two miles per second and thus a kinetic energy of 56×10^{15} foot-pounds. "Hence the energy consumed in shearing and crushing the cylinder of rock is inappreciable in comparison with that required to give the rock the terminal velocity of the meteorite, a conclusion quite in harmony with that of Professor Webster."

The main point of this discussion, however, was one which Webster "has passed with general comments, without going to numerical results." The kinetic energy of the meteorite at impact had been 13×10^{17} foot-pounds; the kinetic energy of the 8,000,000 tons of meteorite, and the 16,000,000 tons of rock "it had given its own velocity," at two miles per second, was 84×10^{15} foot-pounds. "Hence we still have to account for 1216×10^{15} foot-pounds of kinetic energy, or 93.6% of that with which the meteorite struck the earth. Even if we had made liberal allowances for the cost of shearing and crushing rock, and losses in other ways, this percentage would not be reduced materially."

This unaccounted for energy, he pointed out, was sufficient to pulverize 62×10^9 tons of rock, "or 150 times more than the entire rock displaced in making the crater." It could not have been dissipated "through a cubic mile of rock, for we are considering the matter at a time when the velocity of the meteorite was so great that these waves could not escape from it. . . . We are considering the situation within one-twentieth of a second after the meteorite touched the surface of the earth. We have lost track of something like 1216×10^{15} foot-pounds of energy which still must be in the neighborhood of the meteorite."

Webster, he noted, had referred to the energy used in ejecting the rock flour, and "energy was undoubtedly used up this way." But the 1216×10^{15} foot-pounds of energy would blow the 16,000,000 tons of rock shattered by the meteorite to a height of 40×10^7 feet, or about 80,000 miles. "Suppose the entire 400,000,000 tons of the crater were ejected, though I do not see how the energy under consideration could have been imparted to it at this time. It would blow the whole mass to a height of 1,500,000 feet, or 300 miles. Nothing approaching this happened. . . . This line of thought affords no escape from our difficulty."

FURTHER MASS AND ENERGY CALCULATIONS

The problem of the energy required to produce Meteor Crater, which figured so prominently in the debate over Moulton's reports, provides a case in point. Magie seems to have been the first to try to estimate this basic datum in 1910, concluding that 12×10^{16} foot-pounds of energy, or in modern terms, 38.8 megatons TNT equivalent, would be sufficient to account for the phenomena, a figure he admitted was no better than a guess.[2] Moulton, curiously enough, in his hurriedly compiled first report, derived this same value by applying a multiple of the "stamp mill factor" to a very liberal estimate of the amount of rock crushed and pulverized by the meteorite.[3] In his second report, however, he decided that this was much too high and settled on 9×10^{15} foot-pounds, or 2.91 megatons.[4] Russell felt that Moulton's energy allowance was "very low," while Bridgman thought it was not low enough.[5]

Over the ensuing years, values ranging from about eight kilotons to 64 megatons TNT equivalent have been suggested, the most recent estimate to find favor being about 15 megatons.[6] These later estimates, in the main, represent extrapolations, or "scaling" as it is called, from craters produced by underground, point-source explosions of high explosives such as nitroglycerine, or of nuclear devices, the assumption being that because the effects are similar, the processes that produce them also are similar. This assumption is familiar enough in science and, indeed, is the same assumption that William Herschel made in 1787, albeit not without caveat, in describing "volcanoes" on the moon. But whether the energy generated by such point-source explosions acts in precisely the same way as the energy generated by "100,000 (or more) tons of iron crashing into rock" at cosmic velocities is at least problematical.

As might be expected, estimates of the mass of the Arizona meteorite also have varied widely. Magie in 1910 suggested 400,000 tons, a figure he soon abandoned under Barringer's persuasion; Moulton's range in his first report of 50,000 to 3,000,000 tons was narrowed to 100,000 to 500,000 tons in his second, depending, of course, on the impact velocity. In 1936 Ernst J. Öpik, using energy estimates only slightly lower than Moulton's and a straight energy equation, found a range of from 7,000 to 60,000 tons.[7] But then, by applying aerodynamic and hydrodynamic theory to considerations of the penetration and of the crater volume, he derived ranges of from 3.9 to 4.8×10^6 tons, and from 1.7 to 5.1×10^6 tons, respectively. As both these methods yielded a mass of about four million tons for a velocity of 25 kilometers (15.6 miles) per second, he felt it was "safe to conclude that the total mass of the Arizona meteor may have been from two to five million tons."

In 1943 meteoriticist C. C. Wylie, extrapolating from World War I mine craters and explosion craters in Iranian oil fields, concluded that 125,000 tons of nitroglycerine, yielding 9×10^{21} ergs of energy, or roughly 0.21 megatons, would be sufficient to produce the crater, computing a range for the meteorite's mass of from 5,000 to 25,000 tons.[8] His fellow meteoriticist Lincoln LaPaz promptly challenged this estimate as too low, arguing that the nitroglycerine analogy "fails . . . to give any trustworthy indication of the mass of the meteorite."[9] But Wylie's range was accepted as "approximately correct" by selenologist Ralph B. Baldwin in 1949, although he felt Wylie's energy estimate was too high, offering his own value of 3.3×10^{21} ergs, or 0.08 megatons. By 1963, however, with the further development of scaling techniques from high-yield explosion craters, Baldwin "readily granted" that these estimates had been too low and that both he and Wylie had been in error.[10]

In 1950 LaPaz, director of the Institute of Meteoritics at the University of New Mexico, offered his own estimate of the meteorite's minimum mass. From partial data on the penetration of ultra-high velocity steel jets produced by shaped explosive charges aimed at steel targets, he derived a lower limit of 1.1 million tons.[11] That same year, however, John S. Rinehart of the Smithsonian Astrophysical Observatory used an energy estimate of 10^{15} foot-pounds and an impact velocity of 9.5 miles per second to make a straightforward calculation of a mass of only 12,500 tons, a figure which, he noted, was "in fair agreement" with Wylie's value.[12]

In 1953 the Carnegie Institute's Norman Rostoker pointed out that the experimental data on shaped-charge jets seemed to support Öpik's hydrodynamic analysis.[13] Rostoker himself then reworked Öpik's method to find for comparable velocities a somewhat higher range of from 2.6 to 7.8×10^6 tons. Rinehart nonetheless stuck to his guns, repeating his 12,500-ton estimate in 1958 on the basis of his estimate, made as a result of his survey of the crater in the summer of 1956, of the amount of minute meteoritic material contained in the soil within a few miles' radius of the rim.[14]

In 1956, Rand Corporation scientists J. J. Gilvarry and J. E. Hill estimated the energy expenditure at the crater at 2×10^{23} ergs, or 4.8 megatons, and applied scaling techniques to crater relationships derived by Baldwin in 1949 to calculate a mass range of from 80,000 to 400,000 tons at velocities of 70 km/sec (43.75 miles) and 10 km/sec (6.25 miles), respectively.[15] And in 1958 Öpik, noting this and other recent work, revised his earlier work and now concluded that at a velocity of 16 km/sec (10 miles/sec), its mass would be 2.6×10^6 tons, still, of course, within his earlier range.[16] His energy allowance, however, was now 2.7×10^{24} ergs, or just over 64 megatons, the highest figure yet advanced, excepting only Magie's wild 1929 guess of 30×10^{25} ergs, or more than 7,100 megatons.

In 1959–60, the U.S. Geological Survey's Eugene M. Shoemaker, the preeminent investigator at the crater for the next quarter-century, suggested a mass for the meteorite of 63,000 tons, using a "conservative" energy estimate of 1.7 megatons, an impact velocity of 15 kilometers (9.3 miles) per second, and basing his calculations on depth-of-burst scaling from the small, but "structurally similar" Teapot Ess nuclear explosion crater at the Nevada nuclear test site.[17] Within the next few years, somewhat higher values for the mass were advanced, notably by Baldwin in 1963, who now used an energy factor of 3.4×10^{23} ergs, or about 8.1 megatons, and scaling techniques to derive a mass of 288,000 tons.[18]

By 1974 Shoemaker, scaling from the larger Sedan nuclear test crater in Nevada, increased his energy estimate to 4 to 5 megatons,[19] and he has since accepted a value of about 15 megatons obtained by his Survey colleague David J. Roddy and others in 1980 from computer code simulations of the Meteor Crater event.[20] He now informally suggests a mass of about 300,000 tons.[21] Thus, estimates of the meteorite's mass are once again falling below the 500,000-ton limit beyond which Moulton was willing "to put up a dinner" for all concerned on Meteor Crater's rim.

Moulton's only practical recommendation for reducing the uncertainties in the Meteor Crater problem was his repeated suggestion, strongly seconded by Russell and Bridgman, that geophysical methods be used to determine whether any meteoritic mass was buried beneath the crater and, if so, then its location, approximate dimensions, and depth. Shaw and the board of directors of the Meteor Crater Exploration and Mining Company now opted to take his advice.

Thus, in March 1930, a crew from The Radiore Company of Los Angeles, California, in the charge of engineer James E. Dick, spent three days at the crater making a preliminary survey with a vertical magnetometer. The results, Dick reported to Colvocoresses, indicated a "considerable disturbance of an unusual character" within the crater.[22] While the anomalies were of "considerable magnitude," however, "the horizontal extent of the unusually high or low spots is small," leading Dick to conclude that "the disturbance is not due to any large buried meteoric mass, but rather to scattered fragments close to the surface, and is, therefore, not of any great significance except to note that such fragments may be more numerous within the crater than outside of it." The survey, he added, "brings out very little except that no important magnetic disturbance occurs under the south rim where the meteorite is believed to be buried."

To this, Radiore's chief engineer, W. R. Nelson, added that the "most interesting anomaly" was a "well-defined 'high' in the northwest corner of the Jupiter placer," that is, in the southeast quadrant of the crater and in an area tested by the drill back in 1906–08.[23] "This, presumably, represents meteoric material of indeterminate size laying at an indeterminate depth," he wrote. "The size is probably not large and the depth is not great since the sphere of influence is only noticeable in an area whose diameter is about 500 feet." Both Dick and Nelson suggested electrical conductivity studies.

Over the next six months, such studies would be undertaken by International Geophysics, Inc., another Los Angeles area firm, and would also produce an interesting anomaly, this time in the Saturn claim in the unexplored southwest quadrant of the crater. Late in May a crew spent four days making preliminary measures. The firm's consulting engineer, J. J. Jakosky, informed Colvocoresses in June that this brief survey "indicates that definite and practical results in reference to the structure, water level and approximate horizontal and vertical location of meteoric material if present, may be expected from a properly conducted geophysical survey."[24] Shaw and the directors, no less than Colvocoresses, were encouraged to press the investigation further, undoubtedly in part because of Jakosky's willingness to take a substantial part of his fee in exploration company stock.[25]

A comprehensive geological, electrical, and magnetic survey of the crater began on July 30 and continued through September 7. Jakosky would detail the results in a fifty-four-page report with four large maps and sixty-two plates, and subsequently in several lengthy publications.[26] But he summarized them in advance early in October in a brief letter to Colvocoresses. The geological phase of the survey proved to be largely confirmatory; it showed that the crater had been "formed by the impact of a meteorite, or a swarm" of meteorites and that "the explanation previously advanced by some investigators that a steam explosion is responsible for the crater, does not seem probable." The crater, he added, "is of considerable age, possibly 50,000 years."[27]

Of far greater interest to Colvocoresses, however, were Jakosky's conclusions from the electrical and magnetic surveys. The former indicated "the presence of an area of high conductivity in the southwest quadrant . . . between the center of the crater and the south rim," the engineer reported. "This material lies at an effective depth of approximately 700 feet. A careful study of the original and altered materials found in the area indicates that this zone . . . cannot be due to rocks in place or their products, or fill material. Our conclusions are that this material is of metallic character and meteoric in origin." The magnetic study, he added, also indicated "an area of magnetic material in the southern portion of the crater at depths of 200 feet or more, and probably concentrates with depth." Jakosky, although reluctant to be specific, suggested that "the mineralized area is at least six hundred feet in length and possesses appreciable breadth and depth." All together, he concluded, the results "present evidence sufficiently strong to warrant drill hole explorations."

Colvocoresses quickly wrote Jakosky urging him to state his conclusions in his final report more specifically; and the engineer, in reply, assured him that "we will . . . make our conclusions as complete and definite as possible," adding, however, that "overdrawn conclusions which are not justified by field data will actually weaken our entire report when it is critically examined by your associates, or their technical advisors."[28] Jakosky's final report, indeed, merely repeated his earlier conclusions.[29]

Late in November, Colvocoresses forwarded lengthy quotations from Jakosky's report to Shaw, declaring that the report itself, "in so far as the present development of Geophysics permits . . . is an extremely thorough study of the Crater problem. . . ."[30] He then gave his assessment of the salient points in the report, going somewhat further than Jakosky had been willing to go. The survey "has changed our previous ideas regarding the location of the meteorite, but in a manner which on the whole is advantageous from a mining and pumping standpoint," he wrote. "Otherwise it has tended to confirm and substantiate the essential portions of the theories on which our exploration was originally based and also our calculations of possible profit. . . . In other words, it gives further ground for the belief that the minimum tonnage of commercial ore cannot be less than two million and that probably this minimum will be substantially exceeded; therefore, there is futher justification for continued exploration and our prospects of final success have been strengthened." He recommended that "the next exploration should take the form of drilling to substantiate and amplify their findings. . . ."

As Jakosky surmised, his report was indeed submitted to "technical advisors." Geophysicist Hans Lundberg of the Swedish American Prospecting Corporation reported to Shaw early in January that the "procedures adopted for the electrical exploration were not entirely suitable" and that the "existence of a meteorite or other highly conductive body cannot be deduced from the data presented."[31] Jakosky's resistivity curves, he argued, merely measured differing densities at the contact planes of the crater fill and the Coconino sandstone of the lower rim. Results of the magnetic survey, he added, were "the only ones which might possibly indicate the existence of a meteoric mass," but he contended that the maximum anomalies Jakosky had found were only one-fourth as large as those that might be expected from theoretical considerations. "The investigation carried on by the International Geophysics at the Meteor Crater," he declared finally, "does not give any positive proof of the existence of a meteoric mass."

But Shaw now sent a copy of Jakosky's report to W. R. Rooney of the Carnegie Geophysical Laboratory, one of the scientists who had devised the so-called Gish-Rooney alternating current apparatus Jakosky had used in his electrical survey, and Rooney disagreed with Lundberg. "In my opinion the resistivity method, as employed by Mr. Jakosky, was particularly suited to the problem," Rooney wrote. "I am also of the opinion that the results indicate the presence of a considerable body of conductive material, rather than any variations in the sedimentary rocks or debris."[32]

HITTING ROCK BOTTOM

Colvocoresses, of course, accepted the conclusions of Jakosky and Rooney. Early in May he sent a lengthy report to Shaw in which he discussed in detail his proposal for processing the crater "ore" at the Humboldt smelter, as well as the comparative economics of the operation vis-à-vis those of the International Nickel Company which, as the principal producer of nickel in North America, would be the crater company's chief competitor. His predictions, not surprisingly, were highly optimistic.[33] In June he and Shaw briefly explored the problem of "dewatering" the crater and subsequently controlling the inflow of water during mining operations with, among others, Frank R. Dravo, whose contracting firm had unsuccessfully attempted to stem the flow of water in the final shaft in the spring of 1929.[34] And no more prone to miss a bet than Barringer had been, he initiated tests of the crater silica sand, conducted that summer by consulting engineer James B. Girand at the American Concrete Pipe Company in Phoenix, which showed that "the use of Meteor Crater Sand will reduce the volume cost, increase the quality, and make a more water-tight concrete."[35]

In mid-July, in a long report to the exploration company's stockholders, Shaw declared that "the other directors still retain their faith in the original theories and conclusions of Mr. Barringer, and believe these to have been strengthened by the majority of the participants in the recent scientific discussion and by the geophysical survey. . . . The directors agree that it will be unfortunate to have the company close its career after an expenditure of over $300,000 without having obtained some definite answer to the scientific enigma of the Meteor Crater, as well as the possible commercial profits from mining operations."[36] However, he noted, the company was now $68,000 in debt, and implementing Jakosky's recommendation for additional drilling would cost, at Colvocoresses's estimate, at least $10,000. A new stock subscription had been suggested, and the directors, he added, proposed that the price for a unit of one share each of preferred and common stock be reduced to $21.

The stockholders, however, apparently lacked the faith of the directors. A few weeks later Colvocoresses conceded ruefully to Lowell astronomer C. O. Lampland that "Unfortunately we have now reached a crisis where this work must either be supported by additional financial assistance or in all probability be discontinued entirely and very likely for a long period of time. Our stockholders . . . appear either to have lost interest in the project or to have been discouraged by the calculations of Dr. Moulton and others and altho we are only trying to raise $10,000 for drilling . . . it seems doubtful if we are going to obtain that comparatively small amount."[37]

Colvocoresses's efforts to raise new money probably were not helped by two minor, short-lived controversies over the crater's origin that had recently been thrashed out in the pages of *Science*. In November 1930 geologist Herman L. Fairchild, long a Barringer supporter, revived the old "plums in a pudding" idea, suggesting that not only were there at least two kinds of Canyon Diablo meteorites—the permanent irons and the decomposable, chlorine-bearing "iron shale"—but that these had been "inclusions or nodules of resistant nature, inclosed in some kind of perishable material. . . . That substance could have been only the stony materials of which most known meteorites are composed." All of the facts, he declared, "support the view that the Arizona visitor was a very large stony mass with metallic inclusions."[38] Reau Barringer, now Standard Iron's president, quickly replied that Fairchild "has presented no new evidence" to support this idea, and "his conclusions from the old evidence do not warrant, to my mind, a change from the more accepted picture of the comet." Reau did note, as Fairchild had not, the single small stony aerolite that his father had found in the plain in 1905, but added that it could "hardly be thought of as a chip from a larger mass. . . ."[39]

Fairchild's article, however, in January 1931, inspired Frederick S. Dellenbaugh, who had been a member of Major John Wesley Powell's second river expedition through the Grand Canyon in 1871 and had since made something of a career by writing about it, to challenge the crater's impact origin. "Inasmuch as no evidence is discoverable of a meteoric body capable of excavating such a large basin, notwithstanding long, competent and diligent examination," he wrote, "and as my poor intelligence sees nothing reported that substantiates in the slightest degree the meteor theory, I venture to disbelieve that theory in toto."[40] Instead, he pointed to the numerous sinkholes known to occur in the Kaibab limestone of the region and declared that "These sinkholes . . . being in the same limestone that forms the upper structure of Meteor Butte would seem to offer a perfectly reasonable explanation of the origin of Meteor Butte. That is to say: *Meteor Butte is entirely the work of erosion* and no meteor has had anything to do with its formation. . . . Meteor Butte, then, seems to be merely the reverse of a solitary mesa which preserves itself by a hard roof against erosion. The Meteor Sink had a soft spot where its hat ought to have been" [Dellenbaugh's emphasis].

Dellenbaugh's article in turn was quickly answered by Yale University geologist Chester R. Longwell, who took the occasion to dispute Fairchild's paper as well. If Dellenbaugh's suggestion was addressed to geologists only, he declared, "there would be no need for a reply. The geologic facts speak for themselves; they are not merely unfavorable to Mr. Dellenbaugh's idea—they disprove it conclusively." These facts, he noted, included the presence of once-buried Coconino sandstone fragments on the outer rim and plain around the crater, the upward tilt of the strata of the rim that "indicates a powerful lifting force acted inside the pit," and the occurrence of fused quartz, or lechatelierite, in the Coconino sandstone, "astonishing in view of the extremely high melting point of quartz. . . ."[41] As to Fairchild, while his idea of a stony meteorite "has interest to the geologist, it appears to be wholly speculative, and creates difficulties more serious than those it purports to remove. There is strong evidence, however, in favor of the theory involving metallic meteorites."

By October 1931 Colvocoresses's efforts seem to have borne at least some fruit, and he now confided to Lampland that "after a lot of hard work the chances now seem good that we will be able to continue some exploration."[42] Indeed, a few weeks later, the directors authorized a modest drilling program.[43]

The first hole, begun late in November in the center of the anomaly delineated in Jakosky's report, encountered "meteoric" material at 414 feet, and progressed laboriously to 675 feet where, Colvocoresses's report notes, "we again hit one very large and solid block or more probably the top of the compact cluster. Further progress was impossible as the stellite teeth were worn entirely off the drill bit with an advance of only a few inches." In sending a copy to Lampland, he added that "If we are able to continue our exploration (which depends entirely upon the amount of money available), it now appears probable that we may develop a very substantial mass which would definitely contradict Moulton's conclusions and I suppose necessitate a substantial revision of some of his theories. . . ."[44] To Shaw he reported that "We must conclude that scattered meteoric fragments lie in the debris . . . below 414 ft. It is, therefore, evident that the fragments constitute the material of 'high' electric conductivity and possessing magnetic attraction described by Jakosky. . . . It appears to me that this definitely contradicts the conclusions of Lundberg. . . . The confirmation of the results of the geophysical survey down to 675 ft. tends to substantiate its further indications that the very compact cluster will be found around 700 ft. and below, in fact we probably struck the top of it when the drill was stopped."[45]

A second hole was begun early in January 1932 and reached a depth of 650 feet. From 540 feet to the bottom, Colvocoresses informed Shaw a month later, "meteoric" fragments were encountered and the drill sludge tested positively for nickel.[46] To Lampland, he later explained that "We were obliged to suspend our drilling . . . as our water supply had frozen up tight and I am not sure that we will be able to continue further, but our second hole penetrated among meteoric fragments, and the conditions were very similar to the first hole so I feel that our work has given altogether favorable results. . . ."[47]

Colvocoresses's doubts about continued exploration were well taken. The economic depression that held the nation and much of the world in its grip took its toll at Meteor Crater as well. In April 1932, Shaw resigned as president of the exploration company. Colvocoresses was now seeking new ways to raise money; and in May he prepared a report in which he proposed to mine, package, and market the deposits of rock flour on the crater's rim as a household scouring powder to be called "Star Dust."[48] Production costs, he estimated, including containers, labels, shipping, and advertising, would be ten cents per two-pound box, which would sell for twenty-five cents at retail and net a seven-cent profit for the exploration company. "Star Dust" would be promoted first in the communities around the crater, but eventually in national markets. Colvocoresses felt that public interest "will be attracted by the unusual source of STAR DUST, by the name itself and the story of its origin which is roughly depicted on the label of the container."

But in October, the company's vice president, Ledyard Hecksher, reported to the stockholders that "The Company is now entirely without funds for taxes and other essential expenses" and urged them to contribute 10 percent of their original investment, to be paid in monthly installments over a year's time. "Subscriptions as low as $10.50 per month will be gladly accepted," he declared.[49] In March 1933, soon after the inauguration of Franklin D. Roosevelt as the nation's new President, Reau Barringer confided to Lampland in Flagstaff that there was nothing new to report regarding the crater. "Like everything else dependent upon finances these days," he opined, "it is sitting awaiting better times."[50] By November, Colvocoresses advised the stockholders that he had sold all the drilling equipment at the crater and still needed new subscriptions at $21 per stock unit to provide the $700 per year required to maintain the company's lease of the crater property.[51]

TIME AND TRAJECTORY

While there would be no further drilling at the crater, in later years more than half-a-dozen geophysical surveys were undertaken, no two of them, seemingly, giving quite the same result. In 1937 Lundberg made a magnetic survey of his own in which he recorded two "pronounced anomalies," the larger one centered 1,000 feet southwest of the crater rim, and a smaller one in the crater itself some 500 feet southwest of the center. Both anomalies, Lundberg concluded, "could be caused by a magnetic body buried at relatively great depth."[52] In 1947, however, a magnetometer survey run on three east-west lines across the area south of the crater by A. W. Whelan for meteoriticist H. H. Nininger's American Meteorite Museum failed to find the anomaly that Lundberg had reported there.[53]

In 1948 the U.S. Geological Survey apparently surveyed the crater with an aerial magnetometer. The results were not published, but Reau Barringer noted later that "Mr. Balsley of the U.S.G.S. told me that they reveal a rather large regional magnetic anomaly extending southeastward from the crater. He doesn't feel (and neither do I) that the anomaly can be meteoritic in origin, because of its size and location."[54]

In 1951 a gravimetric survey, financed by Standard Iron, was conducted at the crater by Roswell Miller III of Princeton and Norman Harding, a graduate student at the University of Wisconsin. Harding later reported that the survey "bears out the indications of the earlier magnetic and electrical resistivity studies which have been corroborated in part by drilling, that there is an anomalous mass of probable meteoric iron in the southwest quadrant of the crater. . . . On a conservative basis there would appear to be somewhere in the neighborhood of two million tons of meteoric iron in the anomaly area . . . at a depth of probably no greater than 600 feet."[55]

In 1953 geologist Dorsey Hager, who had proclaimed the crater a limestone "sink" in 1926, published a Fairchild Aerial Surveys aeromagnetic map in the course of developing his thesis of a geologic origin of the crater, which showed no appreciable magnetic anomaly at all at the crater.[56] In 1965, however, Roy G. Brereton of the California Institute of Technology, from an airborne magnetometer survey, reported a regional magnetic high marginal to the crater and centered some seven miles to the southeast. This, he thought, might be related to a dike or system of dikes associated with the crater, or to an extensive, horizontal sill-like mass in the deep basement rock. It "cannot possibly be related to a buried mass of extraterrestrial material," he declared; "however, its presence could lend credence to a terrestrial origin for this feature."[57]

In the early 1970s both gravity and magnetic studies were made at the crater by Robert D. Regan of the U.S.G.S. and William J. Hinze of Purdue University.[58] In the gravity survey, they found a small negative anomaly in the center of the crater, which they attributed to the underlying breccia, and an encircling negative anomaly centered on the crest of the rim, which they explained by the low-density rim debris and the underlying fractured and uplifted rock. They found no evidence of Harding's positive anomaly, but conceded that they had made no observations in the immediate area he had designated. Their magnetic survey failed also to confirm the positive anomaly in the southwest quadrant reported by Jakosky. It did, however, show a "subtle" negative anomaly in the southeast quadrant which extended some 2,000 feet beyond the rim and which, they felt, resulted from the "disturbance to the underlying formations, particularly the Supai. . . ." In sum, they declared, the two surveys revealed "an approximately symmetrical simple bowl-shaped breccia lens centered beneath the crater. There is no evidence for a buried massive meteoritic body, and there is no gross asymmetry to the subsurface structure that would indicate the trajectory of the impacting meteorite."

Barringer and Tilghman had thought originally that the meteorite had had a north-northwest to south-southeast trajectory and that it had struck at a near-vertical angle, reasoning from the circularity of the crater. By 1909 and thereafter, however, Barringer concluded, on the basis of the apparent bilateral symmetry of the crater, that its angle of impact had been steeper—perhaps 45°—and thus its mass would be found under the southern cliffs. This interpretation was not questioned for nearly fifty years.

But in 1957, Rinehart, from the results of his soil survey at the crater the previous year, disputed this view. He too found a bilateral symmetry, but he interpreted it as being centered along a line running "somewhat north of east," where he had found a concentration of meteoritic material in the soil.[59] "Thus, if we assume that the debris now lies where it originally fell, then there is little doubt that the meteorite approached the earth along an axis of asymmetry of our pattern: roughly north of east or south of west. . . . A highly reasonable hypothesis is that the meteorite approached the earth from a southwesterly direction and, when it struck, pitched forward large quantities of meteoritic material to the position where it now rests," he declared. Meteorites recovered on the rim and plain to the east and northeast had been "severely altered by heat and deformation," he noted, while those found to the west and southwest "are in their virgin state," and there were more large boulders on the eastern rim and plain than to the west. Moreover, on his interpretation, the magnetic, electrical, and gravimetric surveys up to that time "seem, in spite of some contradictory evidence, to favor a southwest to northeast direction. . . ."

Reau Barringer quickly challenged Rinehart's view, pointing out that his father's studies had indicated that the axis of symmetry "starts somewhat west of north, and ends a little east of south."⁶⁰ Exploration by the drill had also shown "a considerable concentration of meteoritic material, both metallic and oxidized, deep below the southern and southwestern portions of the Crater floor," and this evidence was "reinforced" by the Jakosky and Miller-Harding geophysical surveys. And now, he revived an idea that Thomson had broached to the elder Barringer many years earlier—the effect of the earth's rotation on the meteorite as it penetrated through the rock. The motion of the earth's surface, in traveling eastward at some 800 feet per second, Reau argued, would tend to make the meteorite "curve to the right as it penetrated the ground, logically ending up in the southwest quadrant rather than in the southeast toward which it was originally headed."⁶¹

Still another direction of approach was suggested in 1962 by C. H. Roach and his colleagues in the U.S.G.S. from shock-induced thermoluminescence characteristics of the Moenkopi sandstone and Kaibab limestone exposed on the crater's rim.⁶² "Comparison of the thermoluminescence of samples of the three marker beds taken from a measured section of the north rim of the crater with those from a corresponding section on the southeast rim shows that rocks exposed in the southeast quadrant of the crater appear to have received the strongest shock, and rocks along the north part of the crater appear to have been shocked least," they reported. "At greater depths rocks also seem to have been more strongly shocked in the southeast part of the crater than elsewhere." Consequently, they suggested that "the point of impact and upper part of the path of penetration of the meteorite may have lain southeast of the center of the crater. As the crater rim is probably roughly centered about the center of gravity of the total energy delivered during penetration of the meteorite, the results of this study suggest that the meteorite was moving southeast to northwest. Additional study is needed, however, before definite conclusions can be drawn."

In 1975 H. D. Ackerman and R. H. Godson of the U.S. Geological Survey and J. S. Watkins of the University of Texas reported a seismic survey of the crater which did indeed indicate a subsurface asymmetry which, in turn, led them in still a different direction. From data on the velocity of shock waves generated by underground dynamite explosions in rock in and around the crater, they found greater time delays in the rock under the south and west walls than on the north and east, and thus that the subsurface structure appeared to be "skewed slightly toward the south and west."⁶³ They also reported that a subsurface asymmetry had been found during the partial excavation of a small missile crater at the White Sands, New Mexico, missile test site, with the skew occurring opposite to the known direction of the missile's approach.

"The asymmetry in the Meteor Crater data," they noted, "is much less than that in the missile impact data, probably because the mechanism for disrupting rock with hypervelocity impact differs from the mechanism of the slower missile impact. . . . Therefore, some uncertainty is present in our assumption of the trajectory of the meteorite. Nevertheless, two lines of evidence in our data suggest that the meteorite came from the north or northeast. . . . Travel time anomalies observed in data from the crater floor and the greater intensity of brecciation and fracturing beneath the south rim support a hypothesis that the meteorite came from a northerly direction."

Shoemaker more recently has favored a south-easterly approach and has suggested that the Odessa, Texas, craters, some six hundred miles to the southeast, were a part of the same fall and were formed by smaller companions which, being subject to greater atmospheric retardation, plunged to earth a few tens of seconds before the Arizona mass. Examination of the lacustrine deposits and other debris beneath "silica hill" in the northern part of the crater floor, unprobed so far by shaft or drill, might well shed additional light on the problem, he has pointed out.[64]

These deposits have figured prominently in attempts to fix Meteor Crater's age. Barringer's belief that the crater was only two or three thousand years old did not survive him, and Tilghman's 10,000-year upper limit also soon went by the boards, with Merrill and later others preferring an estimate of about 20,000 years. Colvocoresses, as noted earlier, thought it might be about 100,000 years old; and in 1953 Dorsey Hager, arguing for a geologic rather than a meteoritic origin for the crater, assigned it an age "of the order of 200,000 years," with the fall of the Canyon Diablo meteorites occurring "considerably later."[65]

In 1930 two geological studies addressed the question, one, of course, by Jakosky, and the other by a well-known geologist, Eliot Blackwelder of Stanford University. Both reasoned from the accumulation of more than 100 feet of sediments and debris on the crater floor and the probable slow rate of erosion in the arid environment of the high northern Arizona plateau. Blackwelder, who spent several days at the crater in the spring, suggested an age range of from 40,000 to 75,000 years, arguing that evidence for changes from a wetter climate to a drier one, as shown by the interlaced fossil-bearing deposits in the fill, indicated that the crater had existed during the last interglacial or post-Tahoe epoch of the Pleistocene.[66] Jakosky was less sure of Pleistocene chronology, but concluded its age to be "measurable in terms of tens of thousands of years and very probably to be in the neighborhood of 50,000 years."[67]

Both men discussed the crater's minimum age, long set at about 700 years by the growth rings of the ancient junipers Holsinger found on the rim in 1903, and both cited a thin layer of volcanic lapilli and ash in the top six feet of the fill, first reported by Fairchild in 1907, and its implication that the crater must predate the most recent volcanic eruption in the region. Jakosky quoted a pioneer study by H. H. Robinson in 1913, who had suggested that the most recent volcanic activity in the eastern part of the San Francisco Peaks volcanic field around Flagstaff "may be (in round numbers) 1,000 years old."[68] Blackwelder, however, apparently unaware of Robinson's earlier work, declared that "No such volcanic eruption is known to have occurred since Pleistocene times," a statement challenged as soon as it appeared in print. F. Martin Brown quickly called attention to a recent report that three-ring dating of burned beams in prehistoric pithouses buried in cinders and ash from Sunset Crater, forty miles west-northwest of Meteor Crater, suggested its eruption in A.D. 793.[69] If this eruption produced the volcanic ash layer at Meteor Crater, he added, then Barringer's dating of 2,000 to 3,000 years age "is probably the most acceptable of all."

Nonetheless, a 50,000-year age was widely accepted over the next thirty years, Hager being the most notable dissenter. Harvey Harlow Nininger, the indefatigable collector and student of meteorites who spent a number of years at the crater during this period, concluded in 1939 that it was between 20,000 and 30,000 years old on the basis of the "corrosion" of the Kaibab limestone of its upper walls. But after Blackwelder, in 1947, pointed out the siliceous character of the limestone and the probability of a slower rate of corrosion, Nininger adopted the 50,000-year figure.[70]

In 1961, however, another meteoriticist, John David Buddhue, in a paper presented to a meeting of the Meteoritical Society in Nantucket, Massachusetts, suggested that the "true age is around 22,500 years."[71] Buddhue correlated a "conjectural chronology" of the various advances and retreats of the final Wisconsin phase of the Pleistocene glaciation with the layers of sediment and debris in the crater fill as determined by Jakosky, who, in 1930, had descended one of the early development shafts in the northern part of the crater and just west of "silica hill" and had recorded a 35-foot, 8-inch-deep stratigraphic section.[72] The lake bed deposits in this section, Buddhue reasoned, must have formed during a pluvial period when conditions were wetter; and a pluvial period during the Wisconsin glaciation recently had been dated by radiocarbon methods at Searles dry lake in California as occurring from 24,000 to 10,500 years ago. In his correlation, he equated the bottom six feet of Jakosky's section to a period 18,000 to 20,000 years ago, suggesting the latter figure as a minimum age for the crater. To derive its "true" age, however, proved to be much simpler; here he pointed to a nine-foot-thick outcrop of

Moenkopi sandstone under the north rim which, he suggested, was a remnant of a formation that in less protected areas elsewhere had eroded away since the crater was formed. Then, using Hager's estimate of the erosion rate for this sandstone of one foot per 2,500 years,[73] and multiplying by nine, he arrived at an actual age of 22,500 years.

Buddhue's estimate seems to have been accepted for more than twenty years;[74] Shoemaker, for example, published an age of 25,000 ± 5,000 years as recently as 1983,[75] basing it on a correlation of the stratigraphy of sediments and soils in the crater with the last major Pleistocene pluvial event as dated by radiocarbon methods. But almost at the same time, Steven R. Sutton, a doctoral candidate at Washington University in St. Louis, suggested a tentative age of 40,000 years on the basis of thermoluminescence dating techniques applied to shock-heated limestone and sandstone at the crater.[76] He has since concluded from additional study and analysis that the crater's age is 49,900 ± 2,900 years.[77]

"METEOR CRATER" IT IS

In the years after Barringer's death, only two recognized geologists—Dorsey Hager and the U.S. Geological Survey's Nelson Horatio Darton—publicly insisted on a geologic rather than meteoritic origin for the crater. In 1932 Darton won a small point when the U.S. Board of Geographical Names officially adopted his favored name of "Crater Mound" for the crater. However, writers of scientific papers and popular articles persisted, somewhat perversely Darton must have felt, in calling the feature "Meteor Crater." In December 1945 a frustrated Darton formally deplored this usage before a joint meeting of the Geological Society of America and the American Association of Geographers in Pittsburgh, reminding his listeners that "Crater Mound" was the crater's official name.[78] "The words 'Meteor Crater' are not admissable," he argued. "I am convinced that no meteorite is present . . . Extensive shafting and drilling have given entirely negative results as to the presence of buried meteoric iron, and a geophysical survey verified the magnetic tests made for Gilbert." Still, "Crater Mound" was indeed an unusual feature, and thus it was his hope "to have it made a national monument."

Darton's paper brought some immediate reactions. Blackwelder, who had been in his audience, pointed out in *Science* that the majority of scientists "disagree with Mr. Darton and are convinced that the crater was made by the impact of a large meteorite." To Darton's point that no such meteorite had been found, he cited Moulton's suggestion that such a meteorite would explode and vaporize. Darton, he added, apparently was "unaware" of the presence of quartz powder and silica glass at the crater.[79] Privately, he advised Harold S. Colton, director of the Museum of Northern Arizona in Flagstaff, that "Apparently, the old man has learned nothing in the last thirty years, so far as the Crater is concerned. In the discussion which followed his statements, several of us took the opportunity to differ with him and pointed out the overwhelming evidence in favor of the meteoric theory."[80]

Meteoriticist Frederick C. Leonard of the University of California at Los Angeles, after reprinting Darton's abstract, noted that there were "two points . . . to which we can agree: (1) that the name 'Meteor Crater' should be changed, albeit not to 'Crater Mound' but to "Meteorite Crater'; and (2) that the crater should be made a national monument or, we may add, a national or state park. The amazing thing is, however, that any geologist should still disbelieve in the meteoritic origin of what is undoubtedly the finest example of a meteorite crater known on the earth!"[81] Leonard would continue his campaign to substitute the term "meteorite" for "meteor" in the crater's name; and in a few years, the Meteoritical Society would resolve that the crater be known as "Barringer Meteorite Crater," thus combining linguistic purism with a tribute to Barringer's pioneering work at the crater.[82] But while a few meteoriticists went along, just about everyone else has continued to call it "Meteor Crater," apparently agreeing with Fairchild, who had proposed the name in the first place, on its "brevity" and "euphony."[83]

Within a year of Darton's well-publicized complaint, "Meteor Crater" became its official name, largely through the efforts of Blackwelder. Early in March 1946 he began corresponding with Meredith F. Burrill, director of the U.S. Board of Geographical Names, and the matter was brought before the board in July.[84] Blackwelder later wrote Colton that "I am informed that the United States Board on Geographical Names reversed itself on July 15, 1946, and adopted the name Meteor Crater. This action was apparently taken in response to representations which I made to the director . . . last spring."[85]

Blackwelder, however, was more intent on changing the crater's status from private to public ownership, a point on which Darton and his critics agreed. He had written Colton shortly after the Pittsburgh meeting to inquire about the situation at the crater in view of reports he had received that the crater was closed to the public and that silica mining operations were soon to begin on the crater's south rim.[86] Colton, interested in monument or park status for several other features in the area as well, immediately wrote to National Park Service director Newton B. Drury and to Arizona's governor, Sidney P. Osborn, to sound them out on the idea.[87] Blackwelder also wrote to Drury, urging him to consider "the ways and means for the Government to acquire the crater and immediately surrounding lands for public purposes. It should be possible to do this without much difficulty," he added, "since the place has very little economic value; and it is certainly in the public interest to do so."[88]

But the political climate in Arizona itself seems to have mitigated against such a move. Osborn quickly replied to Colton that he was "tremendously interested" but that "it will take an act of the legislature and some money to get the thing going." He hoped that the matter could be presented to the legislature the following year "by whoever may be governor when that time arrives."[89] Osborn died later that same year, however, and no subsequent legislative action was taken. Blackwelder also received a reply to his letter to Drury which, he advised Colton, "alludes to 'the opposition of certain interests to the establishment of additional national monuments in Arizona.'"[90] Colton quickly confirmed "a very strong local opposition.... Certain short-sighted taxpayers as well as the [Coconino County] Board of Supervisors will fight any effort to establish new monuments...."[91]

Standard Iron, of course, strongly opposed any such move, and Barringer's son Richard sent out a "publicity release" detailing the company's position on the "nationalization of its property." In this, the attitude of the U.S. Geological Survey over more than forty-five years was a major point. Gilbert, he noted, had concluded that the crater originated in a volcanic steam explosion; and although "incapacitated" in his later years, "he was a big enough man, had he been active, to have admitted his mistake.... But his successors were of no such caliber.... Resolutely they continued to ignore the facts, until by many years of uphill effort, Barringer succeeded in swinging the weight of scientific thinking of the whole world to his side.... How," he asked, "would the public's knowledge, or the advancement of science, have been served if Meteor Crater remained government property throughout those years?"[92]

Richard, incidentally, also played on the rising public emotions in the burgeoning post–World War II "Cold War" with Russia. Private ownership, he pointed out, was the "American Way" as against the "fetish of state ownership" that prevailed in Russia and other communist nations. Under government ownership, Meteor Crater would "share the fate of the Siberian fall of 1908 and be owned by the state and wholly barred to all scientists and all visitors."

Oddly enough, Blackwelder's attitude was ambivalent when, in the autumn of 1946, the Meteor Silica Corporation of Los Angeles actually began mining the rock flour on the south rim on a lease from Standard Iron for use by glass manufacturers. While mining "might disfigure the place rather badly even if it did not destroy things of scientific importance," he confided to Colton, "there may be some compensation in the new facts which may be brought to light during the excavating."[93] Colton agreed.[94] Later Richard Barringer pointed out that the mining operation was "out of view" to visitors and that the mining firm was obligated by the lease "not to deface the crater" and "to record the geology."[95]

The mining operation soon ended, after several hundred tons of rock flour were shipped from the crater, but the ownership issue was brought up sporadically in later years. Standard Iron, however, renamed "Barringer Crater Company" in 1951, successfully maintained its position, in part by arranging to install many improvements, such as a paved access road and a new, spacious museum facility and viewing point on the crest of the crater's north rim. By 1955 the Meteoritical Society, on a motion by Lincoln LaPaz, recognized these efforts. Noting the "criticisms of the private ownership of the Barringer Meteorite Crater that have appeared in the press and elsewhere from time to time," the Society resolved to "go on record as approving the manner in which the Barringer Meteorite Crater is being handled, and reiterates its conviction that this is *not* the appropriate time for nationalization of the Crater, or for its inclusion in the Arizona State Park System [emphasis in original]."[96]

Hager's postulate of a geologic origin for the crater also stirred a brief controversy, and again Blackwelder was involved. Hager's thesis now was an expanded and refined version of his early "sink" theory and was developed, he said, through study not only of the crater but of the region in which it lies and through discussions with numerous geologists who "made suggestions and raised questions, especially members of the United States Geological Survey in Salt Lake City." He first promulgated it in 1949 in talks before the Meteoritical Society[97] and to regional scientific societies in Utah and New Mexico. In 1953 he published a lengthy exposition of his hypothesis, concluding that the "cumulative evidence for a geologic origin seems to have been overwhelming." His paper, he noted, "can justly be said to represent the opinions of numerous geologists and others interested in an objective explanation of the phenomenon."[98] Incidentally, he used Darton's name for the crater, explaining that, while the official name had been changed to "Meteor Crater," the feature was nonetheless "a crater surrounded by a mound."

"The crater and the mound are two separate, though interrelated, geologic features," he declared. "The present mound is the remnant or skeleton of an elliptical dome with a graben-like sink in its apex. Originally the dome was far higher, in the order of 1,000 feet or more above its present height." The boulders and blocks of limestone and sandstone on the rim and surrounding plain had not fallen upward, as Reau had suggested sarcastically in criticizing Hager's 1926 sinkhole theory, but had fallen downward from the steep walls of the dome. Much of the "white sand so prominent on the rim," he declared, "is principally derived by weathering action from the sandy Kaibab limestone." Moreover, the silica glass, or lechatelierite, found at the crater "is probably derived from old explosive volcanoes northwest and southwest of Crater Mound."

The dome or mound, he argued, was formed by a folding of the strata, perhaps 5,000,000 years ago, along a northwest-trending anticline known as "Sunshine Nose" whose axis coincided with the natural faulting in the area and with the major axis of the elliptical dome itself. The gradual removal by underground water of underlying limestone and evaporites in the still lower Supai red beds had caused the dome to break into blocks which "settled downward to finally reach a state of equilibrium on the lower floor of the crater." Two of these downthrown blocks, or graben, he identified in the southeast and northwest corners of the crater. Such grabenlike structures with associated sinkholes, he added, occurred over a broad band stretching from some forty miles southeast of the crater to Cameron, Arizona, sixty-five miles to the northwest. Furthermore, "Evidence exists from sink holes both in sandstone and limestone at many other places away from Arizona to justify a sink of the proportion found here."

Blackwelder quickly demurred, focusing his criticisms on Hager's explanations of the occurrence of lechatelierite and the "white sand" at the crater. Lechatelierite was not found in the ejectamenta of volcanoes, he declared, and its formation required temperatures "far higher than afforded by volcanic eruptions." Nor was the "white sand" derived from the Kaibab limestone. "Careful microscopic study shows . . . that the two kinds of material have little in common."[99] Hager, in reply, cited Dana's *Textbook of Mineralogy* to the point that lechatelierite "is reported in volcanic rocks," and declared that "the white sand is clearly a weathered product." Blackwelder, he complained, had not considered other matters of greater importance raised in his paper.[100]

NININGER'S CONTRIBUTIONS

A more strongly worded critique was offered by H. H. Nininger, certainly a leading authority on the crater at this time. Hager's paper, he declared, was "an elaborate presentation of geological features spread over an area of some 10,000 square miles, most of which can be considered as related to the crater only on the basis of a forbidding list of assumptions."[101] Hager "assumes that four of the thirteen faults in the crater rim are connected through the floor of the crater by concealed faults, thus making it a graben"; that the crater "is underlain by a bed of evaporites," and that there once existed a "1,000-foot high dome where the crater now is." He also "asserts that the 'rock flour' is mainly just sand" but "cites no proof." And Hager's discussion of the distribution of meteoritic material in and around the crater "reveals a notable lack of familiarity with the behavior of meteorites as they land on the Earth." In making these assumptions, he added, Hager "apparently saw no reason for making a detailed examination of the area under consideration."

To this, Hager retorted that Nininger "has entirely missed the significance of the geologic observations as he dismisses the geologic features quite cavalierly."[102] In particular, he wrote, Nininger ignored the crater's orientation on the axis of Sunshine Nose. As to specifics, there were not thirteen faults in the rim, but only six; other "slight displacements" being "purely superficial features." To back his postulate of evaporite beds beneath the crater, he now pointed to evaporites encountered in the Supai at a depth of 2,000 feet in test wells drilled thirty-five miles to the southeast of the crater. Nor was his postulated height for the original dome out of line, he contended, calling attention to an 1,800-foot uplift in southeastern Utah known as Upheaval Dome. As to the "white sand," Hager referred to a 1948 analysis by Buddhue showing that only 14 percent of the sand was fine enough to pass through a 200-mesh screen.[103] What little evidence of fracturing there was in the grains could be accounted for by weathering and "release of tension from folding." On the distribution of meteoritic material, Hager quoted Barringer's 1909 paper, italicizing passages pertinent to his point that the ordinary Canyon Diablo irons, at least, were always found "on or very near the surface," a fact that Barringer had been unable to explain. "I do not claim that a geologic hypothesis is final," he concluded, "but the evidence for one of several geologic hypotheses of origin seems far stronger than for meteoritic origin. Certainly a meteorite or meteoritic shower fell in the near vicinity, but the crater was already present and not formed by the meteoritic fall."

Nininger now offered some specifics of his own. As to Hager's evaporite beds, he noted that extensive exposures of the Supai formation in Oak Creek Canyon thirty miles west of the crater "fail to indicate evaporite beds. Thus this postulate seems purely arbitrary." Buddhue, he also pointed out, had been "dealing with rock flour only incidentally" and had merely analyzed "a sample of sandstone in the crater wall that had been 'crushed but not dispersed.'" Moreover, if weathering and folding produced such rock flour, he wondered, "why do we not find comparable deposits of rock flour at other sites? I have examined a dozen large sinks 30 miles south of the crater where the entire section of the Kaibab is exposed. Not even a small sample of rock flour was in evidence."[104]

That the crater was located on the axis of the Sunshine Nose anticline, he also noted, he had learned by reading a copy of Hugh M. Roberts's 1919 report, but there were "thousands of such structures on the land surface of the Earth, and the coincidental association of any feature with one of them does not require explanation involving generic relationship." Hager's Sunshine Nose, Nininger added pointedly, was "much more conspicuous on the Hager map than it is in the field," and by Hager's own statement, "every line on that map represents 'inferred conditions' except that which outlines the crater pit. . . ."

> One has difficulty following the Hager logic. On the one hand he considers as very significant the simple fact that the crater lies on the axis of an anticline, of which there are literally thousands. Yet he sees no necessary significance in the association of a startlingly distinctive collared pit with some 40,000 heavy nickel-iron meteorites, billions of metallic spheroids also of nickel-iron, millions of slaggy bombs containing more millions of nickel-iron particles and, except for these particles, showing approximately the same composition as the sediments in which the crater appears. This pit is unlike known volcanoes. It is unlike known sinkholes. It lacks certain essential features of steam or gas blowouts. There are few other rimmed pits similar in structure and in each case where they have been adequately explored meteorites have been found associated. All these facts are regarded as coincidental by Hager and are given very small consideration in his paper, and some are not even mentioned.

Hager, incidentally, made three subsequent visits to the crater, later publishing "several corrections" on points in his paper raised by his critics.[105] But while he still urged the crater's geologic origin, few if any scientists were persuaded.

Quite recently, it may be added, Hager's argument by analogy to Upheaval Dome, in what is now Canyonlands National Park, has begun to erode. Shoemaker and K. E. Herkenhoff of the California Institute of Technology now strongly suggest that it, too, is meteoritic in origin. From a reinvestigation of the structure in 1983, they have concluded that Upheaval Dome is probably a remnant of an impact crater that may originally have been eight or nine kilometers in diameter and 1.3 to 1.4 kilometers deep, from which at least one kilometer of rock strata has been stripped off by erosion since it was formed, perhaps some sixty million years ago.[106] Their evidence includes the presence of shock-metamorphosed sandstone that "corresponds approximately" to some of the metamorphosed Coconino sandstone at Meteor Crater.

Nininger's remarks about "billions of metallic spheroids" and "millions of slaggy bombs" containing nickel-iron particles refer to his own major contributions to the study of Meteor Crater. Nininger took up the study of meteorites in the early 1920s and over the next three decades became widely known as a meteoriticist. In 1933 he was founding secretary of the Society of Meteorite Research, precursor to the Meteoritical Society, and later served as its president.[107] But he was also, by his own later admission, "regarded by certain geologists as somewhat of a radical in respect to meteoritical questions."[108] His interest in the crater itself first appeared early in 1929, when he inquired about current developments there, and Barringer put him off tersely by noting that a new shaft was being sunk but that he could give him no further information.[109]

In 1930 Nininger joined the staff of the Colorado Museum of Natural History which, for the next sixteen years, became the base for his extensive meteorite collecting expeditions around the country. During this time, through collecting and purchase, he amassed perhaps the largest and most representative meteorite collection in the world, including numerous specimens of the Canyon Diablo meteorites.[110] In 1933 he excavated a meteorite crater, a small 36-by-55-foot depression with associated irons he had found four years earlier on a farm near Haviland, Kansas.

In 1939, with the aid of a small grant from the American Philosophical Society, he spent several weeks at Meteor Crater combing twenty-three acres of the surrounding plain with a large magnetic rake of his own devising, garnering more than 12,000 small meteorites weighing, in all, just over forty-two pounds. Among these, he later found some anomalous specimens which contained more nickel and showed different Widmanstätten patterns than the ordinary Canyon Diablo irons. Thus he suggested that at least three kinds of meteoritic irons were present and that the crater might have been formed by "a *composite* fall consisting of an undetermined number of *components* [Nininger's emphasis]."[111] In 1964, Dieter Heymann of the Enrico Fermi Institute for Nuclear Studies confirmed that the anomalous irons were indeed different. From the ratios of cosmogenic rare gases they contained, however, he concluded, although with reservations, that all the Canyon Diablo irons had probably originated in a single mass, but that "substantial structural and composition variations existed in the nickel-iron phase of the parent body. . . ."[112]

In 1946 Nininger moved his collection, weighing 16,000 pounds, from Denver to a vacant towered structure known locally as the "Observatory," six miles north of the crater and on heavily traveled U.S. Highway 66. There he established his American Meteorite Museum, hoping that the twenty-five cent fee he charged curious tourists to view his collection would not only provide a living for himself and his family, but would finance his continuing researches. He had obtained another small Philosophical Society grant, as well as a permit from Standard Iron to collect specimens in and around the crater. But he had barely begun his researches there before events began to conspire against him.

Complaints about his collecting and surveying activities came from the Tremaine Cattle Company of Phoenix, which had held grazing rights there since before World War II and which apparently viewed Nininger's museum as a threat to its growing post-war tourist business. In 1948 Standard Iron terminated Nininger's permit, assigning exclusive collection rights to Lincoln La Paz's Institute for Meteoritics.[113]

Now, the Tremaine Company expressed an interest in constructing a modern tourist facility on the crater's rim, and the Barringer Company agreed to lease a portion of the north rim and its land to the north on the condition that a paved access road, a museum, and suitable tourist facilities be provided. Two prominent architects, Frank Lloyd Wright and Philip Johnson, submitted plans, with Johnson's being accepted. The crater company also stipulated that there be no other construction or activity which might deface the crater or exclude any portion of it from scientific exploration and research.

By 1951 the new access road as well as major realignment of busy U.S. 66 effectively isolated the American Meteorite Museum from the main traffic flow and drastically reduced the number of paying visitors there. Bound by his lease, Nininger remained until 1953, continuing to make his surveys and collect meteorites and soil samples whenever possible from scattered state-owned sections of land around the crater. He then moved to Sedona, thirty miles south of Flagstaff, where he built a new museum and operated it for many years until his retirement.

To Nininger, the minute "metallic spheroids" as well as slightly larger "metal-centered pellets" that he began to find in great abundance in the soil around the crater in 1946, all of which reacted strongly for nickel, were "condensation products."[114] Indeed, he had undertaken his soil survey on the premise that if the explosion-and-vaporization hypotheses of Merrill, Moulton, and others were valid, evidence of condensation should be present. Barringer and Tilghman, of course, had reported such particles in 1906, but Barringer, unwilling to accept extensive vaporization, had thereafter ignored them.

From quantitative measures at sixty locations around the crater, Nininger estimated that "from 4,000 to 8,000 tons of the little spheroids are now present in the upper four inches of the soil within an average radius of two-and-a-half miles from the crater's rim." Their distribution was asymmetric, extending out more than six miles on the northeast. Assuming 50,000 years of erosion since impact, he suggested that it would "not be unreasonable" that this residual amount "represents an original deposit of 100,000 to 200,000 tons!"

He also speculated that these tiny particles had been formed "in the heart of a large cloud of metallic vapors and other gases from the core of which oxygen was excluded," that had been thrown up several miles in the air by the impact and had fallen to earth as droplets of metal which became cemented to the soil grains through oxidation. The concentration of the spheroids on the northeast of the crater, he thought, might be explained if the strong prevailing southwesterly winds there had been blowing when the cloud was formed.

Rinehart, incidentally, after a more extensive survey than Nininger was able to make, later disputed this last idea in arguing for a southwesterly approach of the meteorite which, on impact, had "pitched forward" meteoritic material on the northeast rim and plain. Noting the similarity of distribution of the "bits" and the "ponderable chunks" of meteoritic material on the northeast, he declared it "not easy to imagine a wind strong enough to propel large chunks and minute pieces of the same material in exactly the same direction and for the same distance. . . . Many of the bits are probably remnants of ponderable chunks." Shifting of material by the wind over subsequent time was "a reasonable thing to postulate," but it did not explain the asymmetrical distribution of the larger meteoritic material. "If wind were a factor we might expect that the bits would be sorted by weight; no evidence of such sorting was found."[115]

Nininger's discovery of the "slaggy bombs" of what he called "impactite" is of considerably greater significance, for it demonstrated that there had been more fusion and melting of rock at the crater than anyone had previously thought. Barringer, of course, had used what he believed was the meager evidence of fusion to argue that the impact had not generated high temperatures, pointing to the Kaibab limestone as an ideal source of slag, and thus the meteorite had struck at a low velocity. But the slaggy "bombs," "bomblets," and twisted "blobs" of fused rock that Nininger now found mixed in with the debris around the crater rim were, in fact, derived from the Kaibab limestone, as his laboratory tests showed. Moreover, as he had quickly discovered, tiny metallic particles that tested positively for nickel were embedded in every piece of this "melted country rock" that he examined.[116]

This slaggy material, he suggested, was formed as the meteorite had plunged through the Kaibab limestone, splashing melted rock in all directions. The embedded metallic particles also were splash droplets "from those parts of the meteorite which touched the rock through which it was passing." As the meteorite penetrated deeper, it progressively lost energy and, reaching the Coconino sandstone and the water table, generated superheated steam which exploded. The relative scarcity of silica glass, or lechatelierite, in the Coconino then was "due to the reduced energy of the meteorite and premature explosion, together with the cooling action of water volatilization." The absence of fusion products in the Moenkopi sandstone, above the Kaibab and on the surface, was due to the meteorite's greater energy at the moment of impact, which had been sufficient to volatilize rather than merely fuse the rock.

While Nininger's work at the crater was cut short by events over which he had no control and about which, indeed, he was quite bitter, he was well aware that much research remained to be done.[117] In summing up his own work in 1956, he listed twenty-eight specific projects that he thought should be carried out before any really reliable conclusions about the mechanics of the Meteor Crater event could be reached.[118] The last of these is of particular interest. In 1953 an industrial chemist named Loring Coes Jr. had discovered a previously unknown high-pressure polymorph of silica in his laboratory by applying pressures in excess of 20,000 kilobars (atmospheres) to quartz.[119] The new form was named coesite, and Nininger suggested that it might well be present at Meteor Crater, and that perhaps someone ought to look for it.

Four years later Eugene Shoemaker sent samples of the sheared and compressed Coconino sandstone that Barringer had dubbed Variety A to Washington, where his U.S.G.S. colleague Edward C. T. Chao, using x-ray diffraction techniques, found that they contained coesite.[120] Two years later, another high-pressure polymorph of silica, discovered in 1961 by Russian scientists S. M. Stishov and S. V. Popova and named stishovite, was also found in the crater sandstone by another U.S.G.S. team.[121]

By this time, as is clear from what has been said, Survey geologists, for the first time in more than sixty years, were once again actively investigating Meteor Crater, and in 1959 Shoemaker's first paper on its impact origin was published with the authorization of the Survey's director.[122] In subsequent years, it should be added, Shoemaker and his Survey colleagues have contributed significantly not only to our understanding of what happened many thousands of years ago at Meteor Crater, but to our understanding of the processes of meteoritic impact and impact cratering in general.

FOURTEEN

IMPACT: A COSMIC PROCESS

> From our knowledge of these terrestrial impact structures and craters, we have been able to establish with confidence the impact origin of most large lunar craters. From the combined evidence from the Earth and Moon we have learned to recognize impact craters on such diverse celestial bodies as Mercury, Mars and the icy satellites of Jupiter and Saturn.
>
> EUGENE MERLE SHOEMAKER, 1984

Progressively over the half-century and more since Barringer's death, scientists have come to the realization that impact has been a fundamental process in the origin and evolution of the solar system.

The evidence that impact has been operating over the 4.5-billion-year history of our system is now quite massive. Scores of meteoritic impact structures, analogous to Meteor Crater but in far more advanced stages of erosional decay, are known to exist on all the continental areas of the earth except ice-sheathed Antarctica. Some of these crateriform structures are comparable in size to the larger craters on the moon, and some of them are believed to be more than a billion years old. America's historic manned exploration of the moon by the Apollo astronauts in the late 1960s and early 1970s has produced indisputable evidence that the impact of bodies large and small began shaping the face of our satellite at least four billion years ago. Through the 1960s and into the early 1980s too, a succession of unmanned spacecraft missions to both the inner and outer planets has revealed that Mercury, Mars, Mars' two tiny moons Phobos and Deimos, and the satellites of giant Jupiter and Saturn are pockmarked with impact craters. Even cloud-shrouded Venus, as both earth- and space-based radar observations show, bears circular scars, strongly suggesting that the impact process has been at work there as well.

A REVOLUTION IN THOUGHT

That impact is an ongoing process also seems quite certain. As H. H. Nininger has pointed out, twice in the current century the earth has encountered large meteoroids in its annual course around the sun with what, in terms of human experience, at least, can only be considered catastrophic effects.[1] His references, of course, were to the Tunguska and Sikhote-Alin events in central and eastern Siberia in 1908 and 1947, respectively.

More recently, Eugene Shoemaker has stressed the potential for even more devastating impact events represented by the uncounted thousands of small asteroids, cometary nuclei, and other meteoric debris whirling about the sun in none-too-stable orbits that closely approach or actually cross that of the earth. Systematic searches, mainly with an 18-inch Schmidt telescope atop Palomar Mountain in California, have to date turned up about fifty of these objects with diameters of one kilometer (0.62 mile) or more, but Shoemaker estimates that there must be at least a thousand of them altogether.[2] Smaller bodies with similar earth-approaching or earth-crossing orbits—too small to be observed astronomically—are certainly far more numerous.

Using estimated cratering rates for the earth and moon, Shoemaker has calculated the statistical probabilities of a collision between one of these objects and the earth. He finds that an object 0.5 kilometer (0.31 mile) in diameter, capable of producing a crater 10 kilometers (6.2 miles) in diameter, should strike the earth, on the average, once in about 100,000 years, and that the impact of a body 10 kilometers in diameter, which would produce a crater comparable to the larger lunar craters, probably occurs once in about 40 million years. These odds perhaps sound reassuring, but it is, of course, possible that such an event could happen at any time.

Moreover, it is now suggested that the impacts of large extraterrestrial objects have influenced the course of biological evolution on the earth. Nobel laureate physicist Luis Alvarez and his colleagues at the University of California at Berkeley have presented impressive evidence, in the form of an apparently ubiquitous iridium anomaly in a centimeters-thin layer of ancient claystone found in many parts of the world, to argue that the drastic environmental changes wrought by the impact of an asteroid 10 kilometers in diameter were responsible for the extinction of the great dinosaurs and of numerous other lifeforms at the end of the Cretaceous period of geologic time, some 65 million years ago.[3] More recently, this hypothesis has received strong support from the serendipitous discovery of shock-metamorphosed grains of quartz in this same iridium-rich claystone layer which contain traces of the high-pressure polymorph of silica, stishovite, indicative of pressures of 150,000 bars (atmospheres) that could have been produced only by the hypervelocity impact of a massive asteroid.[4] Impact now has also been invoked to explain other episodes of mass extinction that seem to appear periodically in the paleontological record of the earth's rocks.[5]

Given the attitudes that prevailed in Barringer's day, all this amounts to a revolution in scientific thought. This revolution, as we have seen, was already underway at Barringer's death; it was sustained in the 1930s largely by the discovery of a number of other indubitable impact craters in widely separated parts of the earth and by the recognition that certain other enigmatic features, long described as "cryptovolcanic" by geologists, might be explained at least equally as well by impact. By the 1940s a few scientists had begun anew to explore the implications of impact for the lunar craters. The development of atomic weapons in World War II provided them with a new analog, the nuclear explosion crater, and soon sophisticated mathematical methods were devised for comparing these to the crateriform structures both on the earth and the moon. By the mid-1950s the list of known or suspect meteorite craters had been substantially increased, in part by the use of aerial photography by astronomers at Canada's Ottawa Dominion Observatory; also advances in chemistry, physics, and geology had equipped geologists not only with more accurate dating techniques, but with new and more definitive criteria for identifying meteoritic impact structures.

That the seeds of this revolution were firmly rooted by the early 1930s is evident from the immediate acceptance, on the basis of criteria developed by Barringer and others at Meteor Crater, of the meteoritic origin of two clusters of craters discovered at this time in remote desert areas of the earth—the Henbury craters in central Australia and the Wabar craters in the Rub' al-Khali, or "Empty Quarter," of southern Arabia.[6] L. J. Spencer, the keeper of minerals at the British Museum and a leading authority on meteorites, quickly pronounced these to be of impact origin from the meteoritic irons and silica glass, or lechatelierite, that were found associated with them, and his conclusion has not since been questioned.[7] Spencer later determined that the silica glass at the Wabar craters contained up to two million microscopic nickel-iron spherules per cubic centimeter and proposed that these had condensed from a cloud of metallic vapor and had fallen into boiling silica produced from the pure quartzose sand of the desert by the intense heat generated by the impacts.

For Spencer, the Henbury and Wabar craters, along with Meteor Crater and the Odessa, Texas, crater,* brought the list of proven impact sites to four,

---

*Three other small, shallow craters have since been identified as being associated with the Odessa fall.

sufficient to constitute a distinct class of terrestrial topographic features. To this list he now added a fifth example—the Campo del Cielo craters of the Argentine Gran Chaco. These were an undetermined number of shallow, roundish depressions scattered over a wide area of the Argentine pampa where meteoritic irons had been found since 1576. Recent excavations in one of the craters, he noted, had yielded "white volcanic ashes," although there were no volcanic rocks in the vicinity and the nearest Andean volcanoes were some 500 miles to the west, along with a "transparent glass" which he correctly assumed to be silica glass. Some fragments of "rusted" iron and of what Spencer called "typical" iron shale were found in a pit dug into the rim of the crater.

Spencer, however, was skeptical of claims for the meteoritic origin of two other features because they lacked one or more of the necessary criteria for impact structures. These were the craters on the Estonian island of Ösel and the Ashanti crater, 6.5 miles in diameter and containing Lake Botsumtwi, in the jungles of Africa's Gold Coast (now Ghana). These, it may be noted, are now considered to be meteoritic; in 1937 Reinvaldt finally found tiny fragments of metoritic iron at Ösel, and in the early 1960s coesite was identified in the rock at the Ashanti crater.[8] Spencer reserved judgment on the "craters" Kulik had reported at the site of the Tunguska event, noting that photographs suggested that they were simply "a series of small pools in a swamp" that formed seasonally atop the perennially heaving permafrost of the central Siberian taiga. "Unfortunately," he opined, "no good and connected account has yet been given, but sensational reports appear periodically in the newspapers."[9]

In discussing meteorite craters in general, Spencer concluded that they were formed by explosive forces. "It seems therefore that meteorite craters are not merely dents in the ground made by the percussion of a meteorite; but that they are explosion craters due to the sudden vaporization of part of the material, both of the meteorite and of the earth. . . . When a large mass of iron traveling with planetary velocity is suddenly stopped, the kinetic energy ($\frac{1}{2}mv^2$) is transformed into heat at a localized spot with the development of a very high temperature. Simple calculations give very high figures."

Spencer would expand on this point four years later in advocating the impact origin of the craters on the moon.[10] Historically, he pointed out, the objections to this theory had been the circularity of the lunar craters and the fact that the earth itself was not pockmarked with such features. "A study of meteorite craters and of meteorite showers on the earth's surface," he now declared, "shows that both these objections are without justification." The Henbury and Wabar craters in particular were explosion craters, and such craters would "be circular whatever the angle of approach of the projectile."

Spencer conceded that the idea of explosion craters had been "anticipated" by Gifford in 1924 "in a remarkable paper which has only now come to my notice and which appears to have been generally overlooked."

Moreover, although the presence of meteoritic iron was diagnostic of a meterorite crater, the absence of such iron did not necessarily deny an impact origin, Spencer now decided; and he pointed out that 95 percent of all observed meteorite falls involved stony meteorites. On the moon then, he suggested, the rayed craters might have been produced by iron meteorites, their rays consisting of "tiny spheres of shining metal" such as he had found in the silica glass from Wabar, while the far more numerous rayless craters had been formed by the impacts of stony meteorites.

For Spencer, too, the paucity of terrestrial impact craters could be explained easily enough by the erosive forces at work on the earth over time. The few known impact craters on earth were all apparently relatively young, and the evidences of their origin "will be destroyed by weathering and all traces of them obliterated by denudation long before the next geological period."

BOON AND ALBRITTON'S HYPOTHESIS

By this time, however, two American geologists, John D. Boon and Claude C. Albritton, Jr., in a brace of landmark papers, had already proposed on theoretical grounds that the impact of large meteorites would indeed leave long-lasting scars that should be recognizable even after erosion, acting over many millions of years, had removed all surficial evidence of the event, including the crater itself. Criteria for identifying such ancient impacts, they further proposed, should be found in the underlying rocks.[11] "When a large meteorite strikes the earth," they pointed out in 1936, "it deals a terrific blow to a medium which has a limited degree of freedom and a high degree of elasticity of volume. . . . Brittle substances are not shattered by pressure, if the pressure be applied to all sides, but by tension. Hence after compression they all rebound."

> Therefore, as a result of impact and explosion, a series of concentric waves would go out in all directions, forming ring anticlines and synclines. These waves would be strongly damped by the overburden and by friction along joint, bedding, and fault planes. The central zone, completely damped by tension fractures produced in rebound, would become fixed as a structural dome.
>
> The general and simplest type of structure to be expected beneath a

large meteorite crater would, therefore, be a central dome surrounded by a ring syncline and possibly other ring folds, the whole resembling a group of damped waves.

There was evidence for this even at Meteor Crater, they noted, in the radially outward dip of the strata in the walls which "strongly suggests" that the underlying beds were domed. "Evidently the bolide which produced Meteor Crater formed also a faulted, domical structure which is likely to persist as a 'meteorite scar' long after the crater, the ejectamenta, and meteorite fragments have been removed by weathering and erosion."

The appearance of such structures at a given time, then, was dependent upon the depth to which the surrounding country had been eroded away since the impact. "It is only in the initial stage that the crater clearly reflects its origin in the rim of ejected material, silica glass, and meteorite fragments distributed about it," they wrote. With time, the crater would become increasingly inconspicuous, but as denudation progressed still further, the underlying structures would begin to appear, and eventually the central uplift and ring folds would be revealed. Only when erosion in the area reached a level below the deepest rock affected by the impact would all the evidence of the nature of the event be entirely obliterated.

As possible examples of such meteorite scars, Boon and Albritton pointed to a class of structures which had puzzled geologists for more than three decades and which were called "cryptovolcanic," the designation having been first applied to the Steinheim Basin in southern Germany in 1905.[12] These were marked by subcircular domed uplifts composed of chaotically jumbled, brecciated rocks indicative of violent explosive action, but lacking any volcanic materials or evidence of thermal action. Geologist Walter H. Bucher, who had been studying examples of such structures since 1920, had recently suggested that they had been formed by "disturbances produced by the explosive release of gases under high tension, without the extrusion of any magmatic materials, at points where there had previously been no volcanic activity."[13] For Boon and Albritton, however, these structures also seemed to meet the criteria that they had outlined for ancient meteorite scars.

"The cryptovolcanic and meteorite hypotheses are strikingly similar," they noted, "in that each supposes tremendous explosions to have entered into the formation of the structures in question. . . . Either hypothesis is adequate to explain the presence of intensely disturbed, subcircular structures with central domical uplifts surrounded by ring folds resembling damped waves. . . . It appears, however, that the meteoritic hypothesis can account for two features which are unsatisfactorily explained by the alternate mechanism." These were the "distinctly bilateral symmetry" found in several examples of cryptovolcanic structures, and the lack of "volcanic materials and signs of thermal activity" associated with them. It was, they contended, more difficult to explain how an upwardly directed explosion could produce bilateral symmetry that it was to explain how an obliquely impinging meteorite could produce a radially symmetrical structure.

Boon and Albritton applied their criteria to several cryptovolcanic structures, including the so-called Flynn Creek disturbance near Gainesboro, Tennessee, which was "suggestive of a meteorite scar which retains evidence of the original crater"; the Sierra Madera dome in Texas, which "suggests a scar from which the crater and its filling have been completely removed by erosion, so that the underlying structures are revealed"; and the huge Vredefort ring in South Africa, 75 miles in diameter and thus far larger than the lunar crater Copernicus. "Impact and explosion of a gigantic meteorite could account for the salient structural features of the Vredefort area: a great circular dome with a core of non-intrusive granite surrounded by a girdle of tilted and overturned strata." The meteorite hypothesis, they added, could also account for transverse and oblique faults around the margins of the core, which were similar to those at Sierra Madera; the absence of volcanic materials, and "the evidence for the operation of unprecedented pressures in association with updoming."

The Boon and Albritton hypothesis cannot be said to have taken geology by storm; and, indeed, even a half century later many geologists remained skeptical, although the evidence for an impact origin of such structures has steadily mounted while there has been no further elucidation of the volcanic gas explosion explanation advanced by Bucher back in 1933. Bucher himself, in his last paper written shortly before his death in 1965, still favored the gas explosion idea.[14]

Nonetheless, the Boon and Albritton hypothesis has made progress over the years, and many of the cryptovolcanic structures recognized by the mid-1930s, including the type locality of the Steinheim Basin, as well as others discovered since, are now considered to be of impact origin. This has come about primarily because of the development of additional criteria for identifying meteorite craters or their remanent structures over and above those derived from Meteor Crater, or suggested by Boon and Albritton. These criteria, all related to shock stress, are: Shatter cones, which are conical, striated slip surfaces in rock that were first described in 1905 at the Steinheim Basin and are indicative of intense pressure; the high-pressure polymorphs of silica, coesite, and stishovite, which, as noted earlier, were discovered in the laboratory in 1953 and 1961, respectively; and the lamellar deformation of the crystalline structure of quartz under sudden, intense shock, reported in 1965 by Neville L. Carter in rock specimens from Meteor Crater, the Vredefort dome, and the Clearwater Lakes craters in Quebec, Canada.[15] Of these criteria, it may be noted, only indubitable shatter cones have not yet been found at Meteor Crater.[16]

That shatter cones might be indicative of impact was first suggested by geologist Robert S. Dietz in 1946 from his study of these rock forms at the Kentland, Indiana, cryptovolcanic structure. Shatter cones were well known at such structures, but Dietz now recognized that the apexes of the Kentland cones were nearly all pointed upward.[17] "The common orientation of the apexes indicates that the deforming stress was unidirectional," he wrote. "A consideration of the mechanics of this type of deformation shows . . . that the shock was applied from the direction in which the apexes point. Therefore, the orientation of the shatter cones suggests that, assuming that the beds were essentially horizontal prior to deformation, the shock force resulted from some type of explosion above the beds rather than from a cryptovolcanic explosion below the beds." Dietz also suggested that, in view of the uncertainty surrounding these structures, they might better be referred to as "cryptoexplosion" structures, a term that in time found some favor with geologists and was eventually accepted by Bucher himself.[18]

Over the years, Dietz and others have found shatter cones in a number of other cryptoexplosion structures, or "astroblemes" (star scars), as Dietz later called them. In cases where it has been possible to make a determination, the cones have tended to show a "preferred" upward orientation, but it has also been found that the interpretation of these features is complicated by the fact that they may be formed by reflected shock waves resulting from an impact as well.[19] In 1961 notably, Shoemaker and his colleagues demonstrated that shatter cones can indeed be formed by high velocity impact by firing a 3/16-inch aluminum sphere at a velocity of 5.61 kilometers (3.5 miles) per second into a block of sandy dolomite from the Kaibab limestone at Meteor Crater and producing "three easily recognized shatter cones and a number of striated slip surfaces of irregular or less definite form" at the bottom of the resulting scar.[20]

But shatter coning, as Dietz himself stressed in 1968, is not in itself definitive of meteoritic impact and "still remains only an indication of shock; and hence of impact only by inference." The same objection has been raised against coesite, and some other effects of shock metamorphism, on the contention that the shock might be produced by some other mechanism. What this mechanism might be, however, has remained a mystery, for volcanism is not known to generate the pressures required to produce these features and forms, and Dietz has pointed out that a viable alternative to impact is "unlikely" as it must necessarily involve "some intraterrestrial explosion triggered by a geologic process of an as yet unspecified nature."[21]

The classification of terrestrial features as known or probable meteorite craters or structures, however, does not depend upon the presence of one criterion alone, but on the cumulative evidence of a multiplicity of them; the more criteria there are, the more certain it is that a specific crater or structure is of meteoritic origin. In 1949, on the basis of the criteria then available, Baldwin listed ten craters or groups of craters known to be of meteoritic origin, and a dozen others as probably having been formed by impact. Of these latter, he noted that impact theory "seems to fit the observed facts better than any other. It alone explains the bilateral symmetry so often found, and it alone seems capable of supplying the vast amounts of energy which are needed to give the observed results."[22] By 1963, after coesite and stishovite had been discovered, his list had grown to forty-eight.[23] Twenty years later, Shoemaker put the total at nearly eighty, noting that a large majority of these were larger than three kilometers (1.86 miles) in diameter.[24] Most of them, too, had been found in North America, western Europe, the Soviet Union, and Australia, areas which "constitute the major portion of the world most thoroughly investigated by geologists." The search, of course, goes on.

BEYOND THE EARTH

The steadily growing list of known or probable impact craters on earth during these years also provided a broader base for extending the impact analogy beyond the earth to the moon. In 1930 the volcanic analogy for the origin of the lunar craters still held its own among selenographers and selenologists as well as those astronomers who considered the problems of the moon at all; but it was being severely challenged. Even Walter Goodacre, long the dean of English selenographers, conceded as much in his book *The Moon,* published in 1931. "Until quite recent times," he wrote, "the cause which has led to the present appearance of the lunar surface was universally ascribed to volcanic energy, but now a rival theory has appeared, by which it is held that the impact of meteoric bodies upon the once plastic surface, sufficiently explains what we see, and quite a number of protagonists who hold this view or modifications of it, have set forth their reasons for its acceptance."[25] But Goodacre then proceeded to attack the impact idea not as conceived by Ives, Gifford, or even Barringer, but by Proctor, Shaler, and See, and in terms of the traditional volcanic analogy.

In 1937, as already noted, Spencer lent his considerable prestige to the idea of an impact origin for the lunar features, stressing that their circularity was explainable by explosive forces. Nininger, while not disputing Spencer's major thesis, took issue with him on this latter point. He not only protested Spencer's exclusion of the oval-shaped Haviland, Kansas, crater, which Nininger had earlier excavated, from his list of terrestrial meteorite craters; but he further contended that not all meteoritic craters need be either circular or formed by an explosion. Some meteorites, he argued, must approach the earth at a very low angle and eventually strike at a comparatively low velocity, and thus "we may have a wide range of applicable results. We may conceive of a situation in which a very large mass would produce no explosion at all."[26]

In 1938, too, geologist Herman L. Fairchild, one of the first to support Barringer's claim of an impact origin for Meteor Crater, advocated an impact origin not only for the lunar craters but for the moon itself, and all other solar system bodies as well, in a lead article in *Science*. "The basins and pittings on the lunar surface are impact craters and are ocular confirmation of the view that the planets and satellites were built by cold accretion," he declared. "This implies acceptance of the Planetesimal hypothesis of Chamberlin and Moulton."[27] Like Spencer, he also pointed out that the great majority of observed falls involved stony meteorites, and that this would be the case on the moon.

By 1941 the impact theory of lunar origins had made such progress that Roy K. Marshall of Philadelphia's Franklin Institute felt obliged to reassert the volcanic analogy in a brief paper read before the sixty-sixth annual meeting of the American Astronomical Society at Williams Bay, Wisconsin. "Despite the evidence which has led most professional astronomers to conclude that the lunar formations are the product of igneous activity," he noted, "textbooks and popular treatments devote so much space to the meteoric theory that it is only by strong arguments that the volcanic theory can be kept in the field." Marshall contended that if the impact theory was valid, then impact phenomena should be observed with relative frequency on the moon and that the presence of "laccolith-like formations" on the moon "lends credibility to the suggestion of the late H. G. Tomkins that a slow igneous formation and not explosive vulcanism is responsible for the formation of the craters."[28]

But almost immediately, Ralph B. Baldwin, soon to become a leading figure in the impact revolution, disputed Marshall's contentions in two articles published in 1942 and 1943 advocating the impact origin of the lunar features.[29] Baldwin, then at the Dearborn Observatory in Illinois, noted that the Tomkins-Marshall hypothesis "does not explain the existence of a central mountain peak, unless later vulcanism built up the peak within the crater formed by the laccolith, nor the great quantities of material ejected radially from so many craters. Each is an effect to be expected if the craters are of meteoritic origin," he declared. And after discussing some specific lunar features in this context, he proceeded to reconstruct a portion of the moon's probable history.

"We shall assume, in accordance with a widely accepted hypothesis, that the moon was once part of the earth. When it broke away there was undoubtedly a good deal of material from the fission which did not immediately coalesce with either major body. The moon very quickly formed a solid crust. Its gravitation, acting like a gigantic vacuum sweeper, gradually caused numerous small bodies to plunge to the lunar surface. These impacts were the sources of the craters."

Subsequently, the vast Mare Imbrium was formed, probably by a body "of asteroidal dimensions" and "of the order of 75 miles in diameter. Upon impact a tremendous amount of rock would have been liquified. In addition, the surface layers of the moon would have been smashed so thoroughly as to expose lower layers which, presumably, were still molten. Hydrostatic pressure alone, from the weight of the newly arrived material as well as that from the neighboring lava level, forced molten rock to the surface where it practically filled up Mare Imbrium and then burst its bonds" to spill over into other maria. Many other craters were formed later as their superposition on these lavas showed.

Volcanism, Baldwin added, played a role in all this, but only a minor one. "Volcanic action alone probably caused some of the smaller craters," and "Perhaps some of the central mountain peaks in craters are to be considered as volcanic. The great majority of lunar craters, however, were not formed by this method, but were produced by a violent explosion attendant upon meteoritic action."

Marshall now returned to the fray, publishing an expanded version of his earlier paper to refute Baldwin in his turn, again complaining that in "textbooks and popular discussions there is a degree of preference, or at least a lack of dislike for the meteoric impact hypothesis. . . . We find in textbooks the most amazing statements which attempt circuitously to destroy any confidence in an igneous theory at the very beginning."[30] His citations here were to Moulton's *Astronomy* (1931), H. Spencer Jones's *General Astronomy* (1923), and Robert H. Baker's *Astronomy: An Introduction* (1930). He again argued the failure to observe actual impacts on the moon, and now added the shallowness of the lunar craters to his case. To make this point, he invoked Meteor Crater, declaring that its depth-to-diameter ratio of about 1:7 "makes this object quite different from any except perhaps the smallest lunar craters. At all events," he added, "it is intuitively very difficult to imagine an impact crater which is 100 miles or more in diameter, yet only two miles deep."

Marshall also discussed briefly the question of a lunar atmosphere, citing William Henry Pickering and A. C. D. Crommelin and opining that "it is certainly no stretch of the possibilities that the Moon is possessed of an atmosphere which is capable of destroying meteoroidal bodies which enter it." And he again turned back to Tomkins, to laccolith-like structures on the moon, and to Wargentin and several other craters more or less filled to their brims with lava. "Until the laccoliths and Wargentin-like craters are otherwise accounted for, it would seem that we must consider the igneous theory of the origin of the lunar craters, as outlined by Tomkins, to be the true one. The hypothesis of meteoroidal impact must be considered seriously challenged because of the failure to observe impact phenomena on the Moon today, probably due to the presence of a thin, but effective lunar atmosphere."

On this latter point, parenthetically, there is an intriguing account of an extraordinary observation of the moon on June 18, 1178, that is highly suggestive of a meteoritic impact. It was recorded by the medieval chronicler Gervase of Canterbury and reported in 1976 by planetary scientist Jack Hartung. The observation was made, Gervase wrote, "after sunset when the moon first became visible" and the "marvelous phenomenon was witnessed by some five or more men who were sitting there facing the moon." The moon was new, he noted, "and suddenly the upper horn split in two. From the midpoint of this division a flaming torch sprang up, spewing out, over a considerable distance, fire, hot coals, and sparks. Meanwhile the body of the moon which was below writhed, as it were, in anxiety, and to put it in the words of those who reported it to me and saw it with their own eyes, the moon throbbed like a wounded snake."

This phenomenon was "repeated a dozen times or more, the flame assuming various twisted shapes at random and then returning to normal. Then after these transformations the moon from horn to horn, that is along its whole length, took on a blackish appearance." His informants, Gervase added, were "prepared to stake their honor on an oath that they have made no addition or falsification in the above narrative."

From this description, Hartung estimated the lunar latitude and longitude of the event and then examined the area on photographs made by lunar orbiting spacecraft. He found an "impressive" crater 20 kilometers in diameter with a very prominent ray system indicative of its recent origin—the lunar crater Giordana Bruno. "We concluded that this crater was the one formed on June 18, 1178," he wrote.[31]

In 1944 the volcanic analogy, in a somewhat more traditional form than the Tomkins-Marshall version, was again advanced in two monographs by Josiah Edward Spurr, whose long career included not only the editorship of the *Engineering and Mining Journal,* but service as geologist to the Sultan of Turkey, the U.S. Geological Survey, and as a consulting economic geologist.[32] These contained the results of his exhaustive investigation of the Mare Imbrium and various other regions based on visual observations of the moon and a detailed study of telescopic photographs—sometimes, apparently, more detailed than their resolution would allow.[33] In summarizing his findings, he declared that the surface of the moon "displays phenomena which are related to phenomena we describe as volcanicity on the earth," a conclusion which, he noted, "of course, has been the dominant one offered by astronomers, both earlier and at the present time."

Spurr mentioned the impact hypothesis only in brief addenda to his monographs, and reluctantly, for "I have encountered nothing that would lead me to interpret any feature as the result of meteoric infall. . . . The inclination to explain the surface of the moon in the simple and picturesque terms of meteoric accumulation doubtless had one of its main stimuli in Coon Butte in Arizona, interpreted, no doubt justly, as due to the fall of a great meteor. However, if meteoric bodies had been an important factor in geologic history we should find a record of it in the rock strata accumulated during a billion years. There is no such record" and the "argument that the record of abundant meteors in the past would be obliterated by shattering, disintegration, and oxidation is not sound. . . ."

Spurr took issue particularly with Gilbert and Elihu Thomson that the central hill in lunar craters "is entire," as Gilbert had put it, and "shows no volcanic vent," as Thomson had written, citing rather Campbell's statement that the presence of "craterlets in the summits of central crater peaks is absolutely fatal to the impact theory." In an addendum to his second monograph, he again found "no occasion to mention again the meteor or bolide theory," but he disputed Marshall's contention that meteors would be consumed by a lunar atmosphere. "The general evidence," he noted, "indicates that the moon has no atmosphere."

But now, another geologist entered the controversy over the lunar craters. Dietz, in 1946, in proposing shatter cones as a criterion for terrestrial meteorite craters, had done so in the course of suggesting that cryptovolcanic, or cryptoexplosion structures were possibly related to the craters on the moon. That same year, he published perhaps the most comprehensive summary up to that time of the arguments in favor of the meteorite origin of the lunar features and against the volcanic analogy.[34] He made several pertinent points. Gilbert, he noted, had been obliged to postulate a terrestrial ring to assure that his "moonlets" would strike the moon at a high angle of incidence, but this was unnecessary "because an explosion crater, in contrast to a percussion

41 P. L. Webster

42 Percy W. Bridgman

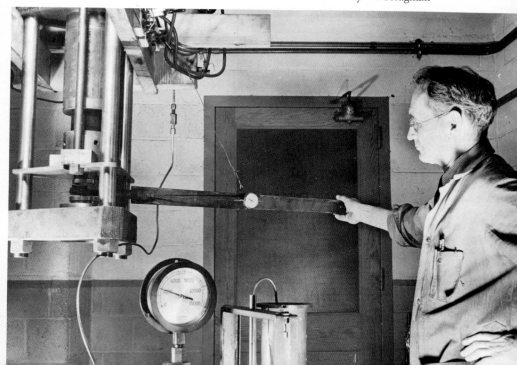

43 H. H. Nininger

44 Eugene M. Shoemaker

Meteorite approaches ground at 15 km/sec.

Meteorite enters ground, compressing and fusing rocks ahead and flattening by compression and by lateral flow. Shock into meteorite reaches back side of meteorite.

Rarefaction wave is reflected back through meteorite, and meteorite is decompressed, but still moves at about 5 km/sec into ground. Most of energy has been transferred to compressed fused rock ahead of meteorite.

Compressed slug of fused rock and trailing meteorite are deflected laterally along the path of penetration. Meteorite becomes liner of transient cavity.

Shock propagates away from cavity, cavity expands, and fused and strongly shocked rock and meteoritic material are shot out in the moving mass behind the shock front.

Shell of breccia with mixed fragments and dispersed fused rock and meteoritic material is formed around cavity. Shock is reflected as rarefaction wave from surface of ground and momentum is trapped in material above cavity.

Shock and reflected rarefaction reach limit at which beds will be overturned. Material behind rarefaction is thrown out along ballistic trajectories.

Fragments thrown out of crater maintain approximate relative positions except for material thrown to great height. Shell of breccia with mixed meteoritic material and fused rock is sheared out along walls of crater; upper part of mixed breccia is ejected.

Fragments thrown out along low trajectories land and become stacked in an order inverted from the order in which they were ejected. Mixed breccia along walls of crater slumps back toward center of crater. Fragments thrown to great height shower down to form layer of mixed debris.

crater, has a circular shape and well developed radial symmetry." However, he added, "Such an explosion crater may be somewhat modified by accompanying percussion effects; for example, a circular shape might be altered to a slightly oval shape and the radial symmetry to a slightly bilateral symmetry." Moreover, as an impacting meteorite traveled much faster than the shock waves it produced in rock "there is little pushing aside of material in front of the impinging body" and thus, as Boon and Albritton had suggested, "instead of deforming plastically, rock becomes highly compressed and then 'backfires' elastically," forming a "damped wave structure, consisting of a central dome with a surrounding ring graben. . . ."

As to craterlets on the summits of central peaks in lunar craters, he pointed to the crater Timocharis, which appeared to have such a hilltop craterlet. "Yet," he noted, "if one examines a photograph of this same feature with the light falling from the opposite direction, no craterlet is seen, so that the supposed crater may be the shadow cast by an isolated peak. Certain other central prominences appear to have a group of isolated peaks with a saddle or possibly a slight depression or 'pseudo-crater,' which might be misinterpreted as a true crater." The ray systems of the lunar craters, he also noted, "may be finely pulverized rock, meteoritic fragments, and solidified, recrystallized, or glassy material ejected in a molten mass," similar to the small, drop-shaped tektites found in strewn fields in many areas of the earth. These, he added, contained lechatelierite or fused quartz "which requires too high a formation temperature to be a product of volcanism, although such material is produced by meteoritic impacts." Finally, he argued, aside from a dearth of volcanic features on the moon, the lunar surface "lacks features resulting from large-scale folding, faulting and intrusive activity which indicate crustal instability and usually accompany volcanism."

BALDWIN'S "MANIFESTO"

In 1949 Baldwin published an exhaustive study of the lunar surface features which not only broke new ground, but clearly influenced the course of future investigations of the problem.[35] His thesis was the explosive meteoritic impact origin of the craters on the moon; and in marshaling his arguments, he also built up a strong case against the volcanic analogy. His methods were analogical and statistical.

He began with a brief survey of the history of the problem, a description of the lunar surface features, and a summary of the major hypotheses that had been advanced to explain them. He then discussed the known terrestrial meteorite craters, i.e., those where associated nickel-iron fragments had been found, and "fossil meteorite craters," i.e., cryptovolcanic structures, and proceeded to analyze their dimensions vis-à-vis manmade explosion craters and the craters on the moon. "If definite relationships are found," he pointed out, "one may reasonably expect that these structures were each formed by variant applications of the same forces."

Baldwin's data on manmade explosion craters came largely from the "tremendous amounts of knowledge" accumulated in World War II "concerning the properties of mortar shells and artillery shells and bombs." He drew his data on the lunar craters primarily from measurements made by half-a-dozen early selenographers, going back to Beer and Mädler, compiled in 1931 by Thomas L. MacDonald, an English observer and a Fellow of the Royal Astronomical Society, and published in a series of papers under the general heading of "Studies in Lunar Statistics."[36] MacDonald had divided the lunar features into two main types—normal craters and walled plains—each further subdivided on the basis of whether they occurred in the "continental" areas or the lunar maria, or "seas." For each of these groups, he derived equations relating diameter to depth, and to height of the rim above the surrounding plain.

Expanding on this, Baldwin added manmade explosion craters and known meteorite impact structures to his analysis and found that the logarithmic ratios between the diameters and depths of all these various craters could be represented by a quadratic equation which, when solved for each individual crater and plotted on a graph, yielded a smooth curve showing a progression from the small pits made by mortal shells, through larger shell, bomb, and mine craters, up to the lunar craters. Terrestrial meteorite craters fell neatly on this curve and served to fill the gap between the largest manmade craters and the smallest observable craters on the moon.

"The only reasonable interpretation of this curve," he declared, "is that the craters of the moon, vast and small, form a continuous sequence of explosion pits, each having been dug by a single blast. No available source of sufficient energy is known other than that carried by meteorites." The observed relationship, he added, is "clearly of the type to be expected for explosion pits. It can also be shown that there exist other correlations between crater forms and dimensions which may be extended from manmade craters through terrestrial meteoritic craters and the lunar craters. . . . The craters of the moon fulfill every logical extrapolation of the known explosion pits and the terrestrial meteoritic craters and cannot be correlated successfully with any known type of vulcanism. The case for the explosive origin of the moon's craters," he concluded, "is unassailable. The probability is very great that the explosions were caused by the impact and sudden halting of large meteorites."

Baldwin found the volcanic analogy wanting on every score. There was simply "no similarity" between terrestrial volcanoes and the lunar craters and the "divergences are so great as to make it certain" that the lunar craters were not collapse calderas either. "Therefore it is apparent that further to pursue the will-o'-the-wisp of a dominant-process lunar volcanism of the type of any known terrestrial vulcanism is futile. . . . To claim that the moon's craters are volcanic is tantamount to postulating an entirely new, entirely hypothetical mode of origin and to fly in the face of the fact that a known process is completely able to explain the vast majority of the observed lunar features."

Baldwin also calculated the diameters of meteorites required to form various lunar craters, basing his computations on energy requirements for man-made explosion craters, and lunar crater volumes, and finding that the resulting dimensions were "rather surprisingly small. . . . The calculated meteoritic diameters bring forcibly to mind such names as Apollo, Hermes, Amor, Adonis, and Albert; names of tiny asteroids which in recent years have paid fleeting visits to our neighborhood; tiny asteroids, each of which could, in some future year, entirely devastate an American state or a European country; tiny asteroids which might wipe out local species of flora and fauna. Sudden disappearances of long-established groups of contemporary life have been reported in past geologic history," he noted. "Is it not possible that the causes of these occurrences were meteoritic impacts?" Finally, it may be added, Baldwin confidently suggested that impact craters should exist on the surfaces of the planets Mercury and Mars.

Like others before him, going back more than a century to Gruithuisen, Baldwin outlined his conception of the impact process, a conception quite reminiscent of the ideas that Moulton privately advanced in 1929.

> At first, the meteorite would plunge into the earth, moving faster than the shock waves and pushing ahead of it an ever increasing plug of compressed rock and probably a similar plug of compressed air. When the speed of the meteorite becomes less than that of the elastic waves, the vast amount of compression produced finds a shoulder against which to push, and the mass is soon stopped.
>
> As long as the velocity is greater than that of the shock waves, very little heat is generated, as heat is a measure of the random motion of molecules and atoms, and these random motions are temporarily stopped during the initial phases of the impact. . . . It is only after the velocity drops below that of the shock waves that the phenomena of heat enters the picture. . . .

With the stoppage of motion, the meteorite is sitting on top of a tremendously compressed, tremendously hot plug of matter. Naturally, an explosion of the utmost violence follows.

Baldwin's monograph was, in effect, the "manifesto" of the impact revolution, and its influence is evident in the research subsequently reported by a host of investigators. The Rand Corporation's J. J. Gilvarry and J. E. Hill, for example, in offering confirmation that "bodies impinging on the moon with astronomically possible velocities actually are subjected to temperatures and pressures of explosive magnitude," declared in 1956 that Baldwin "has put the hypothesis on a strong foundation. . . . The prescription, within a certain error, of diameter as a function of depth for all these craters by one analytic function, corresponding to the Baldwin curve, leaves little doubt of the common genesis in explosion. For the terrestrial meteoritic and the lunar craters, the explosion arises from the high pressures and temperatures behind the shock waves generated on meteoritic impact."[37]

SHOEMAKER'S CHALLENGE

Baldwin's thesis that impact craters are explosion craters has had, as we have seen, a long history dating back to the early decades of the twentieth century. Unquestionably, the explosion idea was primarily responsible for the growing acceptance over the intervening years of impact as a structure-forming process on both the earth and the moon. Nonetheless, the explosion idea has been challenged by Shoemaker who, since 1959, has been in the forefront of the continuing study of terrestrial and extraterrestrial impact phenomena. The shock of impact alone, acting through purely mechanical processes, is sufficient to form a crater, he has suggested.

"In the case of meteorite impact at sufficiently high velocity, the meteorite will penetrate the ground by a complex mechanism which may be described in terms of three phenomena, 1) compression of the target rocks and meteorite by shock, 2) hydrodynamic flow of the compressed material, and 3) dispersal of the meteoritic material in the flowing mass. These phenomena occur simultaneously or overlap each other in time, but their relative importance changes during the course of the penetration process."[38]

Shoemaker has outlined the mechanics and developed the mathematics of this process in several publications since 1959, and a more detailed description is not needed here; his diagrammatic sketches of his hypothesized sequence of events are shown following page 356. In describing these sketches, he emphasizes shock propagation and its ground surface reflected rarefaction together with hydrodynamic flow of material as crater-forming mechanisms, rather than an explosion which implies vaporization. His conclusions were based primarily on his own exhaustive investigation of Meteor Crater, begun in 1957, and on "scaling" from the 1.2-kiloton Teapot Ess underground nuclear explosion crater produced in 1955 at the Nevada test site. The Teapot Ess test, he found, reproduced nearly all of the major structural features of the Arizona crater.

In 1961, in a key paper discussing the lunar craters as impact structures, Shoemaker elaborated on the idea that an explosion, as such, is not involved in the formation of meteorite craters.[39] "From the nuclear explosion craters it can be seen that the surface and structural features of impact craters can be produced by strong shocks originating at shallow depths . . . and also that the structures formed depend upon the depth of the origin of the shock. . . . It may be anticipated that the structure of large meteorite craters will depend upon the depth of penetration of the meteorite, the total energy released, and the nature of the target rocks."

But, he added, it would be "erroneous" to conclude from the structural similarities to explosion craters "that meteorite craters are produced by the explosion of the meteorite. . . . So widely has this concept been described that craters of inferred high-velocity impact origin are now commonly referred to as explosion craters. In the form developed by Ives and Gifford, the concept of explosion of the meteorite was derived by computing the specific kinetic energy of the meteorite travelling in the known range of geocentric velocities of meteors and equating this kinetic energy to specific internal energy in the meteorite at the moment of impact. The specific internal energy was thus found to exceed the enthalpy of vaporization for any solid at atmospheric pressure, and it was concluded that the meteorite would explode."

The error in this, Shoemaker contended,

> lies in the neglect of the partition of energy in the shock waves generated by the impact and in the neglect of the equations of state of the shocked materials. Very high specific energies are produced by hypervelocity impact, but these are the consequence rather than the cause of the shocks which produce the craters. In fact, the fraction of energy which is retained thermally by material engulfed by the shock is unavailable for further propagation of the shock. For the same total energy, the higher the initial specific energy the smaller will be the crater. Vaporization of the meteorite

or target rocks would not, therefore, facilitate the opening up of an impact crater, except possibly where the rocks are rich in volatile constituents.

Indeed, only a small fraction of the kinetic energy of the meteorite remains in the meteorite itself, the major part being transferred to the shocked rock ahead of it, and only a fraction of this remanent internal energy is trapped thermally, the rest being released by expansion of the meteorite behind the rarefaction wave to contribute to the further propagation of the shock into the rock. Thus, there is little heat available for vaporization.

There is, indeed, a touch of semantics in all this, as Shoemaker himself has conceded. "We may properly speak of meteorite impact craters as explosion craters in the sense that materials fly out of the crater. But in this sense the pits formed by raindrops on soft mud are also explosion craters. It would be better if the term 'explosion' were dropped with reference to hypervelocity impact mechanics." In short, while no true explosion was involved, the effect was nonetheless explosive.

Baldwin, in expanding in 1963 on his earlier thesis, granted Shoemaker one point.[40] "In effect, Shoemaker is saying that the production of a hypervelocity crater differs from an ordinary explosion crater in the sense that most of the velocity is imparted to the rocks in the former case by the shock mechanism rather than by the production of a great amount of vapor. This is very probably true." But he took issue with Shoemaker's statement that "No evidence has been recognized . . . by which it could be shown that more than a small fraction of the meteorite or rocks at Meteor Crater ever behaved as vapor." Baldwin here pointed to the samplings of the soil around Meteor Crater by Nininger and Rinehart which "indicate that at least 12,000–15,000 and probably considerably more tons of meteoritic materials were vaporized, then condensed, and *still exist* in the general neighborhood of the crater in the form of tiny spherical drops [Baldwin's emphasis]. How much more was immediately oxidized or was condensed into droplets too small to fall near the crater or has been oxidized in the thousands of years which have elapsed is not known, but in all probability the amount was very large."

· Moreover, he declared, the "percentage of nickel-iron still remaining as droplets beneath the crater must be considered to be a relatively small part of the original body. Droplets of this type have been found at the Wabar craters and at Odessa and others. It is extremely doubtful that all these objects struck within such a narrow velocity range that the impinging mass liquified but did not vaporize. The nature of the droplets and their wide distribution clearly point to the fact that a moderately high-velocity meteorite will almost completely vaporize and that the craters produced by such impacts owe much of their size to a true explosive action supplemented by a shock mechanism of ejection."

Shoemaker, however, disagrees and argues on the basis of laboratory studies by Survey colleagues of the spheroids at Meteor Crater and of other material brought up from the early drill holes and shafts,[41] that the spheroids or "droplets" are a result of shock-melting rather than condensation from a vapor. Moreover, he suggests that possibly up to one-third of the initial mass of the meteorite, or as much as 100,000 tons, remains buried in the crater in the form of these tiny droplets embedded in shock-melted rock forming a "liner" up to fifty feet thick coating the bottom and lower walls of the crater.[42] The controversies over Coon Mountain, it seems, are not entirely over yet, and the research goes on.

If there is disagreement and uncertainty surrounding the precise nature of the impact process, there is no longer doubt that impact has been and is of fundamental importance in the cosmic scheme of things. What to Proctor was a "wild and fanciful" idea in 1873 became by 1973, when the historic Apollo manned landing missions to the moon were terminated, a tenet in planetary science. This, it has been suggested here, amounts to a revolution.

But it is a revolution only in the sense that an idea once scorned and ridiculed has been accepted into the body of scientific thought. It is a revolution, indeed, which did not depose or dispose, but rather modified and expanded the conventional wisdom of sience. If impact is now admitted as a cosmic process, vulcanism, while yielding its former position of dominance, is still admitted at court, so to speak, and retains a place in the hierarchy of concepts that govern the continuing efforts of scientists to understand the origin and evolution of the solar system.

This revolution seems to have reached its peak in the 1950s and early 1960s when the impact concept became, as Baldwin would later note, "a glittering toy, and many played with it."[43] During these years, the scientists who were concerned at all with such matters now directed their researches primarily to the elucidation and elaboration of the impact idea, rather than to the validity of the idea itself. This subtle but significant shift in emphasis developed progressively after 1906 when Barringer and Tilghman compiled and published the first hard physical evidence that a massive meteorite impact had actually occurred on the earth. Notably, Fairchild, Merrill, and the few others who investigated the evidence at Coon Mountain for themselves quickly turned their attention to the corollary problems of the nature, dimensions, and disposition of the impacting mass.

In 1957, with Sputnik and the advent of the space age, the problems of lunar origins in particular were catapulted abruptly into the mainstream of science. Thousands of scientists began to work on problems that for many years had been the concern of only a very few. The moon itself, for a few years at least, came under the most intense, systematic telescopic scrutiny in history. The U.S. Air Force's Aeronautical Chart and Information Center (ACIC), using telescopes at the Lowell Observatory in Flagstaff, began producing the most detailed and accurate lunar maps ever made to be used in this new exploration of the moon. A new discipline—planetary science—embracing scientists from many fields, quickly began to evolve and in a few years became institutionalized, notably as separate divisions of the American Astronomical Society and the Geological Society of America, but also in independent departments or branches in many of the larger research-oriented universities and institutions of the world. The U.S. Geological Survey organized its Branch of Astrogeology in 1959, in part as a result of Shoemaker's work at Meteor Crater and Germany's Rieskessel and his insistence that much could be learned about terrestrial geology from the study of the geology of the moon and the planets.

In 1959, and for ten years thereafter, a remarkable series of unmanned spacecraft missions to the moon—Russia's Lunas and Zonds and America's Rangers, Lunar Orbiters, and Surveyors—produced massive amounts of data for this vastly expanded research effort, including hundreds of thousands of photographs of the moon made from near-moon space and later from the surface of the moon itself. In 1964 the Mariner 4 spacecraft flyby of the planet Mars returned the first close-up photographs of the Martian surface which revealed, to the surprise of not a few, a number of large, lunarlike craters. Ten years later, imagery from the Mariner 10 spacecraft probe of the inner planets showed that the planet Mercury, like the moon, is pitted with crateriform structures.

Such data, however, differed from that obtained through earth-based observations only in degree, and thus also proved to be amenable to varied interpretations; impact theorists most certainly found in them what they were seeking, but the vulcanists also gleaned new grist for their mill. Several events gave the vulcanists encouragement during these years. In 1958 the Russian astronomer N. A. Kozyrev visually and spectroscopically observed what appeared to be an emission of hydrogen, and perhaps some other gases, from the slopes of the central peak in the lunar crater Alphonsus, and Kozyrev himself and others took this to be a manifestation of lunar volcanic activity.[44] More widely publicized, perhaps, were the observations of ACIC moon mappers James Greenacre and Edward Barr at the Lowell Observatory in the fall of 1963. Twice in one month they found a portion of the rim of the bright lunar crater Aristarchus and its immediate environs glowing ruby red, the display lasting for almost two hours before the color faded back into the normal chiaroscuro aspect of the lunar surface. These observations were claimed by some to be indicative of volcanic action, although Greenacre and Barr, as well as several Lowell astronomers who also witnessed the phenomenon, remained noncommittal.[45]

The resurgence of a volcanic explanation for the craters and other surface features on the moon, perhaps, came to a head in 1965 at a symposium of scientists from many nations of the world convened by the New York Academy of Sciences to discuss what was known, or what was thought to be known, about the moon. Here the vulcanists and the impact theorists both had their say, neither seemingly persuading the other to their point of view. But concessions were indeed made. Baldwin, for instance, argued strongly that there were phenomena enough on the moon to support both analogies. "The key point here," he declared, "is that those formations which can best be explained as of igneous nature should be so considered. Those objects which can best be explained as of meteoritic impact origin must be considered to have an external origin. The great majority of lunar craters cannot be anything but impact pits. The statistics of the various phases of the problem demand this. More problems would be raised on earth, in nearby space, and on the moon by denying this than could possibly be solved by claiming the lunar craters to be primarily volcanic. Conversely, a great many lunar volcanic and igneous processes clearly have operated on the moon in major fashion."[46]

The first hard physical evidence for the operation of impact on the moon, and thus elsewhere in the solar system beyond the earth itself, did not become available for another four years, until July 20, 1969, when Apollo 11 astronauts Neil Armstrong and Edwin Aldrin became the first men to set foot on the lunar surface. Exhaustive analyses of the samples of lunar rocks and soils that they and their counterparts in five subsequent Apollo lunar landings returned to earth left no doubt that impact had played a dominant role in bringing the moon to its present condition.[47] Yet these same samples also showed that vulcanism, although not perhaps in the oversimplified form of the long-entrenched volcanic analogy, has also been operative through lunar history.

The case for impact on other planets and satellites in the solar system is still founded largely on analogies drawn from known terrestrial meteorite craters, and preeminently Meteor Crater, and the craters on the moon. Ironically, perhaps, in view of all that has been said above, the case for vulcanism is on a somewhat firmer basis. Late in 1969 the Mariner 9 spacecraft arrived in orbit around Mars and over the next two years telemetered thousands of high-resolution photographs of the planet's surface back to earth for study by planetary scientists. These revealed, along with many undisputed impact craters, a number of huge shield volcanoes, long inactive, analogous to Hawaii's Kilauea volcano, and including the 17-mile-high Olympus Mons, the largest volcano yet known in the solar system. A decade later, the spectacular Voyager I and II missions to Jupiter, Saturn and beyond sent back imagery recording volcanic eruptions actually in progress on Jupiter's Galilean satellite, Io. These are the first active volcanoes known anywhere in the solar system beyond the earth itself.

Nonetheless the impact revolution, if not a total revolution, has changed the course of scientific thought, and thus meets the principal criterion for revolutions as such. That extraterrestrial masses large enough to form vast craters could impact on the earth or any other solar system body was, at the turn of this century, an incredible idea; today, meteoritic impact is widely recognized as a fundamental cosmic process.

NOTES TO THE CHAPTERS

INTRODUCTION

1. W. G. Hoyt, "Meteor Crater: Historical Note on Nomenclature," *Meteoritics*, 18(1983):159–63.

2. B. Barringer, "Daniel Moreau Barringer (1860–1929) and His Crater (The Beginning of the Crater Branch of Meteoritics)," *ibid.*, 2(1964):183–99.

3. D. M. Barringer, "Coon Mountain and Its Crater," *Proceedings of the Academy of Natural Sciences of Philadelphia*, 57(1906):861–86.

4. H. L. Fairchild, "The Origin of Meteor Crater (Coon Butte), Arizona," *Bulletin of the Geological Society of America*, 18(1907):493–504.

5. Decision Lists Nos. 4607–4609, U.S. Department of Interior Board of Geographical Names, July–September, 1946.

6. H. L. Fairchild, *op. cit.*

7. M. S. Wilkinson and S. P. Worden, "On Egregious Theories—The Tunguska Event," *Quarterly Journal of the Royal Astronomical Society*, 19 (1978):282–89.

8. H. H. Nininger, "Geological Significance of Meteorites," *American Journal of Science*, 246(1948):101–08.

9. L. W. Alvarez, W. Alvarez, F. Asaro and H. V. Michel, "Extraterrestrial Cause for the Cretaceous-Tertiary Extinction," *Science*, 208(1980):1095–1108; and R. Ganapathy, "Evidence for a major meteorite impact on the Earth 34 million years ago: Implication for Eocene extinctions," *ibid.*, 216(1982):885–86.

10. G. K. Gilbert, "The Moon's Face; a Study of the Origin of Its Features," *Bulletin of the Philosophical Society of Washington,* 12(1893):241–92.

11. G. K. Gilbert, "The Origin of Hypotheses, Illustrated by the Discussion of a Topographic Problem," *Science,* n.s., 3(1896):1–13.

12. B. Barringer, *op. cit.*

CHAPTER ONE. THE CRATERS OF THE MOON

Epigraph: W. Herschel, "An account of three volcanoes in the Moon," *Philosophical Transactions of the Royal Society,* 77(1787):229–32.

1. This definition is still emphasized in many modern dictionaries; see *Webster's New Twentieth Century Dictionary,* unabridged, 1976.

2. B. M. Middlehurst and J. M. Burley, "Chronological Listing of Lunar Events," National Aeronautics and Space Administration #X-641-66-176, Goddard Space Flight Center, Greenbelt, Md., 1966.

3. G. Galilei, *The Starry Messenger (Siderius Nuncius),* in *The Discoveries and Opinions of Galileo,* trans. S. Drake (Garden City: Doubleday Anchor Books, 1957), pp. 31–37.

4. G. Galilei, *Dialogue Concerning the Two Chief World Systems,* trans. S. Drake (Berkeley: University of California Press, 1953), p. 63.

5. Z. Kopal and R. W. Carder, *Mapping the Moon* (Dordrecht-Holland: D. Reidel Publishing Co., 1974).

6. Robert Hooke, *Micrographia* (London: J. Martyn and J. Allestry, 1665). A facsimile reproduction was published in 1961 by Dover Publications of New York. The spelling and typographic style in the original have been modernized.

7. R. Grant, *History of Physical Astronomy* (London: Henry G. Bohn, 1852), pp. 44–46.

8. J. H. Schröter, *Selenotopographische Fragmente* (Lilienthal, 1791, Göttingen, 1802).

9. W. F. Ryan, "John Russell, R. A., and Early Lunar Mapping," *Smithsonian Journal of History,* 1(1966):27–48.

10. M. J. Crowe, "New Light on the Moon Hoax," *Sky and Telescope,* 62(1981):428–29.

11. W. Herschel, "Letter to the Reverend Dr. Maskelyne," June 12, 1780, in *The Scientific Papers of Sir William Herschel* (London: Royal Society and the Royal Astronomical Society, 1912), Vol. 1, pp. xc–xci.

12. A. M. Clerke, *A History of Astronomy During the Nineteenth Century,* 2nd edition (London: Adam and Charles Black, 1887), p. 313.

13. F. von P. Gruithuisen, *Entdeckung vieler deutlichen Spuren der Mondbewohner,* Nuremberg, 1824. See also "Notes," *The Observatory,* 42(1919):327.

14. D. S. Evans, "The Great Moon Hoax," *Sky and Telescope,* 62(1981):196–98, 308–11.

15. As suggested by M. J. Crowe, *op. cit.*

16. J. F. W. Herschel, *Outlines of Asronomy,* 8th edition (London: Longmans, Green, and Co., 1865), pp. 288–89.

17. Kopal and Carder, *op. cit.*

18. Clerke, *op. cit.*, p. 313.

19. Z. Kopal, "Topography of the Moon," in Z. Kopal, ed., *Physics and Astronomy of the Moon* (New York: Academic Press, 1962), pp. 244–45.

20. Clerke, *op. cit.*, pp. 314–16.

21. *Ibid.*; see also R. A. Proctor, *The Poetry of Astronomy* (London: Smith, Elder, and Co., 1881), pp. 215–44.

22. Middlehurst and Burley, *op. cit.*

23. J. F. W. Herschel, *op. cit.*, p. 283.

24. J. N. Lockyer, *Elementary Lessons in Astronomy* (London: Macmillan and Company, 1871), pp. 91–92.

25. Summaries outlining early geological interest in the lunar craters can be found in E. M. Shoemaker, "Interpretation of Lunar Craters"; in Kopal, *op. cit.*, pp. 283–359; and in J. Green, "Hookes and Spurrs in Selenology," *Annals of the New York Academy of Sciences*, 123(1965):373–402.

26. J. D. Dana, "On the Volcanoes of the Moon," *American Journal of Science*, 2(1846): 335–53.

27. E. Neison (pseudonym for Edward Neville Nevill), *The Moon and the Condition and Configuration of Its Surface* (London: Longmans, Green, and Co., 1876).

28. J. Nasmyth and J. Carpenter, *The Moon: Considered as a Planet, a World, and a Satellite* (London: John Murray, 1874).

29. R. A. Proctor, *The Moon: Her Motions, Aspect, Scenery, and Physical Condition* (Manchester: Alfred Brothers, 1873).

30. M. Rozet, "Memoir sur le Sélénologie," *Comptes Rendus*, 22(1846):470.

31. Obituary (Richard Anthony Proctor), *Monthly Notices of the Royal Astronomical Society*, 49(1889):164–68; and W. Noble, "Richard Anthony Proctor," *The Observatory*, 11(1888):366–68.

32. For discussions of the attitudes toward meteors and meteorites in this period, see J. N. Lockyer, *The Meteoritic Hypothesis* (London: The Macmillan Company, 1890), pp. 3–10; C. P. Olivier, *Meteors* (Baltimore: Williams & Wilkins, 1925), pp. 1–8; and H. H. Nininger, "Geological Significance of Meteorites," *American Journal of Science*, 246(1948):101–08.

33. A. D. Fougeroux, L. C. Cadet, and A. Lavoisier, "Rapport . . . sur une pierre qu'on prétend être tombée du ciel pendent un orage," *Journal de Physique, de Chimie, et d'Histoire Naturelle*, 2(1777):251–55.

34. Quoted in Lockyer, *The Meteoritic Hypothesis, op. cit.*, p. 141.

35. *Ibid.*, p. 142.

36. E. F. F. Chladni, *Über den Ursprung der von Pallas gefundenen Eissenmassen*, Riga, 1794; abstracted in the *Philosophical Magazine*, Tillock's Series, 2(1798):225. A French translation was published in *Journal des Mines*, 15(1803):286–320.

37. Lockyer, *The Meteoritic Hypothesis, op. cit.*, p. 112.

38. E. F. F. Chladni, *Über Feuermeteore, und über die mit denselben herabgefallenen Massen*, Wein, 1819.

39. J.-B. Biot, *Relation d'une Voyage fait dans le départment de l'Orne, pour constater la realité d'un météore observé a l'Aigle*, Paris: Baudouin, 1803. See also, Biot, "Account of a Fire-ball which fell in the neighborhood of L'aigle," *Philosophical Magazine*, 16(1803): 224–28.

40. W. Herschel, "Observations on two newly-discovered bodies (Ceres and Pallas)," *Scientific Papers, op. cit.*, Vol. 1, pp. 187–98.

41. M. von Bieberstein, *Untersuchungen über den Ursprung und die Ausbildung der gegenwartigen Anordnung des Wellbebaendes*, Darnstadt, 1802.

42. K. E. von Moll, *Über den Zusammenhang der Begirgsbildung mit dem Erscheinen der Jeuerkugeln*, Munich, 1815.

43. F. von P. Gruithuisen, *Analekten Erd-und-Himmels Kunde*, Munich, 1829; and "Geologisches," *Astronomische Jahrbuch*, 1840, 151–60. See also J. F. J. Schmidt, *Der Mond*, 1856, iii–viii. Gruithuisen's reputation was defended many years later by T. G. Elger ["Selenographical Notes," *The Observatory*, 10(1887):419–21], who wrote: "A slur has often been cast upon Gruithuisen's lunar observations on the ground that he was inclined to set down more than he actually saw, and . . . it has been considered that a lively imagination frequently led him astray. The more careful and systematic study of these delicate features [lunar rills] during the last twenty years has, however, sufficed to show that his records are by no means to be depreciated."

44. Nasmyth and Carpenter's illustrations were still being widely used more than thirty years later. See, for examples, Robert Ball, *The Story of the Heavens* (London: Cassell and Co., 1905), and C. Flammarion, *Popular Astronomy*, trans. J. E. Gore (New York: D. Appleton and Co., 1907).

45. R. A. Proctor, *Old and New Astronomy*, A. C. Ranyard, ed. (London: Longmans, Green, and Co., 1895), p. 517.

46. R. A. Proctor, *Our Place Among the Infinities* (New York: D. Appleton and Co., 1876), pp. 18–19.

47. R. A. Proctor, "The Moon's Myriad Small Craters," *Belgravia*, 36(1878):153–71.

48. R. A. Proctor, *The Poetry of Astronomy, op. cit.*, pp. 182–214.

49. H. J. Klein, "On Some Volcanic Formations on the Moon," *The Observatory*, 5(1882): 253–58.

50. W. R. Birt, "Lunar Physics," *The Observatory*, 4(1881):47–51, 108–11.

51. W. Thiersch and A. Thiersch ("Asterios"), *Die Physiognomie des Mondes*, Nördlingen, 1879.

52. A. Meydenbauer, "Die Gebilde der Mondoberfläche," *Sirius*, 15(1882):59–64; "Über die Bildung der Mondoberfläche," *ibid.*, 10(1877):180; and "The Mode of Formation of Lunar Craters," *The Observatory*, 6(1883):153.

53. "The Mode of Formation of Lunar Craters," *The Observatory*, 6(1883):59–60.

54. G. F. Chambers, *A Handbook of Descriptive and Practical Astronomy*, 4th edition (Oxford: Clarendon Press, 1889), Vol. 1, p. 124.

55. J. Ericsson, "The Lunar Surface and Its Temperature," *Nature*, 34(1886):248–51.

56. S. E. Peal, "Lunar Glaciation," *Nature*, 35(1886):100–01; and *A Short Abstract on the Theory of Lunar Surfacing by Glaciation* (Calcutta: Band and Mookerjie, 1886).

57. A. Guillemin, *The Heavens*, 2nd edition (London: Richard Bentley, 1867), pp. 159–61; see also footnote by T. E. Webb, p. 160.

58. R. A. Proctor, *The Moon, op. cit.*, pp. 349–52.

59. H. A. Faye, "Les volcanoes de la lune," *Revue Scientifique*, Series 3, 1(1881):130–38.

60. H. Ebert, "Ein Vorlesungsversuch aus dem Gebiete der physikalischen Geographie (Bildung der Schlammvulkane und der Mondringgebirge," *Annalen die Physik und Chemie*, 4(1890):351–63.

61. J. B. Hannay, "Formation of Lunar Volcanoes," *Nature*, 47(1892):7–8.

62. W. Goodacre, *The Moon* (Bournemouth: Pardy & Son, 1931), p. 48.

63. J. W. Judd, *Volcanoes—What They Are and What They Teach*, 4th edition (London: Kegan Paul, Trench & Co., 1888), p. 367.

64. J. W. Dana, *Manual of Geology*, 4th edition (New York: American Book Co., 1894), p. 11.

CHAPTER TWO. IRON AND DIAMONDS

Epigraph: G. K. Gilbert, "The Origin of Hypotheses, Illustrated by the Discussion of a Topographic Problem," *Science*, n.s., 3(1891):1–13.

1. A. E. Foote, "A New Locality for Meteoric Iron with a Preliminary Notice of the Discovery of Diamonds in the Iron," *Proceedings of the American Association for the Advancement of Science*, 40(1892):279–83.

2. A. E. Foote, "Geological Features of the Meteoric Iron Locality in Arizona," *Proceedings of the Academy of Natural Sciences of Philadelphia*, 43(1891):407. This is a brief report on Foote's presentation at the A.A.A.S., cited above.

3. G. K. Gilbert, "The Origin of Hypotheses, Illustrated by the Discussion of a Topographic Problem," *Science*, new series, 3(1896):4.

4. G. K. Gilbert, "Notes Made in Arizona, Oct. 22–Nov. 19, 1891," Index No. 51, Accession No. 3448, *U.S. Geological Survey Field Records File*, U.S. National Archives; hereafter cited as *Notebook 51*.

5. G. Troost, "Description of three varieties of Meteoric Iron," *American Journal of Science*, second series, 2(1846):356–58.

6. Gilbert, "The Origin of Hypotheses," *op. cit.*, 3.

7. Foote, "A New Locality," *op. cit.*

8. Gilbert, "The Origin of Hypothesis," *op. cit.*, 4.

9. Gilbert, *Notebook 51*.

10. Foote, "A New Locality," *op. cit.*

11. S. J. Holsinger to D. M. Barringer, Feb. 6, 1903, *Daniel Moreau Barringer Papers,* Firestone Library, Princeton University, hereafter cited as *Barringer Papers.*

12. Foote, "A New Locality," *op cit.*

13. Gilbert, "The Origin of Hypotheses," *op. cit.,* 4.

14. The principal biographical source for Gilbert is W. M. Davis, "Grove Karl Gilbert, 1843–1918," *Biographical Memoirs of the National Academy of Sciences,* 21(1926), No. 2. A number of memorials and appreciations by noted geologists shortly after his death, and some of these have been incorporated as well into the most recent biography: Stephen J. Pyne, *Grove Karl Gilbert—A Great Engine of Research* (Austin: The University of Texas Press, 1980).

15. For details of the U.S. Army Corps of Topographic Engineers' work in the West prior to the Civil War, see William Goetzmann, *Army Exploration in the American West, 1803–1863* (New Haven: Yale University Press, 1959). For the Ives expedition, see chapter 10 in Goetzmann.

16. For a brief review of the post–Civil War surveys of the West, see A. Hunter Dupree, *Science in the Federal Government* (Cambridge: Harvard University Press, 1957), chapter 10.

17. *Ibid.,* 202–14.

18. G. K. Gilbert, *Report on the Geology of the Henry Mountains* (Washington: U.S. Government Printing Office, 1877).

19. G. K. Gilbert, *Lake Bonneville,* U.S. Geological Survey Monograph I (Washington: U.S. Government Printing Office, 1890).

20. Gilbert, "The Origin of Hypotheses," *op. cit.,* 2–3.

21. G. K. Gilbert, "The Inculcation of Scientific Method by Example, with an illustration drawn from the Quaternary Geology of Utah," *American Journal of Science,* third series, 32(1886): 287.

22. Gilbert, "The Origin of Hypotheses," *op. cit.,* 1–2.

23. Gilbert, "The Inculcation of Scientific Method," *op. cit.,* 289–90.

24. Gilbert noted that this method "has found favor with the greatest investigators," and cited T. C. Chamberlin, "The Method of Multiple Working Hypotheses," *Science,* first series, 15(1890): 92–96. See Gilbert, "The Origin of Hypotheses," *op cit.,* 2.

25. *Ibid.,* 12.

26. See Pyne, *op. cit.,* 191, who cites an example from Davis, *op. cit.,* 248.

27. Gilbert, "The Origin of Hypotheses," *op. cit.,* 4–5.

28. *Ibid.,* 5–7.

29. *Ibid.*

30. *Ibid.*

31. *Ibid.*

32. *Ibid.*

33. Gilbert, *Notebook 51,* Oct. 22–26, 1891.

34. Gilbert, "The Origin of Hypotheses," *op. cit.,* 6.

35. Gilbert, *Notebook 51*, Oct. 26–31, 1891.
36. *Ibid.*, Nov. 7, 1891.
37. Gilbert, "The Origin of Hypotheses," *op. cit.*, 6–7.
38. Gilbert, *Notebook 51*, Nov. 1, 1891.
39. *Ibid.*, Nov. 4, 6, and 7, 1891.
40. *Ibid.*, Nov. 6, and following Nov. 19, 1891.
41. *Ibid.*, Nov. 6, 8, and 10, 1891.
42. *Coconino Sun* (Flagstaff), Nov. 12, 1891.
43. Gilbert, *Notebook 51*, Nov. 4, 1891.
44. *Ibid.*, Nov. 8, 1891.
45. *Ibid.*, Nov. 12, 1891.
46. *Ibid.*, Nov. 14, 1891.
47. *Ibid.*, Nov. 4, 1891.
48. *Ibid.*, Nov. 14, 1891.
49. *Ibid.*; and Gilbert, "The Origin of Hypotheses," *op. cit.*, 9–10.
50. Gilbert, *Notebook 51*, Nov. 15–19, 1891.
51. G. K. Gilbert, "Notes Made in Arizona, Nov. 21–Dec. 5, 1891," Index No. 52, Accession No. 3449, *U.S. Geological Survey Field Records File, U.S. National Archives*; hereafter cited as *Notebook 52*. See also Gilbert, *13th Annual Report*, U.S. Geological Survey, 1892, Part I, 98.
52. *Coconino Sun*, Dec. 3, 1891.
53. Gilbert, *Notebook 52*, Nov. 20, 21, and 25, 1891.
54. *Coconino Sun*, Dec. 3, 1891.
55. "A Meteoric Crater," *San Francisco Examiner*, Dec. 1, 1891, reprinted in the *Publications of the Astronomical Society of the Pacific*, 4(1892):37. See also E. S. Holden's editor's note, *ibid.*, 263.
56. G. K. Gilbert, *14th Annual Report*, U.S. Geological Survey, 1893, Part I, 187; and "The Moon's Face; a Study of the Origin of Its Features," *Bulletin of the Philosophical Society of Washington*, 12(1893):242–92.
57. G. K. Gilbert, *13th Annual Report*, *op. cit.*, and *14th Annual Report*, *op. cit.*
58. Gilbert, *14th Annual Report*, *op. cit.*
59. H. L. Fairchild, "Grove Karl Gilbert," *Science*, 48(1918):151–54; and "Meteor Crater Exploration," *ibid.*, 69(1929):485.
60. The summary of Gilbert's conclusion regarding Coon Moutain's origin given here is from Gilbert, "The Origin of Hypotheses," *op. cit.*, 8–12.

CHAPTER THREE. MOONLETS AND OTHER MATTERS

Epigraph: G. K. Gilbert, "The Moon's Face; a Study of the Origin of Its Features," *Bulletin of the Philosophical Society of Washington*, 12(1893):241–92.

1. G. K. Gilbert, "Geology," *14th Annual Report*, U.S. Geological Survey, Washington, D.C., 1893.

2. There seems to be some confusion in recent geological writings as to the chronology of Gilbert's Coon Mountain and lunar investigations, with some writers giving his lunar study precedence in time. See, for example, F. El-Baz, "Gilbert and the Moon," *The Scientific Ideas of G. K. Gilbert—an assessment on the occasion of the centennial of the United States Geological Survey (1879–1979)*, Special Paper 183, Geological Society of America, Boulder, Colorado, 1979, 69, 72.

3. Gilbert did, however, later publish a review of a stereoptic method of lunar observation proposed by W. Prinz of Belgium, but without specific reference to his own lunar work. See G. K. Gilbert, "Stereoptic Study of the Moon," *Science*, 13(1901):407–09.

4. H. Jacoby, "Minutes of meeting of Feb. 16, 1893, Astronomy Section, New York Academy of Sciences," *Astronomy and Astro-Physics*, 12(1893):286.

5. "Dr. Gilbert on the Evolution of the Moon," abstract, *Publications of the Astronomical Society of the Pacific*, 4(1892):263–64.

6. G. K. Gilbert, "The Moon's Face; a study of the origin of its features," *Bulletin of the Philosophical Society of Washington*, 12(1893):241–92. Reprinted in *Scientific American Supplement*, 37(1893–94), Nos. 938–40.

7. G. H. Darwin's work on tidal friction will be found in his *Scientific Papers* (Cambridge: The University Press, 1908), Vol. II. His ideas are progressively summarized in his articles on "Tides" in the *Encyclopedia Britannica* from the 9th edition (1887, Vol. 23, 353–81) through the 13th edition (1926, Vol. 25, 938–61).

8. H. Ebert, "Ein Vorlesungsversuch aus dem Gebiete der physikalischen Geographie (Bildung der Schammvulkane und der Mondringgebirge)," *Annalen der Physik und Chemie*, 41(1890):351–63.

9. G. K. Gilbert, "The Moon's Face," *op. cit.*

10. El-Baz, *op. cit.*, credits Gilbert with having "coined the term 'meteoric'" in regard to such theories, although Proctor, for one, had used the word repeatedly twenty years earlier.

11. H. Ebert, "Uber die Ringgebirge des Mondes,"*Sitzungsberichte der physikalischen-medizinischen Sozietät* (Munich: University of Erlangen, 1890), 171.

12. See G. K. Gilbert, "The Moon's Face," *op. cit.*, 267–68, for his mathematical derivation of this expression.

13. This idea, of course, was not new nor was it developed fully for another seventy years. See W. R. Birt, "Lunar Physics," *The Observatory*, 4(1881):47–51, 108–11, who attributed it to J. Chacornac in 1869; and E. M. Shoemaker and R. J. Hackman, "Stratigraphic basis for a lunar time scale," in Z. Kopal and Z. K. Mihailov, eds., *The Moon* (I.A.U. Symposium No. 14) (New York: The Academic Press, 1962), 289–300.

14. G. K. Gilbert, "The Evolution of the Moon," abstract, *American Naturalist*, 26(1892): 1056–57; "Dr. Gilbert on the Evolution of the Moon," *op. cit.*; H. Jacoby, *op. cit.*; and "A Theory of the Formation of Lunar Craters," abstract, *Transactions of the New York Academy of Science*, 12(1893):93–95.

15. E. S. Holden, editor's note, *Publications of the Astronomical Society of the Pacific*, 4(1892):263–64.

16. E. S. Holden, "The Lunar Crater Copernicus," *ibid.*, 114–20.

17. Review of "The Moon's Face," *American Naturalist*, 27(1893): 784.

18. J. F. James, review of "The Moon's Face," *Science*, 21(1893): 305–07.

19. W. M. Davis, "Lunar Craters," *The Nation*, 41(1893): 342–43.

20. W. M. Davis, "Grove Karl Gilbert," *Biographical Memoirs, National Academy of Sciences*, Vol. 21, No. 2, 1926.

21. A. C. Ranyard, "The Great Lunar Crater Tycho," *Knowledge*, August 1, 1893, 149–52.

22. Ranyard completed and edited Proctor's posthumously published *Old and New Astronomy* (London: Longmans, Green, and Co., 1895), in which Proctor cites the volcanic analogy for the lunar craters without reference to his earlier impact hypothesis.

23. E. Miller, "A New Theory of the Surface Markings of the Moon," *Popular Astronomy*, 3(1896): 273–78.

24. S. L. B., "Mr. Miller's Theory on the Surface Markings of the Moon Criticized," *ibid.*, 3(1896): 383.

25. E. Suess, "The Moon as Seen by a Geologist," abstract, trans. C. A. Stetefeldt, *Publications of the Astronomical Society of the Pacific*, 7(1895): 139–48.

26. W. W. Payne, "The Moon," *Popular Astronomy*, 2(1894): 62–72.

27. W. H. Pickering, "Are There at Present Active Volcanoes on the Moon?" *The Observatory*, 15(1892): 250–54.

28. J. B. Hannay, "Formation of Lunar Craters," *Nature*, 47(1892): 7–8.

29. W. H. Pickering, "Lunar Atmosphere and Surface Detail," *Annals of the Observatory of Harvard College*, Vol. 32, Part II, 1900; and "Origin of the Lunar Features," *Popular Astronomy*, 8(1900): 147–52, 181–89.

30. W. H. Pickering, *The Moon; a Summary of the Existing Knowledge of Our Satellite with a Complete Photographic Atlas* (New York: Doubleday, Page and Co., 1903).

31. R. G. Aitken, review of W. H. Pickering, *The Moon*, *Publications of the Astronomical Society of the Pacific*, 16(1904): 42–45.

32. S. A. Saunder, review of *The Moon*, *The Observatory*, 27(1904): 92–97.

33. R. P., "Theory of Lunar Formations," *English Mechanic and World of Science*, 66(1897): 297; and "The Meteoric Theory of Lunar Formations," *ibid.*, 67(1898): 335–36. See also J. S. G., "Canal Gemination and Telescopic Controversy, *ibid.*, 334.

34. E. M. Antoniadi, "Abnormal Planetary Observations," *ibid.*, 67(1898): 547. Other opponents of impact theory included S. M. B. Gemmill, "Variable Stars in Ursa Major and Minor—What Is the Polar Compression of Mars," *ibid.*, 475; W. F. A. Ellison, "Meteoric Theory of Lunar Craters," *ibid.*, 525; and Radix, "Lunar Craters," *ibid.*, 68(1898): 349.

35. E. M. Antoniadi, *op. cit.*

36. G. W. Coakley, "The Probable Origin of Meteorites," *Astronomy and Astro-Physics*, 11(1892): 735–63.

37. R. S. Ball, *The Story of the Heavens* (London: Cassell and Co. Ltd., 1886). See also "new and revised edition," 1905.

376 NOTES TO PAGES 70–76

38. N. S. Shaler, "A Comparison of the Features of the Earth and the Moon," *Smithsonian Institution Contributions to Knowledge,* No. 1438, Vol. 34, 1903.

39. Meteoritic impact has recently been suggested to explain the extinction of the dinosaurs at the end of the Cretaceous Era, some 65 million years ago, and of certain other species at the end of the Eocene Epoch of Tertiary time some 34 million years ago. See L. W. Alvarez, W. Alvarez, F. Araso, and H. Michel, "Extraterrestrial Cause for the Cretaceous-Tertiary Extinction," *Science,* 208(1980): 1095–1108; and R. Ganapathy, "Evidence for a Major Meteorite Impact on the Earth 34 Million Years Ago; Implication for Eocene Extinctions," *ibid.,* 216(1982): 885–86.

40. T. C. Chamberlin and R. D. Salisbury, *Geology* (New York: Henry Holt and Co., 1905), 596, 598.

41. F. R. Moulton, *An Introduction to Astronomy* (New York: The Macmillan Co., 1906), 266.

42. T. J. J. See, *Researches on the Evolution of Stellar Systems* (Lynn, Massachusetts: Thom. P. Nichols & Sons, 1910), Vol. II, 331–36.

43. "Report of meeting of Oct. 26, 1910," *Journal of the British Astronomical Association,* 31(1910–11): 19–20; W. Goodacre, "The Origin of the Lunar Formations," *ibid.,* 25–32; and "Report of meeting of Jan. 25, 1911," *ibid.,* 179. Goodacre, whose paper was discussed at these meetings of the BAA, was then and for many years director of the BAA's Moon Section. *See also* D. P. Beard, "The Impact Origin of the Moon's Craters," *Popular Astronomy,* 25(1917): 167–77; "The Limelight of the Moon—An Alternate Theory," *ibid.,* 32(1924): 325–34; and "Coral Origin of the Lunar Craters," *ibid.,* 33(1925): 74–75.

44. W. M. Davis, "Grove Karl Gilbert," *op. cit.,* 186–88.

CHAPTER FOUR. STAKING A CLAIM

Epigraph: D. M. Barringer, "Coon Mountain and Its Crater," Proceedings of the Academy of Natural Sciences of Philadelphia, 57(1906): 863.

1. D. M. Barringer, "Coon Mountain and Its Crater," *Proceedings of the Academy of Natural Sciences of Philadelphia,* 57(1906): 861–86; and "Historical" note in "Instructions to my executors," typewritten text dated April 1, 1922. *Barringer Papers.*

2. The biographical information is taken from biographical sketches prepared by Barringer himself, copies of which are in the *Barringer Papers;* from B. Barringer, "Daniel Moreau Barringer (1860–1929) and His Crater (The beginning of the Crater Branch of Meteoritics)," *Meteoritics,* 2(1964): 183–99; and from "D. M. Barringer," *Who's Who in America, 1922–23.*

3. "J. W. Mallet," *Who's Who in America, 1902–03.*

4. R. A. F. Penrose, "John Casper Branner," *Biographical Memoirs, National Academy of Sciences,* Vol. 21, No. 3, 1926.

5. B. Barringer, *op. cit.,* 185.

6. D. M. Barringer, "Historical" note, *op. cit.*

7. D. M. Barringer to S. J. Holsinger, January 30, 1903, *Barringer Papers.*

8. S. J. Holsinger to D. M. Barringer, February 6, 1903, *ibid.*

9. The analysis was: Malleable iron, 94%; nickel, 3%; cobalt, 1%; moisture, 1%. Holsinger did not cite the source for this analysis, but it agrees very roughly with an analysis published by Orville A. Derby, "Constituents of the Canyon Diablo Meteorite," *American Journal of Science*, 49(1895): 101–10.

10. Volz subsequently told Barringer that before he, Volz, had arrived at Canyon Diablo, merchants in Winslow (the Williams brothers) had collected and shipped out "perhaps half as much." See D. M. Barringer, "Coon Mountain," *op. cit.*, 875.

11. The earlier claims, located in 1891, apparently had been allowed to lapse. See D. M. Barringer to W. C. Barnes, September 1910, *Barringer Papers*.

12. S. J. Holsinger to D. M. Barringer, February 15, 1903, *ibid.*

13. D. M. Barringer to S. J. Holsinger, February 17, 1903, *ibid.*

14. S. J. Holsinger to D. M. Barringer, February 26, 1903, *ibid.*

15. S. J. Holsinger to D. M. Barringer, March 7, 1903, *ibid.*

16. *Ibid.; see also* B. Barringer, *op. cit.*, 186.

17. B. Barringer, *loc. cit.; see also* D. M. Barringer, "Historical" note, *op. cit.* In 1951, the company's name was changed to The Barringer Crater Company.

18. D. M. Barringer to H. H. Alexander, July 14, 1905, *Barringer Papers*.

19. B. Barringer, *op. cit.*, 187–88; *see also* D. M. Barringer, "Historical" note, *op. cit.*

20. S. J. Holsinger to D. M. Barringer, April 16, 1903, *Barringer Papers*.

21. S. J. Holsinger to D. M. Barringer, March 31, 1903, *ibid.*

22. S. J. Holsinger to D. M. Barringer, April 18, 1903, *ibid.*

23. S. J. Holsinger to D. M. Barringer, April 21, 1903, *ibid.*

24. S. J. Holsinger to D. M. Barringer, May 4, 1903, *ibid.*

25. S. J. Holsinger to D. M. Barringer, April 2, 1903, *ibid.*

26. S. J. Holsinger to D. M. Barringer, April 16, 1903, *ibid.*

27. S. J. Holsinger to D. M. Barringer, April 18, 1903, *ibid.*

28. S. J. Holsinger to D. M. Barringer, May 6, 1903, *ibid.*

29. I. E. Dennis, "Meteorite Mountain," *Coconino Sun*, August 22, 1903.

30. D. M. Barringer to S. J. Holsinger, September 4, 1903, *Barringer Papers*.

31. B. Barringer, *op. cit.*, 186.

32. D. M. Barringer, "Coon Mountain," *op. cit.*, 876–79.

33. *Ibid.*, 882; *see also* D. M. Barringer to J. W. Mallet, August 21, 1905, *Barringer Papers*.

34. D. M. Barringer, "Coon Mountain," *op. cit.*, 869; *see also*, G. K. Gilbert, "Notes made in Arizona, Oct. 27–Nov. 19, 1891," Index no. 52, Accession no. 3449, *U.S. Geological Survey Field Records File, U.S. National Archives*; entry for November 7, 1891.

35. D. M. Barringer, "Coon Mounain," *op. cit.*, 867; *see also* B. C. Tilghman, "Coon Butte, Arizona," *Proceedings of The Academy of Natural Sciences of Philadelphia*, 57(1906): 894.

36. S. J. Holsinger to D. M. Barringer, May 27, 1904, *Barringer Papers*.

37. D. M. Barringer to H. H. Alexander, July 17, 1905, *ibid.*

38. B. C. Tilghman, *op. cit.*, 903–05, 907–10.

39. B. C. Tilghman to S. J. Holsinger, April 7, 1905, with postscript by D. M. Barringer, *Barringer Papers.*

40. D. M. Barringer to J. W. Mallet, June 25, 1905, *ibid.*

41. B. C. Tilghman, *op. cit.*, 905.

42. *Ibid.*, 905–06.

43. D. M. Barringer to American Well Works, July 15, 1905; and to H. H. Alexander, July 7, 1905, *Barringer Papers.*

44. B. C. Tilghman, *op. cit.*, 906.

45. D. M. Barringer to S. J. Holsinger, April 24, 1905, *Barringer Papers.*

46. D. M. Barringer to H. H. Alexander, July 17, 1905, *ibid.*

47. D. M. Barringer to S. J. Holsinger, April 7, 1905; and to J. W. Mallet, April 19, 1905, *ibid.*

48. *See* various calls for assessments in Standard Iron Company letter books, April 1904–July 1905, *Barringer Papers.*

49. D. M. Barringer to S. J. Holsinger, July 1, 1905, *ibid.*

50. D. M. Barringer to S. J. Holsinger, April 7, 1905; to H. H. Alexander, April 19, 1905; and to J. W. Mallet, August 21, 1905, *ibid.*

51. J. W. Mallet to D. M. Barringer, August 17, 1905, *ibid.* This letter, in which Mallet also reported finding "five excellent microscopic diamonds" in the iron, was published by Barringer as a footnote in his "Coon Mountain," *op. cit.*, 862–63.

52. D. M. Barringer to H. H. Alexander, September 7, 1905, *Barringer Papers.*

53. D. M. Barringer to E. J. Bennitt, July 6, 1905, *ibid.*

54. Alexander's analysis is recorded on page 189, Standard Iron letter book 2, *ca.* August 1905, *ibid.*

55. D. M. Barringer to H. H. Alexander, August 9, 1905, *ibid.*; *see also* D. M. Barringer, "Coon Mountain," *op. cit.*, 883; and B. C. Tilghman, *op. cit.*, 913–14.

56. D. M. Barringer to J. W. Mallet, August 26, 1905, *Barringer Papers.*

57. D. M. Barringer to J. W. Mallet, June 25, 1905, *ibid.*

58. D. M. Barringer to E. J. Bennitt, July 6, 1905; *see also* assessment no. 11, July 7, 1905, and assessment nos. 12–16, August 11, 1905, through February 23, 1906, Standard Iron lettter book 2, *Barringer Papers.*

59. D. M. Barringer to S. G. Dixon, September 12, September 15, and September 22 (telegram), 1905, *ibid.*

60. D. M. Barringer to The Academy of Natural Sciences of Philadelphia, September 25, 1905; and to S. G. Dixon, September 26, 1905, *ibid.*

61. D. M. Barringer to H. H. Alexander, September 15, 1905, *ibid.* Barringer here notes that he has "succeeded in raising money for both my exploration company and the gold dredging scheme in northern California—$250,000." A majority of the correspondence in the *Barringer Papers,* in fact, deals with matters other than Coon Mountain.

62. D. M. Barringer to Chalmers & Williams, September 11 and 16, 1905, *Barringer Papers.*

63. D. M. Barringer to S. J. Holsinger, July 1, 10, and August 4, 1905, *ibid.*

64. D. M. Barringer to S. J. Holsinger, July 13 and 15, 1905, *ibid.*

65. D. M. Barringer to S. J. Holsinger, July 1 (two letters), and 10, 1905; to J. Meyer, July 1, 1905; and to Chalmers & Williams, July 6, 10, and August 4, 1905; *ibid.*

66. D. M. Barringer to S. J. Holsinger, July 1 and August 4, 1905, *ibid.*

67. D. M. Barringer to G. Pinchot, July 17, 1905; and to E. J. Bennitt, July 6 and 21, 1905, *ibid.*

68. D. M. Barringer to E. J. Bennitt, January 31, 1906; to S. J. Holsinger, February 6 and 14, and March 21, 1906; and to G. Pinchot, February 19 and March 2, 1906, *ibid.* Pinchot gave Holsinger's health as his reason for not rehiring him.

69. D. M. Barringer to S. J. Holsinger, August 4, 1905, *ibid.*

70. D. M. Barringer to American Well Works, July 17, 1905; and to J. Meyer, August 9, 1905, *ibid.*

71. D. M. Barringer to S. J. Holsinger, December 9, 1905, *ibid.*

72. D. M. Barringer to J. Nevill, October 27, 1905; to E. J. Bennitt, October 26, 1905; and to Chalmers & Williams, November 15, 1905; *ibid.*

73. D. M. Barringer to J. Nevill, November 24, 1905, *ibid.*

74. D. M. Barringer to S. G. Dixon, November 21, 1905, *ibid.*

75. D. M. Barringer to J. W. Mallet, November 22, 1905, *ibid.*

76. D. M. Barringer to S. J. Holsinger, November 23, 1905, *ibid.*

77. D. M. Barringer to E. J. Bennittt, November 23, 1905, *ibid.*

78. S. G. Dixon, "Coon Mountain and Its Crater," *Proceedings of The Academy of Natural Sciences of Philadelphia,* December 5, 1905.

79. D. M. Barringer to S. J. Holsinger, December 9, 1905, *Barringer Papers.*

80. D. M. Barringer to S. J. Holsinger, December 12, 1905; and to J. Nevill, December 26, 1905, *ibid.*

81. D. M. Barringer to J. Nevill, December 14, 1905; and to S. J. Holsinger, telegram, December 12, 1905, *ibid.*

82. D. M. Barringer to S. J. Holsinger, December 12, 1905, *ibid.*

83. D. M. Barringer to J. Nevill, December 22, 1905, *ibid.*

84. D. M. Barringer to S. J. Holsinger, to E. J. Bennitt, telegram; and to Chalmers & Williams, January 19, 1906, *ibid.*

85. D. M. Barringer to Chalmers & Williams, January 23, 1906, *ibid.*

86. D. M. Barringer to F. Breuil, January 26, 1906, *ibid.*

87. D. M. Barringer to Chalmers & Williams, January 23, 1906, *ibid.*

88. D. M. Barringer to J. W. Mallet, February 2, 1906, *ibid.*

89. D. M. Barringer to Chalmers & Williams, January 23, 1906, *ibid.*

90. D. M. Barringer to E. J. Bennitt, February 13, 1906, *ibid.*

91. D. M. Barringer to E. J. Bennitt, March 17, 1906, *ibid.*

92. D. M. Barringer to B. C. Tilghman and E. J. Bennitt, March 8, 1906, *ibid.*

93. D. M. Barringer to J. Nevill, telegram, March 26, 1906, and March 29, 1906, *ibid.*

94. *See* assessment nos. 11–16, Standard Iron letter book 2, *Barringer Papers.*

95. Acknowledgment by survey officials of the impact origin of Coon Mountain/Meteor Crater seems to have begun in 1959 with the publication of the first of a series of studies made there by Survey geologist Eugene M. Shoemaker. *See* Shoemaker, "Impact Mechanics at Meteor Crater, Arizona," prepared on behalf of the U.S. Atomic Energy Commission, Open File Report, July 1959. Earlier publications concerning the crater by Survey members are rare and follow Gilbert's steam explosion hypothesis. *See* N. H. Darton, "The Zuni Salt Lake," *Journal of Geology,* 13(1905):185; "A Reconnaissance of Parts of Northwestern New Mexico and Northern Arizona," *U.S. Geological Survey Bulletin* 435, 1910; "Explosion Craters," *The Scientific Monthly,* 3(1916):417; and "Crater Mound, Arizona," *Bulletin of the Geological Society of America,* 56(1945):1154 (abstract), for the major examples.

96. See, for example, F. El-Baz, "Gilbert and the Moon," in E. L. Yochelson, ed., *The Scientific Ideas of G. K. Gilbert—an assessment on the occasion of the centennial of the United States Geological Survey,* Special Paper 183, Geological Society of America, Boulder, Colorado, 1980, p. 73.

97. D. M. Barringer, "Coon Mountain," *op. cit.,* 863.

98. The quotations in the summary that follows are taken from D. M. Barringer, "Coon Mountain," *op. cit.;* and B. C. Tilghman, "Coon Butte," *op. cit.*

99. G. K. Gilbert, *Notebook 51, op. cit.,* entry for November 7.

100. Gilbert did, of course, refer briefly to a "plentiful black mineral" in his field notes; see G. K. Gilbert, *op. cit.,* entry for November 7, 1891.

101. *See also* D. M. Barringer to S. J. Holsinger, November 28, 1905, *Barringer Papers.*

102. In regard to Rounsville's and Manning's reports, *see also* D. M. Barringer to A. Rounsville, November 24, 1905; and to J. W. Mallet, February 26, 1906, *ibid.* There is still another coincidence here, it seems, for Manning had been a student of Mallet's at Jefferson Medical College, where Mallet had taught prior to joining the University of Virginia faculty.

103. *See also* D. M. Barringer to J. W. Mallet, April 19, 1905, *ibid.*

104. G. K. Gilbert, *op. cit.,* entries for November 1 and November 7, 1891.

CHAPTER FIVE. PROVING THE CLAIM

Epigraph: G. P. Merrill and W. Tassin, "Contributions to the Study of the Canyon Diablo Meteorites," *Smithsonian Institution Miscellaneous Collections,* 50(1907):203.

1. G. P. Merrill, "On a Peculiar Form of Metamorphism in Siliceous Sandstone," *Proceedings of the U.S. National Museum,* 32(1907):547–50.

2. "Arizona's Big Meteor," editorial, *New York Herald,* July 23, 1906.

3. D. M. Barringer to J. C. Branner, October 17, 1906, *Barringer Papers.*

4. H. Shapley to D. M. Barringer, May 29, 1929, *ibid.*
5. G. P. Merrill to D. M. Barringer, April 23, 1908, *ibid.*
6. W. Tassin to D. M. Barringer, April 20, 1907, *ibid.*
7. W. H. Dall to D. M. Barringer, January 14, 1915, *ibid.*
8. D. M. Barringer, "Coon Mountain and Its Crater," *Proceedings of the Academy of Natural Sciences of Philadelphia,* 57(1906):886.
9. D. M. Barringer to B. C. Tilghman, March 9, 1906, *Barringer Papers.*
10. D. M. Barringer to E. J. Bennitt, March 19, 1906, *ibid.*
11. D. M. Barringer to B. C. Tilghman, February 2, 1910, *ibid.*
12. B. C. Tilghman to D. M. Barringer, January 30, 1908, *ibid.*
13. B. Barringer, "Daniel Moreau Barringer (1860–1929) and His Crater," *Meteoritics,* 2(1964):187.
14. D. M. Barringer to S. J. Holsinger, August 31, and October 16, 1906; and B. C. Tilghman to D. M. Barringer, August 22, 1906, *Barringer Papers.*
15. D. M. Barringer to J. C. Branner, and to J. W. Mallet, August 7, 1906, *ibid.*
16. B. C. Tilghman to D. M. Barringer, July 9, 1908, *ibid.*
17. D. M. Barringer to S. J. Holsinger, August 3, 1908, *ibid.*
18. D. M. Barringer to E. J. Bennitt, September 1, 1908, *ibid.*
19. D. M. Barringer to S. J. Holsinger, December 31, 1909, *ibid.*
20. D. M. Barringer to E. J. Bennitt, January 13 and 19, 1910, *ibid.*
21. D. M. Barringer to S. J. Holsinger, March 8, 1911, *ibid.*
22. D. M. Barringer to E. J. Bennitt, March 19, 1906, *ibid.*
23. D. M. Barringer to J. W. Mallet, March 21, 1906, *ibid.*
24. D. M. Barringer to C. D. Walcott, March 27, 1906, *ibid.*
25. D. M. Barringer to H. A. Ward, May 18, 1906, *ibid.*
26. *Ibid.; see also* D. M. Barringer to G. P. Merrill, May 22, 1906; and reply, May 23, 1906, *ibid.*
27. J. C. Branner, "The Policy of the U.S. Geological Survey and Its Bearing Upon Science and Education," *Science,* 24(1906):722–28. Branner, who resigned as a member of the Survey in February 1906 over a dispute over his earlier work for the Arkansas State Geological Survey, here prints Walcott's letter accepting his resignation and his own point-by-point rebuttal to Walcott's statements. *See also* J. C. Branner, "The U.S. Geological Survey," *The Engineering and Mining Journal,* 86(1908):1066, and J. C. Branner to D. M. Barringer, December 5, 1906, *Barringer Papers.*
28. D. M. Barringer to J. C. Branner, April 30, 1906, *Barringer Papers.*
29. J. C. Branner to D. M. Barringer, May 4, 1906, *ibid.*
30. D. M. Barringer to J. C. Branner, September 8, 1906, *ibid.*
31. J. C. Branner to D. M. Barringer, September 18, 1906, *ibid.*
32. D. M. Barringer to T. C. Chamberlin, September 26 and November 15, 1906, *ibid.*
33. J. C. Branner to D. M. Barringer, October 12, 1906, *ibid.*

34. D. M. Barringer to J. C. Branner, October 17, 1906, *ibid*.

35. J. C. Branner to D. M. Barringer, October 22, 1906, *ibid*.

36. E. O. Hovey, report on J. C. Branner, "The Geology of Coon Butte," *Science*, 24(1906):370–71.

37. D. M. Barringer to J. C. Branner, July 3, 1906, *Barringer Papers*.

38. E. O. Hovey, *op. cit.*

39. D. M. Barringer to A. C. Lane, July 12, 1906, *Barringer Papers*.

40. E. O. Hovey to D. M. Barringer, December 12, 1907, *ibid*.

41. E. O. Hovey to D. M. Barringer, October 17, 1911, *ibid*.

42. L. A. Fletcher, "The Search for a Buried Meteorite," *Nature*, 74(1906):490–92.

43. W. P. Blake, "Origin of the Depression Known as Montezuma's Well," *Science*, 24(1906):568.

44. D. M. Barringer to W. P. Blake, October 6 and 23, 1906, *Barringer Papers*.

45. D. M. Barringer to H. L. Fairchild, November 9, 1906, *ibid*.

46. D. M. Barringer to W. P. Blake, November 10, 1906, *ibid*.

47. F. N. Guild, "Coon Mountain Crater," *Science*, 26(1907):24–25.

48. D. M. Barringer to J. W. Mallet, August 13, 1906 (enclosing letter from F. N. Guild), *Barringer Papers*.

49. D. M. Barringer to S. J. Holsinger, February 6, 1906, *ibid*.

50. D. M. Barringer to A. Y. Smith, August 30, 1906, *ibid*.

51. B. C. Tilghman to D. M. Barringer, August 22, 1906, *ibid*.

52. D. M. Barringer to H. H. Alexander, September 14, 1906, *ibid*.

53. D. M. Barringer to P. B. Barringer, May 16, 1906, *ibid*.

54. D. M. Barringer to E. J. Bennitt, July 21, 1906, *ibid*.

55. D. M. Barringer to S. J. Holsinger, August 21, 1906, *ibid*.

56. D. M. Barringer to E. J. Bennitt, March 19, 1906, *ibid*.

57. D. M. Barringer to B. C. Tilghman, April 19, 1906, *ibid*.

58. D. M. Barringer to R. A. A. Penrose, March 28, 1906, *ibid*.

59. D. M. Barringer to E. J. Bennitt, September 9, 1906, *ibid*.

60. D. M. Barringer to S. J. Holsinger, February 26, 1909, *ibid*.

61. G. P. Merrill, *op. cit.*, 550.

62. D. M. Barringer to J. W. Mallet, April 13, 1906, *Barringer Papers*.

63. J. W. Mallet to D. M. Barringer, telegram, April 27, 1906, *ibid*.

64. D. M. Barringer to J. W. Mallet, April 28, 1906, *ibid*.

65. J. W. Mallet to D. M. Barringer, April 28, 1906, *ibid*.

66. D. M. Barringer to J. W. Mallet, May 15, 1906; and to S. L. Penfield, May 19, 1906, *ibid*.

67. D. M. Barringer to J. C. Branner, May 28, 1906, *ibid*.

68. D. M. Barringer to J. W. Mallet, May 28, 1906, *ibid.*

69. D. M. Barringer to J. W. Mallet, May 30, 1906, *ibid.*

70. G. P. Merrill, *op. cit.,* 549.

71. A. F. Rogers, "Natural History of Silica Minerals," *The American Mineralogist,* 13 (1928):73–92.

72. D. M. Barringer to O. C. Farrington, May 21, 1906, *Barringer Papers.*

73. O. C. Farrington, "Analysis of 'Iron Shale' from Coon Mountain, Arizona," *American Journal of Science,* 22(1906):303–09.

74. D. M. Barringer to J. W. Mallet, October 25, 1906. Tilghman had actually abandoned Barringer's "flaming drops" theory while his paper was in press but was too far along toward publication for any changes. See D. M. Barringer to J. W. Mallet, February 14, 1906, *Barringer Papers.*

75. D. M. Barringer to J. C. Branner, October 15, 1906, *ibid.*

76. J. C. Branner to D. M. Barringer, August 13, 1906; and reply, August 22, 1906, *ibid.*

77. H. L. Fairchild, "Grove Karl Gilbert," *Science,* 48(1918):151–54.

78. H. L. Fairchild, "A Meteoric Crater of Arizona," *Comptes Rendus,* 1906, 147–51.

79. H. L. Fairchild to D. M. Barringer, October 12, 1906, *Barringer Papers.*

80. D. M. Barringer to J. C. Branner, November 15, 1906, *ibid.*

81. D. M. Barringer to H. L. Fairchild, November 9, 1906, *ibid.*

82. D. M. Barringer to J. C. Branner, November 15, 1906, *ibid.*

83. J. C. Branner to D. M. Barringer, November 20, 1906, *ibid.*

84. D. M. Barringer to H. L. Fairchild, November 20, 1906, *ibid.*

85. H. L. Fairchild to D. M. Barringer, November 23, 1906, *ibid.*

86. D. M. Barringer to H. L. Fairchild, December 1, 1906, *ibid.*

87. B. C. Tilghman to H. L. Fairchild, November 27, 1906, *ibid.*

88. D. M. Barringer to T. C. Chamberlin, December 3, 1906, *ibid.*

89. H. L. Fairchild, "Origin of Meteor Crater (Coon Butte), Arizona," *Bulletin of the Geological Society of America,* 18(1907):493–504.

90. H. L. Fairchild to D. M. Barringer, October 26, 1906, *Barringer Papers.*

91. D. M. Barringer to S. J. Holsinger, February 6, 1906, *ibid.*

92. D. M. Barringer to T. Roosevelt, February 5, 1906, *ibid.*

93. D. M. Barringer to B. Penrose, April 22, 1906, *ibid.*

94. Byrd Granger, Will C. Barnes' *Arizona Place Names,* rev. ed. (Tucson: The University of Arizona Press, 1960).

95. D. M. Barringer to B. Penrose, May 22, 1906. In 1914, long after the name Meteor Crater had come into general use, Barringer declared that he and Holsinger had given this name to the crater. See Barringer to J. M. Gest, November 23, 1914, *Barrington Papers.* The Meteor, Arizona, post office was discontinued on April 15, 1912.

96. G. P. Merrill to D. M. Barringer, May 18, 1906, *ibid.*

97. D. M. Barringer to G. P. Merrill, May 24, 1906, *ibid.*

98. G. P. Merrill and W. Tassin, "Contributions to the Study of the Canyon Diablo Meteorites," *Smithsonian Institution Miscellaneous Collections,* 50(1907): 203–15.

99. G. P. Merrill, "The Meteor Crater of Canyon Diablo, Arizona; its history, origin, and associated irons," *Smithsonian Institution Miscellaneous Collections,* 50(1908): 461–98.

100. D. M. Barringer to J. C. Branner, November 27, 1906, *Barringer Papers.*

101. D. M. Barringer to P. B. Barringer, January 18, 1907, *ibid.*

102. D. M. Barringer to Emma Bennitt, April 19, 1907, *ibid.*

103. D. M. Barringer to J. W. Mallet, April 13, 1906, *ibid.*

104. D. M. Barringer to G. P. Merrill, February 3, 1907, *ibid.*

105. D. M. Barringer to H. H. Alexander, February 25, 1907, *ibid.*

106. W. Tassin to D. M. Barringer, April 20, 1907, *ibid.*

107. W. Tassin, "Note on the Occurrence of Graphitic Iron in a Meteorite," *Proceedings of the U.S. National Museum,* 31(1907): 573–74.

108. G. P. Merrill, "On a Peculiar Form of Metamorphism," *op. cit.*

109. G. P. Merrill and W. Tassin, *op. cit.*

110. G. P. Merrill, "The Meteor Crater of Canyon Diablo, Arizona," *op. cit.*

111. R. Arnold, report of the meeting of the Geological Society of Washington on November 25, 1908, *Science,* 29(1909): 239–40.

CHAPTER SIX. UNDER THE SOUTHERN CLIFFS

Epigraph: E. Thomson, "A Hunt for a Great Meteor—The Commercial Possibilities at Stake," *Scientific American Supplement,* 1912, No. 1896, pp. 282–83.

1. H. E. Wimperis, "Temperature of Meteorites," *Nature,* 71(1904–1905): 81–82. See also G. P. Merrill, "The Meteor Crater of Canyon Diablo, Arizona; its history, origin and associated irons," *Smithsonian Institution Miscellaneous Collections,* 50(1908): 493. Wimperis estimated the minimum mass to be 10 to 20 pounds.

2. R. A. Proctor, "The Moon's Myriad Small Craters," *Belgravia,* 36(1878): 153–71.

3. J. N. Lockyer, *The Meteoritic Hypothesis* (London: Macmillan and Company, 1890).

4. G. H. Darwin, "On the mechanical conditions of a swarm of meteorites and on theories of cosmogony," *Philosophical Transactions of the Royal Society,* series A, 180(1889): 1–69.

5. T. C. Chamberlin, "An attempt to test the Nebular Hypothesis by the relations of masses and moments," *Journal of Geology,* 8(1900): 58–73; and F. R. Moulton, "An attempt to test the Nebular Hypothesis by an appeal to the laws of dynamics," *Astrophysical Journal,* 11(1900): 103–30. Moulton preferred the name "Spiral Hypothesis" for the theory "to keep it in sharp contrast" to Laplace's Nebular Hypothesis. *See* Moulton, *Introduction to Astronomy* (New York: The Macmillan Company, 1906), 463.

6. Merrill, *op. cit.,* cites Lockyer's *The Meteoritic Hypothesis* in a footnote.

7. N. S. Shaler, "A comparison of the features of the Earth and the Moon," *Smithsonian Institution Contributions to Knowledge,* Vol. 34, No. 1438, 1903.

8. Shaler suggested that temperatures on impact might reach 150,000°F. See *ibid.*, 12.

9. D. M. Barringer to J. C. Branner, August 7, 1906, *Barringer Papers.*

10. D. M. Barringer to S. J. Holsinger, August 21, 1906, *ibid.*

11. D. M. Barringer to E. J. Bennitt, September 1, 1906, *ibid.*

12. D. M. Barringer to E. J. Bennitt, April 2, 1907, *ibid.*

13. D. M. Barringer to Emma Bennitt, April 19, 1907, *ibid.*

14. G. P. Merrill to D. M. Barringer, February 13, 1908, *ibid.*

15. G. P. Merrill to D. M. Barringer, April 23, 1908, *ibid.*

16. D. M. Barringer to J. C. Branner, June 30, 1908, *ibid.*

17. D. M. Barringer to J. Douglas, July 30, 1908, *ibid.*

18. J. C. Branner to D. M. Barringer, July 14, 1907, *ibid.*

19. "Mark of Gigantic Meteor," *The New York Times,* June 15, 1908, 2.

20. D. M. Barringer to J. F. Newsom, undated (*ca.* June 1908), *Barringer Papers.*

21. D. M. Barringer to J. W. Mallet, July 1, 1908, *ibid.*

22. "Meteor Crater Discovered by Philadelphians," *Public Ledger* (Philadelphia), June 21, 1908.

23. D. M. Barringer and B. C. Tilghman, "That Meteor Crater," letter to the editor of the *Public Ledger* (Philadelphia), July 19, 1908.

24. D. M. Barringer, "Who Really First Explained Origin of Meteor Crater," letter to editor of the *Public Ledger* (Philadelphia), March 8, 1909; and to editor, *The Iron Age,* April 15, 1909.

25. C. D. Walcott, "Meteor Crater of Canyon Diablo, Arizona," *Smithsonian Institution Annual Report* (for 1908), 9–10.

26. J. C. Branner to D. M. Barringer, March 15, 1909, *Barringer Papers.*

27. D. M. Barringer to C. F. Cox, May 26, 1910, *ibid.*

28. S. J. Holsinger to D. M. Barringer, August 14, 1908; and D. M. Barringer to E. J. Bennitt, September 22, 1908, *ibid.*

29. D. M. Barringer to J. C. Branner, June 30, 1908, *ibid.*

30. D. M. Barringer to H. S. Colton, March 14, 1927, *ibid.*

31. D. M. Barringer to E. J. Bennitt, August 24, 1908, *ibid.*

32. D. M. Barringer, "Meteor Crater (formerly called Coon Mountain or Coon Butte) in Northern Central Arizona," paper read before the National Academy of Sciences, November 16, 1909 (privately printed).

33. D. M. Barringer to G. P. Merrill, September 27, 1907, *Barringer Papers.*

34. D. M. Barringer, "Meteor Crater," *op. cit.*

35. W. F. Magie to D. M. Barringer, June 21, 1909, *Barringer Papers.*

36. D. M. Barringer to S. J. Holsinger, April 18, 1910, *ibid.*

37. D. M. Barringer to H. N. Russell, March 10, April 1 and 3, 1910; and H. N. Russell to D. M. Barringer, April 4, 1910, *ibid.*

38. D. M. Barringer, "Meteor Crater," *op. cit.,* appendix.

39. D. M. Barringer to J. W. Mallet, April 20, 1906, *Barringer Papers.*

40. W. H. Pickering to D. M. Barringer, January 25, 1909, *ibid.*

41. W. H. Pickering, "The chance of collision with a comet, iron meteorites and Coon Butte," *Popular Astronomy,* 7(1909):329–39.

42. W. H. Pickering to D. M. Barringer, October 25, 1909, *Barringer Papers.*

43. D. M. Barringer to H. N. Russell, May 19, 1910. Barringer notes here that the earth "went through the tail of a comet last night," *ibid.*

44. D. M. Barringer to H. N. Russell, May 21, 1910, *ibid.*

45. D. M. Barringer to E. Thomson, February 7, 1912, *ibid.*

46. D. M. Barringer, "A possible partial explanation of the visibility and brightness of comets," *Proceedings of the Academy of Natural Sciences of Philadelphia,* 80(1916):472–75.

47. "Addendum by Elihu Thomson," *ibid.,* 475–78.

48. D. M. Barringer to J. A. Miller, February 24, 1917; and J. A. Miller to D. M. Barringer, March 27, 1917, *Barringer Papers.*

49. D. M. Barringer to E. J. Bennitt, April 4, 1910, *ibid.*

50. W. F. Magie, "Physical Notes on Meteor Crater, Arizona," *Proceedings of the American Philosophical Society,* 49(1910):41–48; see also *American Journal of Science,* 23(1910): 335–36, and *Science,* 31(1910):872–73.

51. D. M. Barringer to E. O. Hovey, August 26, 1911, *Barringer Papers.*

52. D. M. Barringer to W. F. Magie, April 23, 1910, *ibid.*

53. D. M. Barringer to H. N. Russell, May 5, 1910, *ibid.*

54. D. M. Barringer to W. F. Magie, May 7 and 12, August 25, October 24, November 9 and 17, 1910; and February 11 and 21, and May 9, 1911, *ibid.*

55. D. M. Barringer to W. F. Magie, May 10, 1911, *ibid.*

56. W. F. Magie to D. M. Barringer, March 20, 1918, *ibid.*

57. J. M. Davidson, "A contribution to the problem of Coon Butte," *Science,* 32(1910): 724–26.

58. D. M. Barringer to J. M. Davidson, November 18, 1910, *Barringer Papers.*

59. W. C. Barnes, "The Canyon Diablo Meteor," *The Pacific Monthly,* 22(1910):347–56. Barnes later became regionally well known for his book, *Arizona Place Names.*

60. S. J. Holsinger to D. M. Barringer, July 15, 1910, *Barringer Papers.*

61. S. J. Holsinger to D. M. Barringer, August 8, 1910, *ibid.*

62. G. P. Merrill to D. M. Barringer, February 13, 1908, *ibid.*

63. D. M. Barringer to S. G. Dixon, October 26, 1910, *ibid.*

64. D. M. Barringer to E. J. Bennitt, August 9, 1911; and to Mrs. S. J. Holsinger, August 21, 1911, *ibid.*

65. D. M. Barringer to W. C. Barnes, August 24, 1911, *ibid.*

66. N. H. Darton, "The Zuni Salt Lake," *Journal of Geology,* 13(1905):185–93.

67. N. H. Darton, "A reconnaissance of part of northwestern New Mexico and northern Arizona," *U.S. Geological Survey Bulletin* 435, 1910, 27–29, 72–74.

68. D. M. Barringer to J. C. Branner, December 28, 1910, *Barringer Papers.*

69. D. M. Barringer to J. C. Branner, January 30, 1911, *ibid.*

70. D. M. Barringer to S. J. Holsinger, December 28, 1910, *ibid.*

71. D. M. Barringer to N. H. Darton, March 2, 1911, *ibid.*

72. N. H. Darton to D. M. Barringer, March 16, 1911, *ibid.*

73. C. R. Keyes, "Coon Butte and Meteorite Falls in the Desert," abstract, *Bulletin of the Geological Society of America,* 21(1910):773–74; and "Phenomena of Coon Butte Region, Arizona," abstract, *Science,* 34(1911):29.

74. D. M. Barringer to E. Thomson, September 11, 1912, *Barringer Papers.*

75. D. M. Barringer to C. M. Smyth, Jr., January 8, 1914, *ibid.*

76. N. H. Darton, "Guidebook to the Western United States," *U.S. Geological Survey Bulletin 613,* 1915, Part C, 112–13.

77. N. H. Darton, "Explosion Craters," *The Scientific Monthly,* 3(1916):417–30.

78. Decision lists 4607–4609, U.S. Board of Geographical Names, U.S. Department of the Interior, Washington, D.C.

79. N. H. Darton, "Crater Mound, Arizona," abstract, *Bulletin of the Geological Society of America,* 56(1945):1154.

80. D. M. Barringer to H. N. Russell, May 10, 1910; and to S. J. Holsinger, May 12, 1910, *Barringer Papers.*

81. H. C. Wilson, "The fourth conference of the International Union for Cooperation in Solar Research," *Popular Astronomy,* 18(1910):489–503.

82. D. M. Barringer to G. E. Hale, September 14, 1910; and to H. N. Russell, October 19, 1910, Barringer Papers.

83. D. M. Barringer to T. Roosevelt, January 28, 1911, *ibid.*

84. T. Roosevelt to D. M. Barringer, January 31, 1911, *ibid.*

85. D. M. Barringer to T. Roosevelt, February 2, 1911, *ibid.*

86. T. Roosevelt to D. M. Barringer, February 7, 1911, *ibid.*

87. D. M. Barringer to W. M. Davis, August 8, 13, 15, 19, and 22, 1912, *ibid.* The group was known as the Raymond-Whitcomb Party.

88. D. M. Barringer to E. Thomson, September 11, 1912, *ibid.*

89. D. M. Barringer to W. M. Davis, September 16, 1912, *ibid.*

90. D. M. Barringer to W. M. Davis, October 4, 1912, *ibid.*

91. D. M. Barringer to E. Thomson, August 13 and 20, 1912, *ibid.*

92. D. M. Barringer to Harvey House (Winslow), September 10 and 16, 1912; and to E. J. Bennitt, telegram, September 11, 1912, *ibid.*

93. "Scientists Study of Meteor Crater," *The New York Times,* October 6, 1912.

94. D. M. Barringer to G. P. Merrill, October 25, 1912; to W. M. Davis, and to E. Thomson, October 29, 1912, *Barringer Papers.*

95. Barringer had, in fact, informed Holsinger's son, Henry, who became caretaker at his father's death, that "the scientists are coming, so lock up the specimens," and had directed him to dig several trenches "to disclose four to ten shale balls" *in situ*. See D. M. Barringer to H. Holsinger, August 22, 1912, *ibid*.

96. D. M. Barringer to W. D. Johnson, November 7, 1912, *ibid*.

97. D. M. Barringer to E. De Cholnoky, November 8, 1912, *ibid*.

98. D. M. Barringer to W. D. Johnson, November 14, 1912, *ibid*.

99. D. M. Barringer to S. J. Holsinger, July 14, 1911, *ibid*.

100. D. M. Barringer to E. J. Bennitt, July 17, 1911; and to S. J. Holsinger, July 19, 1911, *ibid*.

101. D. M. Barringer to W. F. Magie, July 14, 1911, *ibid*.

102. D. M. Barringer to S. J. Holsinger, July 14, 1911; and to Mrs. S. J. Holsinger, August 21 and 23, 1911, *ibid*.

103. D. M. Barringer to E. Thomson, March 8, 1911, *ibid*.

104. D. M. Barringer to S. J. Holsinger, March 8, 1911, *ibid*.

105. D. M. Barringer to E. J. Bennitt, September 5, 1911, *ibid*.

106. E. Thomson, "The Fall of a Meteorite," *Proceedings of the American Academy of Arts and Sciences,* 47(1912): 721–33.

107. E. Thomson, "A Hunt for a Great Meteor—The Commercial Possibilities at Stake," *Scientific American Supplement,* 1912, No. 1896, 282–83.

CHAPTER SEVEN. COON MOUNTAIN AND THE MOON

Epigraph: F. Meineke, "Der Meteorkrater von Canyon Diablo in Arizona und seine Bedeutung für die Enstehung der Mondkrater," *Naturwissenschaftliche Wochenschrift,* 8(1909): 801–10.

1. A. E. Foote, "Geological Features of the Meteoric Locality in Arizona," *Proceedings of the Academy of Natural Sciences of Philadelphia,* 40(1891): 407.

2. T. C. Chamberlin and R. D. Salisbury, *Geology* (New York: Henry Holt and Co., 1904), 567–69.

3. T. J. J. See, "The Origin of the So-called Craters on the Moon by the Impact of Satellites, and the Relations of these Satellite Indentations to the Obliquities of the Planets," *Popular Astronomy,* 18(1910): 137–44. See also See, *Researches on the Evolution of Stellar Systems* (Lynn, Massachusetts: Thomas Nichol & Son, 1910), 330–50.

4. W. Goodacre, "The Origin of the Lunar Surface Formations," *Journal of the British Astronomical Association,* 21(1910–1911): 25–32. See also Report of Meeting for October 26, 1910, and January 25, 1911, *ibid.,* 19–20, 179.

5. F. Meineke, "Der Meteorkrater von Canyon Diablo in Arizona und seine Bedeutung für die Enstehung der Mondkrater," *Naturwissenschaftliche Wochenschrift,* 8(1909): 801–10.

6. E. H. L. Schwarz, "The Probability of a Large Meteorite Having Fallen on the Earth," *Journal of Geology,* 17(1909): 124–25.

7. R. S. Dietz, personal communication, March 8, 1983.

8. E. Thomson, "A Hunt for a Great Meteor—The Commercial Possibilities at Stake," *Scientific American Supplement,* 1912, No. 1896, 282–83.

9. E. Thomson, "The Fall of a Meteorite," *Proceedings of the American Academy of Arts and Sciences,* 47(1912):721–33.

10. Darwin, however, conceded that the idea of the moon's origin in the earth, which was an extrapolation from his tidal friction theory of the earth-moon system, was "wild speculation, incapable of verification." See G. H. Darwin, *The Tides,* 3rd ed. (London: John Murray, 1911), 287–90.

11. This idea was not new. W. H. Pickering in 1909, for example, suggested that the stony meteorites, at least, "were all of them formed during the great cataclysm that occurred at the time that the moon separated from the earth," and even that these fragments of the earth might have formed a terrestrial ring of "moonlets." See Pickering, "The Origin of Meteorites," *Popular Astronomy,* 17(1909):273–82. A similar idea was advanced in 1910 by Robert Parry; see Parry, "Meteoric Markings on the Moon," *Journal of the British Astronomical Association,* 21(1910–1911):194.

12. D. M. Barringer, "A Discussion of the Origin of the Craters and Other Features of the Lunar Surface," a reprint of "Volcanoes—or Cosmic Shell Holes?" *Scientific American,* 131(1924):10–11, 62–63, 102, 104.

13. D. M. Barringer to E. Thomson, February 7, 1912, *Barringer Papers.*

14. D. M. Barringer to E. Thomson, April 4, 1912, *ibid.*

15. D. M. Barringer to E. Thomson, August 13, 1912, *ibid.*

16. D. M. Barringer to A. Lacroix, October 7, 1913, *ibid.* The Sproul telescope was in fact a 24-inch refractor.

17. D. M. Barringer to G. P. Merrill, October 14, 1913, *ibid.*

18. A. M. Worthington, *A Study of Splashes* (London: Longmans, Green and Company, 1908).

19. D. M. Barringer to A. M. Worthington, January 30, 1914, *Barringer Papers.*

20. D. M. Barringer to W. F. Magie and to E. Thomson, November 12, 1914, *ibid.*

21. A. M. Worthington to D. M. Barringer, December 14, 1914, *ibid.*

22. D. M. Barringer to E. Thomson, December 26, 1914, *ibid.* The idea of vegetation on the moon, however, had been suggested only a month earlier by W. H. Pickering. See Pickering, "The Double Canal of the Lunar Crater Aristillus," *Popular Astronomy,* 22(1914):570–78.

23. D. M. Barringer, "Further Notes on Meteor Crater, Arizona," *Proceedings of the Academy of Natural Sciences of Philadelphia,* 66(1914):556–65.

24. D. M. Barringer to G. P. Merrill, October 14, 1913; to A. Lacroix, October 7 and 18, 1913; and to E. O. Hovey, December 27, 1913, *Barringer Papers.*

25. D. M. Barringer, "Further Notes," *op. cit.*

26. D. M. Barringer to G. F. Kunz, May 6, 1915, *Barringer Papers.*

27. D. M. Barringer to E. J. Bennitt, December 28, 1914 *ibid.*

28. D. M. Barringer, "A Possible Partial Explanation of the Visibility and Brilliancy of Comets," *Proceedings of the Academy of Natural Sciences of Philadelphia,* 68(1916):472–78.

29. There were exceptions, of course. One Stanislaus Meunier, for example, reported to the French Academy of Sciences in January 1916 that his interpretation of the meteoritic material at Coon Mountain "agrees with the view put forth by Barringer and Tilghman that the crater of Coon Butte was excavated by the shock of a meteorite." See "Note on S. Meunier, 'New Observations on the Structure of the Meteoric Irons of the Canyon Diablo (Arizona),'" *Nature*, 96(1916):1697.

30. "A Terrestrial Crater of the Lunar Type," *ibid.*, 595–96. *See also* abstract, *Publications of the Astronomical Society of the Pacific,* 28(1916):99–100.

31. D. M. Barringer to editor, *Nature,* August 11 and October 18, 1915, *Barringer Papers.*

32. O. C. Farrington, "Catalog of the Meteorites of North America," *Memoirs of the National Academy of Sciences,* XIII, 1915.

33. D. M. Barringer to O. C. Farrington, January 22, 1915, *Barringer Papers.*

34. O. C. Farrington to D. M. Barringer, January 25, 1915, *ibid.*

35. O. C. Farrington, *Meteorites—Their Structure, Composition and Terrestrial Relations* (Chicago: 1915), 25.

36. D. M. Barringer to G. P. Merrill, February 22, 1916, *Barringer Papers.*

37. D. M. Barringer to O. C. Farrington, February 25, 1916, *ibid.*

38. D. M. Barringer to O. C. Farrington, February 28, 1916, *ibid.*

39. O. C. Farrington to D. M. Barringer, March 9, 1916, *ibid.*

40. D. M. Barringer to G. F. Kunz, May 6, 1915, *ibid.*

41. D. M. Barringer to E. J. Bennitt, December 15, 1914, *ibid.*

42. Option agreement, dated February 25, 1915, Standard Iron Co. letterbook for October 29, 1912 to July 3, 1918, *Barringer Papers.*

43. D. M. Barringer to E. J. Bennitt, September 3, 1919; to Mrs. S. J. Holsinger, November 21, 1919; and to W. F. Magie, September 26, November 29, and December 15 and 18, 1919, *ibid.*

44. D. M. Barringer to E. J. Bennitt, November 30, and December 2, 1914, *ibid.*

45. D. M. Barringer to Coconino County Treasurer (Flagstaff), December 10, 1914, *ibid.*

46. D. M. Barringer to E. J. Bennitt, December 10, 1914, *ibid.*

47. D. M. Barringer to E. Thomson, September 22, 1914, *ibid.*

48. D. M. Barringer to D. Baugh, January 26, 1915, *ibid.* Baugh was the owner of the Baugh Chemical Co., founder of the Baugh Institute, and a fellow trustee with Barringer on the board of Philadelphia's Jefferson Medical College.

49. D. M. Barringer to E. Thomson, December 15, 1915, *ibid.*

50. D. M. Barringer to E. J. Bennitt, June 18, 1915; and to H. Gray, October 18, 1916, *ibid.*

51. D. M. Barringer to P. M. Sharples, September 26, 1916; and to E. J. Bennitt, September 28, 1916, enclosing Sharples's reply, *ibid.*

52. D. M. Barringer to E. J. Bennitt, and to H. Gray, October 10, 1916, *ibid.*

53. D. M. Barringer to H. Gray, October 18 and 21, 1916; and to P. M. Sharples, October 26 and November 3, 1916, *ibid*.

54. D. M. Barringer to E. J. Bennitt, March 29, 1916, *ibid*.

55. D. M. Barringer to E. J. Bennitt, November 24, 1916, *ibid*.

56. D. M. Barringer, memorandum, dated January 4, 1917, *ibid*.

57. D. M. Barringer to P. M. Sharples, February 5, 1917; May 13, 1919; June 29, October 23 and 29, and December 30, 1917; to H. Gray, April 24 and 28, and October 30, 1917; and to A. P. Anderson, December 30, 1919, *ibid*.

58. D. M. Barringer to S. S. Woods, June 1, 1917, *ibid*.

59. D. M. Barringer to E. Thomson, September 11, 1917, *ibid*.

60. D. M. Barringer to W. C. Sproul, November 8 and 12, 1917; and to A. C. Elkington, December 10, 1917, *ibid*. A major source of high quality silica was, in fact, located at nearby Lewistown, Pennsylvania.

61. A. L. Day to D. M. Barringer, January 5, 1918; and D. M. Barringer to A. L. Day, January 7, 1918, *ibid*.

62. D. M. Barringer to E. Thomson, June 12, 1917, *ibid*.

63. D. M. Barringer to E. Thomson, March 19, 1918, *ibid*.

64. D. M. Barringer to E. Thomson, August 8, 1918; and to W. F. Magie, August 2, 1918, *ibid*.

65. D. M. Barringer to W. F. Magie, August 2, 1917, *ibid*.

66. D. M. Barringer, memorandum, dated September 14, 1917, in letterbook for October 29, 1912 to July 3, 1918, *ibid*.

67. D. M. Barringer to E. Thomson, October 19, 1917, *ibid*.

68. D. M. Barringer to J. F. Newsom, October 23, 1917, *ibid*.

69. D. M. Barringer to J. F. Newsom, December 27, 1917, *ibid*.

70. D. M. Barringer to O. A. Hart, telegram, December 26, 1917, *ibid*.

71. D. M. Barringer to J. F. Newsom, February 5, 1918, *ibid*.

72. D. M. Barringer to E. Thomson, March 8, 1918, *ibid*.

73. D. M. Barringer to W. F. Magie, March 19, 1918, *ibid*.

74. D. M. Barringer to E. Thomson, March 8, 1918, *ibid*.

75. D. M. Barringer to E. Thomson, March 19, 1918, *ibid*.

76. D. M. Barringer, memorandum, dated March 23, 1918, in letterbook for October 29, 1912 to July 3, 1918, *ibid*.

77. D. M. Barringer to S. J. Jennings, March 27, 1918, *ibid*.

78. D. M. Barringer to J. F. Newsom, April 17, 1918, *ibid*.

79. D. M. Barringer to E. Thomson, April 29, 1918, *ibid*.

80. D. M. Barringer to E. Thomson, April 29, 1918 (second letter), *ibid*.

81. D. M. Barringer to E. J. Bennitt, May 3, 1918, *ibid*.

82. D. M. Barringer to W. F. Magie and to E. Thomson, May 9, 1918, *ibid*.

83. D. M. Barringer to W. F. Magie, May 10, 1918, *ibid*.

84. D. M. Barringer to E. Thomson, May 17, 1918, *ibid*.

85. D. M. Barringer to W. F. Magie, June 6, 1918, *ibid*.

86. D. M. Barringer to S. J. Jennings, June 13, 1918, *ibid*.

87. D. M. Barringer to S. J. Jennings, June 18, 1918, *ibid*.

88. D. M. Barringer to H. H. Alexander, July 6, 1918, *ibid*.

89. D. M. Barringer to W. F. Magie, July 9, 1918, *ibid*.

90. D. M. Barringer to C. F. Moore, July 6, 10 (two letters), 11, 16, 19, 27, and 30, 1918; and to H. H. Alexander, July 10, 12, 15, 17, 20, and 22, 1918, *ibid*.

91. D. M. Barringer to O. A. Hart, July 12 and 18 (telegram), 1918; to W. F. Magie, July 6 and 9, 1918; and to A. H. Phillips, August 2, 1918, *ibid*.

92. D. M. Barringer to H. H. Alexander, July 20, 1918, *ibid*.

93. D. M. Barringer to C. F. Moore, August 24, 1918, *ibid*.

94. D. M. Barringer to H. H. Alexander, September 4, 1918, *ibid*.

95. D. M. Barringer to S. J. Jennings, October 11, 1918, *ibid*.

96. D. M. Barringer to C. F. Moore, November 13, 1918; and to S. J. Jennings, December 16, 1918, *ibid*.

97. D. M. Barringer to W. F. Magie, January 10, 1919, *ibid*.

98. D. M. Barringer to E. Thomson, October 5, 1918, *ibid*.

99. D. M. Barringer to E. Thomson, October 11, 1918, *ibid*.

100. D. M. Barringer to W. F. Magie, January 10 and February 19, 1919; to K. Taylor, February 11, 1919; to C. Scribner, February 17, 1919; to M. J. Pyne, February 18, 1919; and to B. Henry, March 17, 1919, *ibid*.

101. D. M. Barringer to E. Thomson, February 26, 1919, *ibid*.

102. D. M. Barringer to E. Thomson, March 5, 1919; and to M. Prescott, April 1, 1919, *ibid*.

103. D. M. Barringer to H. F. Osborn, June 18, 20, 26 (telegram), 30, and July 7, 1919, *ibid*.

104. D. M. Barringer to L. H. Baekeland, February 18 and 19, 1919; and to S. Arrhenius, March 31, 1919, *ibid*.

105. S. Arrhenius to D. M. Barringer, postcard, May 4, 1919, *ibid*.

106. D. M. Barringer to L. D. Ricketts, June 17 and 19, 1919, *ibid*.

107. D. M. Barringer to L. D. Ricketts, June 25, 1919, *ibid*.

108. D. M. Barringer to L. D. Ricketts, June 26, 1919, *ibid*.

109. D. M. Barringer to O. A. Hart, June 27 and 30, 1919, *ibid*.

110. D. M. Barringer to S. J. Jennings, August 5, 1919, *ibid*.

111. D. M. Barringer to S. J. Jennings, August 18, 1919, *ibid*.

112. D. M. Barringer to L. D. Ricketts and to W. F. Magie, August 27, 1919, *ibid*.

113. D. M. Barringer to W. F. Magie, September 6, 1919, *ibid*.

114. D. M. Barringer to S. J. Jennings, telegram, October 6, 1919; and to L. D. Ricketts, telegrams, October 21 and November 11, 1919, *ibid.*

115. D. M. Barringer to O. A.. Hart, telegram, October 22, 1919, *ibid.*

116. D. M. Barringer to S. J. Jennings and to W. F. Magie, November 13, 1919, *ibid.*

117. L. D. Ricketts to D. M. Barringer, November 26, 1919, enclosing a copy of H. M. Roberts, "Report on Coon Crater," E. J. Longyear Co., Minneapolis, Minnesota, *ibid.*

118. H. M. Roberts, "Report on Coon Crater," *op. cit.*

119. D. M. Barringer to C. F. Moore, November 28, 1919, *Barringer Papers.*

120. D. M. Barringer to L. D. Ricketts, December 2, 1919, *ibid.*

121. D. M. Barringer to E. Thomson, and to W. F. Magie, December 1, 1919, *ibid.*

122. D. M. Barringer to G. P. Merrill (two letters), to W. B. Scott, to E. O. Hovey, and to G. F. Kunz, December 2, 1919, *ibid.*

123. E. O. Hovey and G. F. Kunz to D. M. Barringer, December 3, 1919; G. P. Merrill to D. M. Barringer (two letters), December 4, 1919; E. Thomson to D. M. Barringer, December 17, 1919; and W. F. Magie to D. M. Barringer, December 23, 1919, *ibid. See also* D. M. Barringer to C. F. Moore, December 15, 1919; and to S. J. Jennings, December 15 and 17, 1919, *ibid.* Scott did not reply immediately, but he later sent a letter supporting Barringer's ideas; *see* W. B. Scott to D. M. Barringer, February 13 and March 10, 1920, *ibid.*

124. D. M. Barringer to S. J. Jennings, January 7, 1920, *ibid.*

125. D. M. Barringer to W. F. Magie and to E. Thomson, January 16, 1919, *ibid.*

126. D. M. Barringer to E. J. Bennitt, January 30, 1920, *ibid.*

127. Agreement to lease, dated April 28, 1920, in letterbook for July 4, 1918 to March 21, 1921, p. 604, *ibid.*

128. D. M. Barringer to E. Thomson, April 29, 1920; and to J. Hight, April 30, 1920, *ibid.* Hight was an attorney for U.S. Smelting.

129. W. W. Campbell, "Notes on the Problem of the Origin of the Lunar Craters," *Publications of the Astronomical Society of the Pacific,* 32(1920):126–38. *See also* D. M. Barringer to W. W. Campbell, October 6, 1915, *Barringer Papers.*

130. J. C. Branner to D. M. Barringer, November 17 and December 8, 1916; and D. M. Barringer to J. C. Branner, December 1, 1916, *Barringer Papers.* Unbeknownst to Barringer, Campbell had considered inviting N. H. Darton, then in Arizona, to accompany him on this visit. See W. W. Campbell to N. H. Darton, November 3, 1916; N. H. Darton to W. W. Campbell, December 6, 1916; and W. W. Campbell to N. H. Darton, December 15, 1916, *Mary Lea Shane Archives of the Lick Observatory,* University of California at Santa Cruz.

131. D. M. Barringer to W. W. Campbell, telegram, April 30, 1920, *Barringer Papers.*

132. W. W. Campbell to D. M. Barringer, telegram, May 6, 1920, *ibid.*

133. W. W. Campbell to D. M. Barringer, May 6, 1920, *ibid.*

134. D. M. Barringer to S. J. Jennings, May 7, 1920; and to W. W. Campbell, telegram, May 8, 1920, *ibid.*

135. D. M. Barringer to W. W. Campbell, May 13, 1920, *ibid.*

136. W. W. Campbell to D. M. Barringer, May 18, 1920, *ibid.*

137. W. W. Campbell, "The Meteor Crater, Arizona," abstract, *Publications of the Astronomical Society of the Pacific,* 32(1920):197. The paper itself was not published, but Campbell's eleven-page typewritten text, with handwritten corrections, is on file in the Mary Lea Shane Archives of the Lick Observatory.

138. D. M. Barringer to E. Thomson, May 14 and 25, 1920, *Barringer Papers.*

139. D. M. Barringer to W. W. Campbell, October 18, 1920, *ibid.*

140. D. M. Barringer to N. W. Rice, October 28, 1920, *ibid.*

141. D. M. Barringer to E. Thomson, October 28, 1920, *ibid.*

142. D. M. Barringer to W. W. Campbell, telegram, October 28, 1920, *ibid.*

143. W. W. Campbell to D. M. Barringer, November 2, 1920, *ibid.*

144. D. M. Barringer to G. P. Merrill, November 19, 1920, *ibid.*

145. G. P. Merrill, "A Retrospective View of the Origin of Meteor Crater, Arizona," *Publications of the Astronomical Society of the Pacific,* 32(1920):259–64.

146. D. M. Barringer to E. Thomson, October 27, 1920; and W. W. Campbell to D. M. Barringer, February 28, 1923, *Barringer Papers.*

147. D. M. Barringer to W. F. Magie, November 19, 1920; and to E. Thomson, November 24, 1920, *ibid.*

148. W. W. Campbell to D. M. Barringer, July 11, 1921, *ibid.*

149. D. M. Barringer to W. W. Campbell, July 19, 1921, *ibid.*

150. W. W. Campbell to D. M. Barringer, July 26, 1921, *ibid.*

151. H. N. Russell to D. M. Barringer, July 27, 1921, *ibid.*

152. W. W. Campbell to D. M. Barringer, July 11, 1921, *ibid.*

153. D. M. Barringer to W. W. Campbell, July 19, 1921, *ibid.*

154. D. M. Barringer to E. Thomson, April 26 and October 11, 1926; to K. Boutwell, May 18, 1926; and to A. W. Stevens, October 15, 1926, *ibid.*

155. D. M. Barringer to S. J. Jennings, May 1 and 7, 1920; and to O. A. Hart, May 7, 1920, *ibid.*

156. D. M. Barringer to C. F. Moore, July 16, 1920, *ibid.*

157. D. M. Barringer to E. Thomson, July 19, 1920, *ibid.*

158. D. M. Barringer to E. Thomson, July 26, 1920, *ibid.*

159. D. M. Barringer to S. J. Jennings, May 7, 1920; to C. F. Moore, July 16, August 5, 19, 21, 27, September 6, 8, October 21, 1920; to J. Holland, November 4, 1920; and to W. F. Magie, E. Thomson, and N. W. Rice, November 10, 1920, *ibid.*

160. D. M. Barringer to S. J. Jennings, October 11, 1920; and to P. M. Sharples, October 11 and 13, and December 21, 1920, *ibid.*

161. D. M. Barringer to W. F. Magie, October 28, 1920, *ibid.*

162. D. M. Barringer to W. F. Magie, February 25, 1921, *ibid.*

163. D. M. Barringer to W. F. Magie, March 11, 1921, *ibid.*

164. D. M. Barringer to S. J. Jennings, June 3, 1921; and to W. W. Campbell, September 13, 1921, *ibid.*

165. D. M. Barringer to J. C. Branner, December 14, 1921; and to S. J. Jennings, February 28, 1922, *ibid.*

166. D. M. Barringer to W. F. Magie, April 4, 1922, *ibid.*

167. D. M. Barringer to W. F. Magie, April 6, 1922, *ibid.*

168. D. M. Barringer to S. J. Jennings, May 6 and 23, and June 13 and 15, 1922; and to W. W. Campbell, June 8, 1922, *ibid.*

169. D. M. Barringer to S. J. Jennings, August 28, 1922, *ibid.*

170. D. M. Barringer to S. J. Jennings, August 2, 1922; to C. W. Plumb, October 9, 1922; and to W. F. Magie, August 29, 1923, *ibid.* Plumb was the drilling supervisor at the crater.

171. D. M. Barringer to C. W. Plumb, August 24, 1922, *ibid.*

172. D. M. Barringer to W. F. Magie, September 30, 1922, *ibid.*

173. D. M. Barringer to S. J. Jennings, November 4, 1922, *ibid.*

174. D. M. Barringer to S. J. Jennings; and to C. F. Moore, October 27, 1922, *ibid.*

175. D. M. Barringer to C. W. Plumb, November 15, 1922, *ibid.*

176. D. M. Barringer, "Further Notes on Meteor Crater in Northern Central Arizona (No. 2)," text, in letterbook for March 27, 1921 to April 28, 1923, pp. 839–45. *See also Proceedings of the Academy of Natural Sciences of Philadelphia,* 76(1924): 275–78.

177. D. M. Barringer to W. J. Fox, March 16, 1923, *Barringer Papers.*

178. D. M. Barringer to W. W. Campbell, February 16 and March 16, 1923, *ibid.*

179. W. W. Campbell to D. M. Barringer, February 28, 1923, *ibid.*

180. D. M. Barringer to W. W. Campbell, March 9 and 21 and April 4, 1923, *ibid.*

181. D. M. Barringer to E. Thomson, April 9, 1923, *ibid.*

182. D. M. Barringer to E. Thomson, April 12, 1923, *ibid.*

183. D. M. Barringer to E. J. Bennitt, April 19, 1923, *ibid.*

184. D. M. Barringer, "A Discussion of the Origin of the Craters and Other Features of the Lunar Surface," text, in letterbook for April 30, 1923 to February 27, 1925, pp. 96–123, *ibid.*

185. D. M. Barringer to E. Thomson, July 10, 1923, *ibid.*

186. D. M. Barringer to W. F. Magie, November 13, 1923, *ibid.*

187. D. M. Barringer to J. A. Miller, June 6 and 13, 1923; and to E. Thomson, June 29, 1923, *ibid.*

188. D. M. Barringer to W. S. Adams, June 14 and 26, 1923, *ibid.*

189. D. M. Barringer to J. M. Bird, August 2, 9, 16, and 20, 1923, *ibid.*

190. D. M. Barringer, "Volcanoes—or Cosmic Shell Holes?" *op. cit.*

191. R. W. Barringer, "An Argument on the Origin of the Lunar Craters," *Popular Astronomy,* 31(1923): 450–54.

192. D. M. Barringer to E. Thomson, September 20, 1923, *Barringer Papers.*

193. J. M. Bird, "With the Editors," *Scientific American,* 131(1924): 3.

194. Barringer, if not the readers of *Scientific American,* was aware of this. *See* D. M.

Barringer to H. N. Russell, June 23, 1924; and to J. M. Bird, June 26, 1924, *Barringer Papers*.

CHAPTER EIGHT. THE NEW METEORIC HYPOTHESIS

Epigraph: A. C. Gifford, "The Mountains of the Moon," in "Astronomical—Interesting and Instructive Notes," published for the New Zealand Astronomical Society by the *New Zealand Times*, May 1924.

1. R. B. Baldwin, "The Origin of Lunar Features," in *Geological Problems in Lunar Research, Annals of the New York Academy of Sciences*, 123(1965): 543–46.

2. D. M. Barringer, Jr., "A New Meteor Crater," *Proceedings of the Academy of Natural Sciences of Philadelphia*, 80(1928): 307–11. I. Reinwaldt, "Bericht über geologische Untersuchungen am Kaalijärv (Krater von Sall) auf Ösel," *Publications of the Geological Institution of the University of Tartu*, No. 11, 1928; and L. J. Spencer, "Meteorite Craters as Topographical Features on the Earth's Surface," *Smithsonian Institution Annual Report*, 1933, 307–26. Spencer was apparently not familiar with Barringer's and Merrill's papers on Meteor Crater, for he called the finding of "silica glass" at the Wabar Craters in Arabia "a remarkable discovery, for no similar material had ever been found before. . . ."

3. N. T. Bobrovnikoff, "A Remarkable Meteorite," *Publications of the Astronomical Society of the Pacific*, 39(1927): 382–84; and "the Podkammennaya Tunguska Meteorite," *ibid.*, 40(1928): 143–45. These early reports noted that crateriform "funnels" up to 10 yards in diameter were found at the site, but subsequent investigation failed to confirm this, and it is now generally believed that the meteoroid exploded in the atmosphere. These early reports were also sensationalized in the popular media.

4. A. C. Gifford, "The Mountains of the Moon," in "Astronomical—Interesting and Instructive Notes," published for the New Zealand Astronomical Society by the *New Zealand Times*, May 1924; and "The Mountains of the Moon," *New Zealand Journal of Science and Technology*, 7(1924): 129–42. This latter and somewhat longer version was reprinted as a Hector Observatory *Bulletin*.

5. Reports of meetings of May 26 and June 30, 1915, *Journal of the British Astronomical Association*, 25(1915): 326–28, 366–67.

6. E. J. Öpik, "Remarque sur le théorie météorique des cirques lunaires," *Bulletin de la Société Russe des Amis de l'Etude de l'Univers*, No. 3, 21(1916): 125–34; and H. E. Ives, "Some Large-Scale Experiments Imitating the Craters of the Moon," *Astrophysical Journal*, 50(1919): 245–50.

7. W. W. Campbell, "Notes on the Problem of the Origin of the Lunar Craters," *Publications of the Astronomical Society of the Pacific*, 32(1920): 126–38.

8. J. Nasmyth and J. Carpenter, *The Moon Considered as a Planet, a World, and a Satellite*, London, 1874.

9. R. A. Proctor, "The Moon's Myriad Small Craters," *Belgravia*, 36(1878): 153–71.

10. G. K. Gilbert, "The Moon's Face; a study of the origin of its features," *Bulletin of the Philosophical Society of Washington*, 12(1893): 241–92.

11. N. S. Shaler, "A Comparison of the Features of the Earth and the Moon," *Smithsonian Institution Contributions to Knowledge,* No. 1438, Vol. 34, 1903.

12. J. Lankford, "A Note on T. J. J. See's Observations of Craters on Mercury," *Journal for the History of Astronomy,* 11(1980): 129–32.

13. T. J. J. See, "The Origin of the Lunar Terrestrial System by Capture, With Further Considerations on the Theory of Satellites and on the Physical Cause Which Has Determined the Directions of the Rotations of the Planets About Their Axes," *Astronomische Nachrichten,* No. 4343, 181(1909): 365–86; "The Origin of the So-Called Craters on the Moon by the Impact of Satellites, and the Relations of These Satellite Indentations to the Obliquities of the Planets," *Popular Astronomy,* 18(1910): 137–44; and *Researches on the Evolution of Stellar Systems,* Lynn, Massachusetts, 1910, Vol. II.

14. T. J. J. See, *Researches, op. cit.,* 350.

15. *Ibid.,* 695.

16. F. R. Moulton, "Capture Theory and Capture Practice," *Popular Astronomy,* 20 (1912): 67–82.

17. F. J. M. Stratton, "Dr. See's Researches on Cosmogony," *The Observatory,* 34(1911): 347–49.

18. F. W. Henkel, review, *Journal of the British Astronomical Association,* 21(1910–1911): 275–78; and "New Ideas on Planetary Evolution," *The Observatory,* 33(1910): 242–46.

19. W. Goodacre, "The Origin of the Lunar Surface Formations," *Journal of the British Astronomical Association,* 21(1910–1911): 25–32.

20. Report of meeting of October 26, 1910, *ibid.,* 19–20.

21. M. E. Mulder, "De Explosive van Meteoren en hat Onstaan van der Meteorkrater van Canyon Diablo," *Ingenieur,* 26(1911): 880–85.

22. W. M. Foote, "A Remarkable Meteorite Fall," *Popular Astronomy,* 21(1913): 118–20.

23. Mulder quotes from the 14th edition of Meyer's book, which I have been unable to locate. The identical passage, however, appears in the 17th edition (1916).

24. D. M. Barringer to T. J. J. See, November 8, 1911, *Barringer Papers.*

25. W. H. Pickering, "The Double Canal of the Crater Aristillus," *Popular Astronomy,* 22(1914): 570–78.

26. Report of meeting of January 27, 1915, *Journal of the British Astronomical Association,* 25(1915): 163–65.

27. T. J. J. See, "The Origin of the Moon," *ibid.,* 282–84. The moon, of course, is actually receding from the earth due to the acceleration of its orbital velocity by the earth's tides.

28. M. Davidson, "The Origin of the Moon," *ibid.,* 346–48.

29. A. C. Gifford, "A. W. Bickerton—An Appreciation," *The Evening Post,* Wellington, New Zealand, January 26, 1929.

30. Report of meeting of May 26, 1915, *Journal of the British Astronomical Association,* 25(1915): 326–28.

31. M. Davidson, "The Appearance of Lunar Craters," *ibid.,* 380–82.

32. Report of the meeting of June 30, 1915, *ibid.*, 366–67.

33. Report of British Astronomical Association meetings of May 26 and June 30, 1915, *The Observatory*, 38(1915): 384–85, 313–14.

34. E. J. Öpik, *op. cit.*

35. E. J. Öpik, "Researches on the Physical Theory of Meteor Phenomena," *Publications de L'Observatoire Astronomique de L'Universite de Tartu*, 28(1936): 3–27.

36. H. E. Ives, *op. cit.*

37. G. E. Hale, "Lunar Photography With the Hooker Telescope," *Publications of the Astronomical Society of the Pacific*, 32(1920): 112–15; W. H. Pickering, "The Origin of the Lunar Formations," *ibid.*, 116–25; and W. W. Campbell, *op. cit.*

38. Campbell claimed to have observed these craterlets atop some of the central peaks in the 1890s, but "made no notes on the subject, as I supposed these features were well known to students of the Moon." See Campbell, *op. cit.*, 132–33.

39. See "Notes," in *Journal of the British Astronomical Association*, 30(1919–1920): 302–03, and *ibid.*, 31(1920–1921): 90–91.

40. E. O. Fountain, "Dissimilarity Between Lunar Craters and Those Caused by Explosive Bombs," *Journal of the British Astronomical Association*, 31(1920–1921): 42–43.

41. A. C. Curtis, "Lunar Craters," *ibid.*, 87.

42. E. O. Fountain, "Lunar Craters," *ibid.*, 121–23.

43. G. H. Lepper, "Origin of the Lunar Formations," *ibid.*, 120–21.

44. E. O. Fountain, "Origin of Lunar Formations," *ibid.*, 163–64.

45. The genesis of the continental drift hypothesis goes back to 1620 and Francis Bacon, who noted the similar configuration of South America's east coast and Africa's west coast. In the early 1800s, Alexander von Humboldt proposed that the two continents had once been linked. In the late nineteenth century, a number of investigators, including Austrian geologist Eduard Seuss, contributed to the idea, several ascribing the drifting of continents to catastrophes involving the moon, or to tidal action. Wegener's 1912 hypothesis, however, was far more detailed and comprehensive up to its time, and provoked violent opposition among geoscientists. In the early 1960s, through advancements in paleomagnetism, geochronology, and oceanography, the concept led to the modern theory of plate tectonics.

46. A. Wegener, "Versuche zur Aufsturztheorie der Mondkrater," *Nova Acta Leopold-Carolina Deutschen Akademie der Naturforscher*, 106(1920): 109–17; and "Die Aufsturzhypothese der Mondkrater," *Sirius*, 53(1920): 189–94.

47. A. C. Gifford, "The Mountains of the Moon," *New Zealand Times*, *op. cit.*

48. A. C. Gifford, "The Mountains of the Moon," *New Zealand Journal of Science and Technology*, *op. cit.*

49. I. L. Thomsen, "Algernon Charles Gifford (1861–1948)," *Monthly Notices of the Royal Astronomical Society*, 109(1949): 146–48.

50. J. W. Wilson, "Note on the Meteoric Theory of the Origin of the Lunar Craters," *Journal of the British Astronomical Association*, 35(1925): 205–07.

51. Report of meeting, March 25, 1925, *ibid.*, 185–86.

52. Report of meeting of May 27, 1925, *ibid.*, 214–24.

53. E. O. Fountain, "The Origin of Lunar Formations," *ibid.*, 236–37.

54. Report of meeting of June 24, 1925, *ibid.*, 250–56.

55. H. G. Tomkins, "The Igneous Origin of some of the Lunar Formations," *ibid.*, 37(1927):161–81.

56. A. C. Gifford, "Lunar Formations," *ibid.*, 36(1925):32–33; and "The Origin of Lunar Craters, *ibid.*, 84–86.

57. For the volcanic analogy: W. Goodacre, H. G. Tomkins, B. M. Peek, Gavin J. Burns, H. Percy Wilkins, and William F. A. Ellison; favoring the impact theory were: J. W. Wilson, W. H. Steavenson, A. E. Levin, M. A. Ainslie, A. C. D. Crommelin, and A. C. Gifford; for a glacial explanation were: E. O. Fountain, F. J. Sellers, and Arthur Mee. Several other members participated in the debate without, however, indicating their preference for one theory or another.

CHAPTER NINE. BLUE SKY LAWS AND BUREAUCRATS

Epigraph: "Meteor Crater," prospectus, Meteor Crater Exploration and Mining Company, May 1926, *Lowell Observatory Archives*.

1. D. M. Barringer to G. M. Colvocoresses, January 13, 1925, *Barringer Papers*.

2. D. M. Barringer to G. M. Colvocoresses, April 6, 1925 (two letters); and G. M. Colvocoresses to D. M. Barringer, April 6, 1925, *ibid.*

3. D. M. Barringer to H. H. Alexander, May 23, 1923, *ibid.*

4. "Meteor Crater," prospectus, Meteor Crater Exploration and Mining Company, May 1926, *Lowell Observatory Archives*.

5. See correspondence of D. M. Barringer, P. Matthey, P. and C. Handy, June 23, 1923, to February 1924, Standard Iron Letterbook No. 8 (April 30, 1923–February 27, 1925), *Barringer Papers*. Barringer's contact with Matthey was through Parker Handy, a Princeton classmate and partner in the firm of Handy and Harman, bankers and dealers in bullion and specie.

6. D. M. Barrringer to J. S. Douglas, April 4, 1924, *ibid.*

7. D. M. Barringer to D. M. Barringer, Jr., telegrams, May 10 and 13, 1924; and to J. S. Douglas, telegram, May 17, 1924, *ibid.*

8. D. M. Barringer to J. S. Douglas, June 12, 16, and 23, 1924, *ibid.*

9. D. M. Barringer to J. S. Douglas, June 25, 1924, *ibid.*

10. D. M. Barringer to J. S. Douglas, telegram, August 1, and letter, August 7, 1924; and to W. F. Magie, August 6, 1924, *ibid.*

11. D. M. Barringer to L. Holland, June 27, 1924, *ibid.*

12. D. M. Barringer to E. J. Bennitt, September 12, 1924, *ibid.*

13. D. M. Barringer to H. N. Russell, June 23 and 25, July 15 and 19, 1924; to G. P. Merrill, July 1, 3, and 22, 1924; to W. H. Pickering, July 7, 1924; and to E. Thomson, June 30, 1924, *ibid.*

14. D. M. Barringer to F. E. Wright, July 28 and September 2, 1924, *ibid*.

15. D. M. Barringer to T. C. Chamberlin, October 9, 1924, *ibid*.

16. F. E. Wright to D. M. Barringer, December 17, 1928; and F. R. Moulton to D. M. Barringer, June 6, 1929, *ibid*.

17. D. M. Barringer to E. J. Bennitt, telegram, August 15, 1924, *ibid*.

18. D. M. Barringer to G. M. Colvocoresses, September 20 and October 2, 1924, *ibid*.

19. D. M. Barringer to L. H. Baekeland, October 15 and 24, 1924, *ibid*.

20. D. M. Barringer to G. M. Colvocoresses, October 28, 1924, *ibid*.

21. D. M. Barringer to E. J. Bennitt, October 29, 1924, *ibid*.

22. D. M. Barringer to G. M. Colvocoresses, December 3, 1924, *ibid*.

23. D. M. Barringer to G. M. Colvocoresses, December 17, 1924, *ibid*.

24. D. M. Barringer to E. Thomson, December 19, 1924; and to G. M. Colvocoresses, December 27, 1924, *ibid*.

25. D. M. Barringer to W. F. Magie, December 30, 1924; and "Letter to Stockholders," Meteor Crater Exploration and Mining Company, June 1, 1928, *ibid*.

26. D. M. Barringer to W. F. Magie, February 5, 1925, *ibid*.

27. D. M. Barringer to G. M. Colvocoresses, February 19, 1925, *ibid*.

28. D. M. Barringer to G. M. Colvocoresses, March 9, 1925, *ibid*.

29. D. M. Barringer to E. Thomson, March 31, 1925, *ibid*.

30. D. M. Barringer to W. F. Magie, April 1, 1925, *ibid*.

31. D. M. Barringer to T. Barbour, December 1 and 4, 1924, and April 5, 1925, *ibid*.

32. G. M. Colvocoresses to D. M. Barringer, and D. M. Barringer to G. M. Colvocoresses (two letters), April 6, 1925, *ibid*.

33. D. M. Barringer to H. H. Alexander, April 6, 1925, *ibid*.

34. D. M. Barringer to E. Thomson, April 14, 1925, *ibid*.

35. D. M. Barringer to B. L. Miller, January 14, 1925, *ibid*.

36. D. M. Barringer to E. Thomson, April 27, 1925, *ibid*.

37. C. P. Olivier, *Meteors* (Baltimore: Williams & Wilkins Company, 1925). Olivier, however, did not mention Barringer in his book, relying solely on Merrill's papers. *See also* D. M. Barringer to C. P. Olivier, April 29, 1925, *Barringer Papers*.

38. D. M. Barringer to C. P. Olivier, May 1, 1925, *Barringer Papers*.

39. D. M. Barringer to W. W. Keen, June 9, 1925, *ibid*.

40. D. M. Barringer to E. Thomson, July 13, 1925, *ibid*.

41. D. M. Barringer to G. M. Colvocoresses, May 15, and June 12, 1925; and to E. Thomson, June 27, 1925, *ibid*.

42. D. M. Barringer to G. M. Colvocoresses, May 5, 1925, *ibid*.

43. D. M. Barringer to W. F. Magie, June 23, 1925, *ibid*.

44. D. M. Barringer to E. Thomson, June 27, 1925, *ibid*.

45. D. M. Barringer to G. M. Colvocoresses, June 26 and 27, 1925, *ibid*.

46. D. M. Barringer to G. M. Colvocoresses, June 29, 1925, *ibid*.

47. "Meteor Crater," *op. cit. See also* D. M. Barringer to G. M. Colvocoresses, telegram, July 6, 1925, *Barringer Papers*.

48. D. M. Barringer to G. M. Colvocoresses, August 31, 1925, *Barringer Papers*.

49. D. M. Barringer to W. V. Smith, August 18, 1925, *ibid*.

50. Lease to Meteor Crater Exploration and Mining Company, August 19, 1925, Standard Iron Letterbook No. 9 (March 4, 1925–April 13, 1926), pp. 391–400; and Certificate of Incorporation (Delaware) and Bylaws, pp. 401–21, *ibid*.

51. D. M. Barringer to G. M. Colvocoresses, August 31, 1925, *ibid*.

52. D. M. Barringer to T. Barbour, August 14, 1925, *ibid*.

53. D. M. Barringer to G. M. Colvocoresses, December 8, 1925, *ibid*.

54. D. M. Barringer to T. Barbour, August 18, 1925, *ibid*.

55. D. M. Barringer to G. R. Agassiz, August 24, 1925, *ibid*. The only record Lowell himself left was a short note thanking Holsinger for "a most interesting day. . . . Scientifically, it is a place of unique importance; the very difficulty of decipherment enhancing its value. . . . As if to point the day, a fine meteor descended straight in front of us as we neared our return goal." See P. Lowell to S. J. Holsinger, October 15, 1909, *Lowell Observatory Archives*.

56. D. M. Barringer to T. Barbour, August 24, 1925, *Barringer Papers*.

57. D. M. Barringer to T. Barbour, August 26, 1925, *ibid*.

58. D. M. Barringer to T. Barbour, October 8, 1925, *ibid*.

59. D. M. Barringer to E. Thomson, October 9, 1925; and to G. R. Agassiz and T. Barbour, October 13, 1925, *ibid*.

60. D. M. Barringer to E. J. Bennitt, October 21, 1925, *ibid*.

61. D. M. Barringer to G. M. Colvocoresses, October 23, 1925, *ibid*.

62. D. M. Barringer to G. M. Colvocoresses, telegram, October 27, 1925, *ibid*.

63. D. M. Barringer to T. Barbour, October 30, 1925, *ibid*.

64. D. M. Barringer to G. M. Colvocoresses, November 3, 1925, *ibid*.

65. D. M. Barringer to T. Barbour, November 5, 1925, *ibid*.

66. D. M. Barringer to A. B. Johnson, November 7, 1925, *ibid*.

67. D. M. Barringer to T. Barbour, November 7, 1925, *ibid*.

68. D. M. Barringer to T. Barbour, November 20, 1925, *ibid*.

69. D. M. Barringer to E. Thomson, November 11, 1925, *ibid*.

70. D. M. Barringer to C. H. Scott, November 28, 1925, *ibid*.

71. D. M. Barringer to Bureau of Securities, Department of Banking, Harrisburg, Pennsylvania, November 24, 1925, *ibid*.

72. D. M. Barringer to G. M. Colvocoresses, November 27, 1925, *ibid*.

73. W. J. Fallows to D. M. Barringer, December 2, 1925; *see also* D. M. Barringer to P. G. Cameron, December 5, 1925, *ibid*. Cameron was Pennsylvania's Banking Commissioner.

74. D. M. Barringer to G. M. Colvocoresses, December 2, 1925, *ibid*.

75. D. M. Barringer to S. J. Jennings, telegram; to E. Thomson, telegram; and to W. F. Magie, December 4, 1925, *ibid.*

76. D. M. Barringer to W. F. Magie, December 4, 1925, *ibid.*

77. D. M. Barringer to H. H. Alexander, December 10, 1925, *ibid.*

78. D. M. Barringer to S. J. Jennings, December 9, 1925, *ibid.*

79. D. M. Barringer to S. J. Jennings and to T. Barbour, January 9, 1926, *ibid.*

80. D. M. Barringer to G. M. Colvocoresses, January 25, 1926, *ibid.*

81. D. M. Barringer to G. M. Colvocoresses, February 4, 1926, *ibid.*

82. D. M. Barringer to P. B. Barringer, January 25, 1926, *ibid.*

83. D. M. Barringer to G. M. Colvocoresses, December 8, 1925, *ibid.*

84. D. M. Barringer to G. M. Colvocoresses, January 9, 1926; and to T. Barbour, January 25, 1926, *ibid.*

85. D. M. Barringer to H. N. Russell, December 15 and 18, 1925, *ibid.*

86. D. M. Barringer to J. Allen, telegram, December 26, 1925, *ibid.*

87. D. L. Webster to D. M. Barringer, January 1, 1926; and D. M. Barringer to D. L. Webster, January 27, 1926, *ibid.*

88. "Mining for Shooting Stars," *Engineering and Mining Journal,* 120(1925):1001–02.

89. D. M. Barringer, "Exploration at Meteor Crater," *ibid.*, 121(1926):59. *See also* D. M. Barringer to editor, *Engineering and Mining Journal,* December 28, 1925, *Barringer Papers.*

90. D. M. Barringer to editor, *Literary Digest,* February 16, 1926, *Barringer Papers.*

91. "New Device Reported to Have Located Meteorite Near Winslow, Arizona," *Engineering and Mining Journal,* 121(1926):377.

92. "Radio Camera Phone and Its Inventor," *Arizona Gazette* (Phoenix), February 2, 1926.

93. D. M. Barringer to E. Thomson, February 6, 1926, *Barringer Papers.*

94. D. M. Barringer to S. Croasdale and to W. D. Blackmer, telegrams, February 8, 1926, *ibid.*

95. D. M. Barringer to G. M. Colvocoresses, February 8, 1926, *ibid.*

96. D. M. Barringer to J. H. Hammond, February 8, 1926, *ibid.*

97. R. C. Coffin to D. M. Barringer, telegram, February 11, 1926, *ibid.*

98. D. M. Barringer to G. M. Colvocoresses and to R. C. Coffin, telegrams, February 13, 1926, *ibid.*

99. D. M. Barringer to J. H. Hammond Jr., February 17, 1926; to G. M. Colvocoresses, March 1, 1926; to R. C. Coffin, March 12, 1926; to C. F. Rand, March 18, 1926; and to S. Croasdale, March 18, 1927, *ibid.*

100. D. Hager, "Meteor Crater," *Engineering and Mining Journal,* 121(1926):374.

101. D. M. Barringer to E. Thomson, March 19, 1926, *Barringer Papers.*

102. D. M. Barringer, "Exploration at Meteor Crater," *Engineering and Mining Journal,* 121(1926):450–51.

103. But Hager then continued to argue the crater's geological origin. See D. Hager,

"Crater Mound (Meteor Crater), Arizona: Is Its Origin Geologic or Meteoritic?" (abstract) *Popular Astronomy,* 57(1949):457–58; and "Crater Mound (Meteor Crater), Arizona, A Geologic Feature," *Bulletin of the American Association of Petroleum Geologists,* 37(1953): 821–57.

104. "A Worth While Star," *Engineering and Mining Journal,* 121(1926):554–55.

105. D. M. Barringer to J. E. Spurr, April 5, 1926, *Barringer Papers.*

106. D. M. Barringer to H. J. Minhinnick, February 3, 1926; and to W. D. Boutwell, February 5, March 12 and 18, 1926, *ibid.*

107. D. M. Barringer to G. H. Grosvenor, November 12, 1926, *ibid.*

108. D. M. Barringer to H. J. Minhinnick, February 19 and 24, 1926; and to W. D. Boutwell, February 5, 1926, *ibid.*

109. H. J. Minhinnick, "Search for Great Meteor Is to Be Prosecuted at Canyon Diablo by Strong Company," *Verde Copper News* (Jerome), April 20, 1926.

110. "Meteor Mountain, Riddle of Science, One of Many Natural Wonders in Wonder Section Near Winslow," *Arizona Gazette* (Phoenix), April 26, 1926.

111. D. M. Barringer to H. J. Minhinnick, April 9, 14, and 28, 1926, *Barringer Papers.*

112. H. J. Minhinnick, "About That Gigantic Meteor Crater," *Progressive Arizona,* 3 (1926):10–11, 37–38.

113. "The Most Interesting Spot on Earth," *Literary Digest,* 90(1926):21–22.

114. D. M. Barringer to W. D. Boutwell, April 10, 1926, *Barringer Papers.*

115. D. M. Barringer to H. N. Russell, April 19, 1926, *ibid.*

116. D. M. Barringer to J. S. Adams, March 10, 1926, *ibid.*

117. D. M. Barringer to H. L. Fairchild, G. P. Merrill, and W. W. Campbell, March 15, 1926, *ibid.*

118. D. M. Barringer to W. F. Magie, March 15, 1926, *ibid.*

119. D. M. Barringer to T. Barbour, May 10, 1926; and to W. F. Magie, May 13, 1926, *ibid.*

120. H. N. Russell to W. F. Magie, April 15, 1926, *ibid.*

121. D. M. Barringer to E. Thomson, April 29, 1926, *Ibid.,*

122. D. M. Barringer to A. H. Hull, April 24, 1926, *ibid.*

123. D. M. Barringer to A. H. Hull, April 26, 1926; to E. Thomson, April 29, 1926; to C. H. Moore, May 1 and 4, 1926; and to H. H. Alexander, May 10, 1926, *ibid.*

124. D. M. Barringer to E. Thomson, May 27, 1926, *ibid.*

125. D. M. Barringer to G. R. Agassiz, May 27, 1926, *ibid.*

126. D. M. Barringer to C. H. Moore, May 27, 1926, *ibid.*

127. D. M. Barringer to G. R. Agassiz, June 4, 1926, *ibid.*

128. D. M. Barringer to A. H. Hull, June 7, 1926, *ibid.*

129. D. M. Barringer to G. M. Colvocoresses, June 23, 1926, *ibid.*

130. A. B. Bibbins, "A Small Meteor Crater in Texas," *Engineering and Mining Journal,* 121(1926):952.

131. G. P. Merrill, "Meteoric Iron from Odessa, Ector County, Texas," *American Journal of Science,* 203(1922):335–37.

132. Merrill's first opportunity, of course, was his visit to Coon Mountain in 1891 while Gilbert was there.

133. D. M. Barringer to W. D. Blackmer, June 7, 1926, *Barringer Papers.*

134. D. M. Barringer to A. B. Bibbins, June 7 and 10, 1926; and to G. M. Colvocoresses, June 10, 1926, *ibid.*

135. D. M. Barringer to G. M. Colvocoresses, June 10, 1926, *ibid.*

136. D. M. Barringer to E. Thomson, June 8, 1926, *ibid.*

137. D. M. Barringer to G. M. Colvocoresses, June 10, 1926, *ibid.*

138. D. M. Barringer to E. Thomson, June 25, 1926, *ibid.*

139. D. M. Barringer to D. M. Barringer Jr., June 25, 1926, *ibid.*

140. D. M. Barringer to Academy of Natural Sciences of Philadelphia, July 6, 1926, *ibid.*

141. E. Thomson to D. M. Barringer, June 29, 1926, *ibid.*

142. D. M. Barringer to G. R. Agassiz, June 30, 1926, *ibid.*

143. D. M. Barringer, Jr., "A New Meteor Crater," *Proceedings of the Academy of Natural Sciences of Philadelphia,* 80(1928):307–11.

144. D. M. Barringer to G. P. Merrill, December 12, 1928, *Barringer Papers.*

145. D. M. Barringer to D. M. Barringer, Jr., December 4, 1928, *ibid.*

146. D. M. Barringer to G. P. Merrill, December 12, 1928, *ibid.*

147. E. H. Sellards, "Unusual Structural Feature in the Plains Region of Texas," *Bulletin of the Geological Society of America,* (abstract) 38(1927):149.

148. D. M. Barringer to G. M. Colvocoresses, March 1, 1926; to E. J. Bennitt, April 21, 1926; and to E. Thomson, July 7, 1926, *Barringer Papers.*

149. C. H. Scott to I. Heller, July 13, 1926, *ibid.*

150. I. Heller to G. R. Agassiz, T. Barbour, A. Johnson, S. J. Jennings, H. H. Alexander, C. H. Moore, W. F. Magie, and E. Thomson, July 13, 1926, *ibid.*

151. B. Barringer to D. M. Barringer, July 13, 1926, *ibid.*

152. D. M. Barringer to Townsend, Elliott and Munson, July 26, 1926, *ibid.*

153. *See* Standard Iron Letterbook No. 10 (April 13, 1920–March 26, 1927) for correspondence in July and August 1926, *ibid.*

154. D. M. Barringer to Beidelman and Hull, September 21, 1926, *ibid.*

155. C. H. Scott to D. M. Barringer, August 17, 1926, *ibid.*

156. D. M. Barringer to T. Barbour, August 21, 1926, *ibid.*

CHAPTER TEN. THE FINAL SHAFT

Epigraph: F. R. Moulton, "The Arizona Meteorite," typewritten report, dated August 24, 1929, *Lowell Observatory Archives.* The emphasis is Moulton's.

1. D. M. Barringer to E. J. Bennitt, April 25, 1923, *Barringer Papers*.
2. D. M. Barringer to E. Thomson, September 13, 1926, *ibid*.
3. D. M. Barringer to E. Thomson, September 17, 1926, *ibid*.
4. D. M. Barringer to E. Thomson, September 21, 1926, *ibid*.
5. D. M. Barringer to Townsend, Elliot, and Munson, July 26, 1926, *ibid*. The Flagstaff astronomers, of course, had been to the crater before.
6. D. M. Barringer to E. Thomson, August 16, 1926, *ibid*.
7. D. M. Barringer to S. M. Hitchcock, November 10 and December 14, 1925, *ibid*.
8. D. M. Barringer to G. M. Colvocoresses, February 12, 1927, *ibid*.
9. D. M. Barringer to G. M. Colvocoresses, October 20, 1926; and to E. Thomson, October 17, 1926, *ibid*.
10. D. M. Barringer to T. Barbour and P. Kraft, November 15, 1926, *ibid*.
11. D. M. Barringer to P. Kraft, February 16 and 27, 1927; and to G. M. Colvocoresses (telegram), March 8, 1927, *ibid*.
12. D. M. Barringer to F. M. Searls Jr., May 17 and 26, 1927, *ibid*.
13. D. M. Barringer to E. Thomson, September 17, 1926, *ibid*.
14. D. M. Barringer to H. N. Russell, April 30, 1926, *ibid*.
15. E. Thomson to D. M. Barringer, October 25, 1926, *ibid*.
16. E. Thomson to D. M. Barringer, October 25, 1926, *ibid*.
17. D. M. Barringer to G. P. Merrill, October 27, 1926, *ibid*.
18. D. M. Barringer to J. Keedy, May 18, 1926; and D. M. Barringer Jr. to Harvard Club, Boston, June 3, 1926, *ibid*.
19. D. M. Barringer to G. Garceau (Harvard Club), September 24, 1926; and to T. Barbour, September 29, 1926, *ibid*.
20. D. M. Barringer to A. W. Stevens, October 15, 1926, *ibid*.
21. D. M. Barringer to H. T. Stetson, October 29, 1926, *ibid*.
22. D. M. Barringer to W. J. Luyten, November 3, 1926, *ibid*.
23. W. S. Hutchinson to D. M. Barringer, October 29, 1926, *ibid*.
24. D. M. Barringer to C. O. Lampland, October 17, 1928, *Lowell Observatory Archives*.
25. D. M. Barringer to E. Thomson, October 22, 1926, *Barringer Papers*.
26. D. M. Barringer to F. E. Wright, November 8, 1926, *ibid*.
27. For a report of the committee's work, *see* F. E. Wright, "The Surface Features of the Moon," *Smithsonian Institution Annual Report*, 1935, 169–82.
28. D. M. Barringer to T. Barbour, November 15, 1926, *Barringer Papers*.
29. F. L. Thurmond, "Is There a Canyon Diablo Meteorite? Drilling Has Failed to Find Any in the Famous Arizona Crater," *Engineering and Mining Journal*, 122(1926):817–18.
30. D. M. Barringer to E. Thomson, November 22, 1926, *Barringer Papers*.
31. E. Thomson to D. M. Barringer, November 30, 1926, *ibid*.
32. D. M. Barringer to D. M. Barringer, Jr., December 11, 1926, *ibid*.

33. D. M. Barringer to W. F. Magie, December 8 and 13, 1926, *ibid.*

34. D. M. Barringer to E. Thomson, January 4, 1927, *ibid.*

35. D. M. Barringer to G. M. Colvocoresses, January 6, 1927, *ibid.*

36. D. M. Barringer to W. F. Magie, December 27, 1926, *ibid.*

37. D. M. Barringer to W. F. Magie, January 25, 1927, *ibid.* Barringer was disappointed but not surprised. Spurr, he had earlier advised his friend, Alexander, was a member of the U.S. Geological Survey and a "steam explosionist." See D. M. Barringer to H. H. Alexander, December 31, 1926, *ibid.*

38. W. F. Magie, "Meteor Crater," *Engineering and Mining Journal,* 123(1927):293.

39. Russell had just published a noncommittal article on the origin of the lunar craters in which he mentioned Meteor Crater in passing as an impact crater. See "Science Plans a Mass Attack on the Moon's Mysteries—Are the Moon's Craters Volcanic or Are They Meteoric Shell-Holes?" *Scientific American,* 136(1927):93.

40. H. N. Russell to A. G. Ingalls, January 10, 1927, *Barringer Papers.*

41. D. M. Barringer to A. G. Ingalls, January 8 and 13, 1927, *ibid.*

42. D. M. Barringer to E. Thomson, February 16, 1927, *ibid.*

43. D. M. Barringer, Jr., "The Most Fascinating Spot on Earth," *Scientific American,* 137(1927):52–54, 144–46, 244–46.

44. D. M. Barringer to A. G. Ingalls, February 11, 1927; and to W. F. Magie and E. Thomson, February 18, 1927, *Barringer Papers.*

45. D. M. Barringer to G. M. Colvocoresses, January 29, 1927, *ibid.*

46. D. M. Barringer to J. H. Hammond, February 2 and 11, 1927; and to G. M. Colvocoresses, February 16, 1927, *ibid.*

47. D. M. Barringer to G. M. Colvocoresses, December 14, 1926, *ibid.*

48. D. M. Barringer to H. A. Guess, December 23, 1926; and to G. M. Colvocoresses, December 23, 1926, and January 6, 1927, *ibid.*

49. D. M. Barringer to T. Barbour, November 22, 1926; and to H. S. Pritchett, November 29, 1926, *ibid.*

50. D. M. Barringer to H. Hoover, February 7, 1927, *ibid.*

51. D. M. Barringer to G. B. Grinnell, February 7, 1927, *ibid.*

52. D. M. Barringer to G. B. Grinnell, February 11, 1927, *ibid.*

53. D. M. Barringer to G. M. Colvocoresses, March 8, 1927; and to J. Gribbel, March 19, 1927, *ibid.*

54. Q. A. Shaw to Meteor Crater Exploration and Mining Company, March 26, 1926, *ibid.*

55. D. M. Barringer to Q. A. Shaw, February 26, 1927, *ibid.*

56. D. M. Barringer to G. M. Colvocoresses, March 1 and 8, 1927, *ibid.*

57. D. M. Barringer to G. M. Colvocoresses, February 7 and 8, 1927, *ibid.*

58. Prospectus and Stock Subscription Agreement, Meteor Crater Exploration and Mining Company, June 1, 1927, *ibid.*

59. G. M. Colvocoresses, "Report on Meteor Crater as a Mining Prospect," August 30, 1927; *ibid.*

60. G. M. Colvocoresses to V. M. Slipher, August 10, 1927; *Lowell Observatory Archives.*

61. G. M. Colvocoresses to C. S. Lowell, August 10, 1927; *Lowell Observatory Archives.*

62. List of Subscribers, Meteor Crater Exploration and Mining Company, September 16, 1927; *Barringer Papers.*

63. D. M. Barringer to C. P. Olivier, September 10, 1927, *ibid.*

64. List of Subscribers, Meteor Crater Exploration and Mining Company, November 25, 1927, *ibid.*

65. D. M. Barringer to G. M. Colvocoresses, January 19, 1928, *ibid.*

66. G. M. Colvocoresses to D. M. Barringer, February 25, 1928; and Report to Stockholders, Meteor Crater Exploration and Mining Company, June 1, 1928, *ibid.*

67. D. M. Barringer to G. M. Colvocoresses, April 17, 1928, *ibid.*

68. Report to Stockholders, *op. cit.*

69. E. H. Robie, "The Meteor Crater Project," *Engineering and Mining Journal*, 125 (1928):850–52.

70. "Meteor Crater Shaft Now Down 140 Ft.," *ibid.*, 126(1928):350.

71. D. M. Barringer to T. Barbour, September 26, 1928, *Barringer Papers.*

72. A. F. Rogers, "The Natural History of Silica Minerals," *The American Mineralogist*, 13(1928):73–87.

73. D. M. Barringer to A. F. Rogers, November 27, 1926; to J. Allen, January 25, 1927; and A. F. Rogers to D. M. Barringer, January 8, 1927, *Barringer Papers.*

74. J. A. Reinvaldt, "Bericht uber geologische Untersuchengen am Kaalijarv (Krater von Sall) auf Ösel," *Publications of the Geological Institute of the University of Tartu*, No. 11, 1928. Alfred Wegener, who had earlier supported an impact origin for the lunar craters, also favored a meteoritic origin for the Ösel craters at this time. See E. Kraus, R. Meyer, and A. Wegener, "Untersuchungen über den Krater von Sall auf Ösel," *Gerlands Beiträge zur Geophysik*, 20(1928):312–78. Meyer agreed with him, while Kraus thought they were formed from salt domes.

75. D. M. Barringer to D. M. Barringer, Jr., March 11, 1929, *Barringer Papers.*

76. W. D. Boutwell, "The Mysterious Tomb of a Giant Meteorite," *National Geographic Magazine*, 53(1928):720–30.

77. D. M. Barringer to W. D. Boutwell, November 9, 1926, *Barringer Papers.*

78. G. P. Merrill to D. M. Barringer, May 28, 1928, *ibid.*

79. D. M. Barringer to G. P. Merrill, June 8, 1928, *ibid.*

80. G. P. Merrill to D. M. Barringer, June 11, 1928, *ibid.*

81. D. M. Barringer to H. L. Fairchild, October 5, 1928, *ibid.*

82. D. M. Barringer to D. M. Barringer, Jr., December 4, 1928, *ibid.*

83. D. M. Barringer to G. O. Smith, June 15, 1928, *ibid.*

84. G. O. Smith to D. M. Barringer, June 23, 1928, *ibid.*

85. D. M. Barringer to G. O. Smith, July 6, 1928, *ibid.*
86. D. M. Barringer to H. L. Fairchild, October 5, 1928, *ibid.*
87. D. M. Barringer to D. M. Barringer, Jr., September 26, 1928, *ibid.*
88. D. M. Barringer to J. O. LaGorce, July 20, 1928, *ibid.*
89. D. M. Barringer to T. Barbour, September 26, 1928, *ibid.*
90. C. P. Olivier, "Meteor Notes," *Popular Astronomy,* 36(1928):498.
91. D. M. Barringer to G. P. Merrill, December 3, 1928, *ibid.*
92. F. E. Wright to D. M. Barringer, December 27, 1928, *ibid.*
93. D. M. Barringer to D. M. Barringer, Jr., January 15, 1929, *ibid.*
94. D. M. Barringer to H. McClenahan, March 9, 1929. Grosvenor sent his letter to McClenahan on February 5, 1929, *ibid.*
95. H. L. Fairchild to J. J. Edson, October 9, 1928; *Lowell Observatory Archives.*
96. D. M. Barringer to W. F. Magie, March 1, 1929; and H. L. Fairchild, March 3, 1929, *Barringer Papers.*
97. D. M. Barringer to H. L. Fairchild, April 5, 1929, *ibid.*
98. D. M. Barringer to H. L. Fairchild, April 9, 1929, *ibid.*
99. H. L. Fairchild, "Meteor Crater Exploration," *Science,* 69(1929):485–87.
100. D. M. Barringer to C. O. Lampland, May 15, 1929; *Lowell Observatory Archives.*
101. D. M. Barringer to C. P. Olivier, July 16, 1929, *ibid.* But others acknowledged Barringer's claim. "It is a matter of record that some scientists have contended that Meteor Crater was produced by volcanic action," R. G. Skerrett wrote in a two-part series on the crater in the *Compressed Air Magazine* in June [34(1929):2773–78, 2809–13]. "Present knowledge of the true nature of Meteor Crater is due to the labor and research of a small group of experts inspired by the leadership of Daniel Moreau Barringer. . . ."
102. D. M. Barringer to D. M. Barringer, Jr., November 13, 1928, *Barringer Papers.*
103. D. M. Barringer to D. M. Barringer, Jr., December 4, 1928, *ibid.*
104. D. M. Barringer to D. M. Barringer, Jr., December 11, 1928, *ibid.*
105. D. M. Barringer to W. F. Magie, December 24, 1928, *ibid.*
106. D. M. Barringer to D. M. Barringer, Jr., January 5, 1929, *ibid.*
107. D. M. Barringer to E. Thomson, January 3, 1929, *ibid.*
108. D. M. Barringer to C. K. Barnes, December 19, 1928, *ibid.*
109. D. M. Barringer to E. Thomson, January 3, 1929; to E. J. Bennitt, January 21, 1929; and to C. K. Barnes, January 22, 1929, *ibid.*
110. D. M. Barringer to Q. A. Shaw, February 7, 1929, *ibid.*
111. Q. A. Shaw to D. M. Barringer, February 18, 1929, *ibid.*
112. D. M. Barringer to E. Thomson, March 25, 1929, *ibid.*
113. D. M. Barringer to T. Barbour, May 15, 1929, *ibid.*
114. D. M. Barringer to H. Shapley, April 26, 1929, *ibid.*
115. H. Shapley to D. M. Barringer, April 29, 1929, *ibid.*

116. D. M. Barringer to F. R. Moulton, June 3, 1929, *ibid.*

117. F. R. Moulton to D. M. Barringer, June 5, 1929, *ibid.*

118. F. R. Moulton to D. M. Barringer, June 6, 1929, *ibid.*

119. Q. A. Shaw, Report to Stockholders, Meteor Crater Exploration and Mining Company, October 1, 1929; *ibid.*

120. G. M. Colvocoresses to Meteor Crater Exploration and Mining Company, June 17, July 3 and 13, 1929, *ibid.*

121. D. M. Barringer to E. Thomson, May 10, 1929; see also Q. A. Shaw, draft of Report to Stockholders, Meteor Crater Exploration and Mining Company, dated August 19, 1929, *ibid.*

122. Q. A. Shaw to D. M. Barringer, June 14, 1929, *ibid.*

123. Q. A. Shaw to D. M. Barringer, August 6, 1929, *ibid.*

124. G. M. Colvocoresses to D. M. Barringer, June 26, 1929, *ibid.*

125. Q. A. Shaw to D. M. Barringer, telegram, July 19, 1929, *ibid.*

126. Q. A. Shaw to D. M. Barringer, July 25, 1929, *ibid.*

127. Q. A. Shaw to D. M. Barringer, August 6, 1929, *ibid.*

128. Q. A. Shaw, Draft of Report to Stockholders, *ibid.*

129. D. M. Barringer to North American Mines, Inc., August 26, 1929, enclosing extract of minutes of Meteor Crater Exploration and Mining Company board meeting of that date, *ibid.*

130. G. M. Colvocoresses to Q. A. Shaw, September 9, 1929, *ibid.*

131. Q. A. Shaw to D. M. Barringer, September 3, 1929, *ibid.*

CHAPTER ELEVEN. THE GREAT DEBATE

Epigraph: F. R. Moulton, foreword, "Second Report on the Arizona Meteorite," November 20, 1929; *Lowell Observatory Archives.*

1. F. R. Moulton, "The Arizona Meteorite," twenty-two-page typescript dated August 24, 1929; *Lowell Observatory Archives.*

2. F. R. Moulton, "Second Report on the Arizona Meteorite," 127-page typescript with 14-page foreword dated Novmeber 20, 1929, *Lowell Observatory Archives.*

3. F. R. Moulton, "III—The Arizona Meteorite," twenty-one-page typescript with three pages of corrections dated January 20, 1929, *Barringer Papers.*

4. F. R. Moulton, "Some Quantitative Aspects of the Fall of Meteors," abstract, *Popular Astronomy,* 39(1931): 17–18. *See also* "Meteor Crater," *Science,* 72(1930): x; F. R. Moulton, "Swarm of Small Meteors Cause of Vast Crater—The Fragments of a Single Exploded Body Are Now Said to Have Caused a Hole a Mile Wide," *The New York Times,* November 23, 1930; and "Meteor Swarm Produced Famous Crater," *Science News Letter,* 18(1930): 204.

5. E. M. Shoemaker, personal communication, September 14, 1983.

6. F. R. Moulton, "The Arizona Meteorite," *op. cit.;* and F. R. Moulton to W. S. Hutchinson, August 26, 1929, and to R. Livermore, September 6, 1929, *Barringer Papers.*

7. R. Livermore to F. R. Moulton, September 4, 1929; and F. R. Moulton to R. Livermore, September 6, 1929, *Barringer Papers*.

8. Moulton's report is available in the *Barringer Papers* in Princeton University's Firestone Library, the Lowell Observatory Archives, and the *Elihu Thomson Papers* in the Library of the American Philosophical Society in Philadelphia. It has been published, typographical errors and all, as an appendix in H. J. Abrahams, ed., *Heroic Efforts at Meteor Crater, Arizona* (Teaneck, New Jersey: Fairleigh Dickinson University Press, 1983). *See also* F. R. Moulton, *Celestial Mechanics* (New York: The Macmillan Company, 1902).

9. E. Thomson to D. M. Barringer, September 6, 1929; and to Q. A. Shaw, September 19, 1929, *Barringer Papers*.

10. W. S. Hutchinson to F. R. Moulton, August 28, 1929, *ibid*.

11. R. Livermore to F. R. Moulton, September 4, 1929, *ibid*.

12. F. R. Moulton to W. S Hutchinson, August 30, 1929, *ibid*.

13. F. R. Moulton to W. S. Hutchinson, August 30, 1929 (second letter), *ibid*.

14. F. R. Moulton to W. S. Hutchinson, September 5, 1929, *ibid*.

15. R. Livermore to F. R. Moulton, September 4, 1929, *ibid*.

16. F. R. Moulton to R. Livermore, September 5, 1929, *ibid*.

17. Q. A. Shaw to D. M. Barringer, September 3, 1929, *ibid*.

18. D. M. Barringer to E. Thomson, September 4, 1929, *ibid*.

19. D. M. Barringer to Q. A. Shaw, September 5, 1929, *ibid*.

20. F. R. Moulton to Hutchinson and Livermore, September 14, 1929, *ibid*.

21. E. Thomson to D. M. Barringer, September 6, 1929, *ibid*.

22. Q. A. Shaw to D. M. Barringer, September 13, 1929, *ibid*. Shapley and Fisher were nonetheless interested in the problem and later studied all the reports and correspondence exchanged during the debate. *See* Q. A. Shaw to Meteor Crater Exploration and Mining Company Board of Directors, April 10, 1930, *ibid*.

23. E. Thomson to Q. A. Shaw, September 19, 1929, *ibid*.

24. E. Thomson to D. M. Barringer, October 12, 1929, *ibid*.

25. D. M. Barringer to E. Thomson, September 13, 1929, *ibid*.

26. D. M. Barringer to Q. A. Shaw, September 13 and 16, 1929, *ibid*.

27. G. M. Colvocoresses to D. M. Barringer, telegram, September 17, 1929, *ibid*.

28. Financial Report, Meteor Crater Exploration and Mining Company, September 1929, *ibid*.

29. F. R. Moulton to Hutchinson and Livermore, September 14, 1929, *ibid*.

30. Q. A. Shaw to D. M. Barringer, September 19, 1929, *ibid*.

31. D. M. Barringer to Q. A. Shaw, September 13, 1929, *ibid*.

32. D. L. Webster to D. M. Barringer, telegram, September 20, 1929, *ibid*.

33. D. M. Barringer to D. L. Webster, October 16, 1929, *ibid*.

34. D. L. Webster, "The Mass of the Meteorites at Meteor Crater," ten-page typescript dated November 29, 1929, *Lowell Observatory Archives*.

35. Q. A. Shaw to D. M. Barringer, September 23, 1929. *See also Engineering and Mining Journal,* 128(1929):484.

36. Q. A. Shaw to the *Boston Evening Transcript,* October 14, 1929.

37. Apparently only the *Northern Miner* replied, advising Shaw that it could not find the offending item. See *Northern Miner* to Q. A. Shaw, September 26, 1929, *Barringer Papers.*

38. "Arizona's Meteor Crater Mystery Nearer Solution as Miners Report Metal Located," *The Arizona Republican,* September 8, 1929.

39. C. P. Olivier to F. R. Moulton, October 1, 1929, *Barringer Papers.*

40. D. M. Barringer to C. P. Olivier, October 8, 1929, *ibid.*

41. E. Thomson to D. M. Barringer, October 5, 1929, *ibid.*

42. M. B. Huston to D. M. Barringer, October 20, 1929, *ibid.*

43. B. Barringer to Q. A. Shaw, October 8, 1929, *ibid.*

44. Q. A. Shaw to D. M. Barringer, October 3, 1929, *ibid.*

45. Q. A. Shaw to D. M. Barringer, October 30, 1929, *ibid.*

46. C. P. Olivier to Q. A. Shaw, October 7, 1929, *ibid.*

47. F. R. Moulton to C. P. Olivier, October 5, 1929, *ibid.*

48. C. P. Olivier to F. R. Moulton, October 15, 1929, *ibid.*

49. F. R. Moulton to Hutchinson and Livermore, October 7, 1929, *ibid.*

50. E. Thomson to D. M. Barringer, October 2, 1929, *ibid.*

51. E. Thomson to D. M. Barringer, October 5, 1929, *ibid.*

52. D. M. Barringer to Q. A. Shaw, October 11, 1929, *ibid.*

53. Q. A. Shaw to F. R. Moulton, October 22, 1929, *ibid.*

54. B. Barringer to Q. A. Shaw, October 8, 1929; and G. M. Colvocoresses to Q. A. Shaw, September 29, 1929, *ibid.*

55. Q. A. Shaw to D. M. Barringer, October 12, 1929, with six pages of typewritten notes enclosed, *ibid.*

56. B. Barringer to Q. A. Shaw, October 22, 1929, *ibid.*

57. Q. A. Shaw to D. M. Barringer, Ocober 30, 1929, *ibid.*

58. D. M. Barringer to E. P. Adams, October 18, 1929, *ibid.*

59. D. M. Barringer to A. L. Day, October 17, 1929, *ibid.*

60. J. A. Miller to D. M. Barringer, October 21, 1929, *ibid.*

61. J. W. Strutt (Lord Rayleigh), "The Resistance of Fluids," *Philosophical Magazine,* 2(1876):430–41.

62. W. H. Echols to D. M. Barringer, October 26, 1929, *Barringer Papers.*

63. R. F. Riggs to D. M. Barringer, October 31, 1929, *ibid.*

64. E. Thomson to D. M. Barringer, October 25, 1929, *ibid.*

65. E. Thomson to D. M. Barringer, October 15, 1929 (second letter), *ibid.*

66. Q. A. Shaw to D. M. Barringer, October 30, 1929, *ibid.*

67. D. M. Barringer to G. M. Colvocoresses, November 1, 1929, *ibid.*

68. E. P. Adams, "Atmospheric Resistance to a Falling Meteor," twelve-page typescript with covering letter to D. M. Barringer, November 4, 1929, *ibid.*

69. Strutt, "The Resistance of Fluids," *op. cit.*

70. A. Klemin to D. M. Barringer, November 15, 1929, *Barringer Papers.*

71. E. W. Brown to D. M. Barringer, November 21, 1929, *ibid.*

72. C. O. Lampland and V. M. Slipher tot D. M. Barringer, night letter, November 12, 1929, *ibid.*

73. E. Thomson to I. Heller, November 13, 1929, *ibid.*

74. *See,* for example, G. M. Colvocoresses to D. M. Barringer, September 20 and October 3, 1929; and to Q. A. Shaw, September 25 and 30 and October 14 and 26, 1929, *ibid.*

75. G. M. Colvocoresses to D. M. Barringer, November 7, 1929, *ibid.*

76. W. F. Magie to D. M. Barringer, October 25, 1929, *ibid.*

77. D. M. Barringer to W. F. Magie, undated cablegram, *ca.* November 8, 1929, *ibid.*

78. W. F. Magie to D. M. Barringer, November 12, 1929, *ibid.*

79. H. N. Russell, "Note on the Energy Relations Involved in the Meteor Crater," three-page typescript dated October 30, 1929, *ibid.*

80. H. N. Russell to D. M. Barringer, October 30, 1929, *ibid.*

81. D. M. Barringer to W. F. Magie, undated cablegram, *ca.* November 13, 1929, *ibid.*

82. D. M. Barringer to W. F. Magie, undated cablegram, *ca.* November 14, 1929, *ibid.*

83. D. M. Barringer to H. N. Russell, November 16, 1929, *ibid.*

84. D. M. Barringer to W. F. Magie, November 16, 1929, *ibid.*

85. F. R. Moulton to Q. A. Shaw, November 11, 1929, *ibid.*

86. D. M. Barringer to Q. A. Shaw, November 14, 1929, *ibid.*

87. D. M. Barringer, untitled twenty-three-page typescript, dated November 15, 1929, *ibid.*

88. E. P. Adams, untitled, undated two-page typescript, *ca.* November 15, 1929, *ibid.*

89. F. R. Moulton to Q. A. Shaw, November 21, 1929, *ibid.*

90. D. M. Barringer to T. Barbour, November 22, 1929, *ibid.*

91. F. R. Moulton, "Second Report on the Arizona Meteorite," *op. cit.*

92. D. M. Barringer to E. Thomson, November 23, 1929, *Barringer Papers.*

93. W. F. Magie, "Report by W. F. Magie on the Meteor Crater," seven-page holographic manuscript with four-page "Supplementary Statement," dated November 10, 1929, *ibid.*

94. E. Thomson to D. M. Barringer, November 26, 1929, *ibid.*

95. H. N. Russell to D. M. Barringer, November 14, 1929, *ibid.*

96. D. M. Barringer to H. N. Russell, draft of letter dated November 27, 1929, *ibid.*

97. D. M. Barringer to Q. A. Shaw, draft of letter dated November 27, 1929, *ibid.*

98. D. M. Barringer to E. Thomson, draft of letter dated November 27, 1929, *ibid.* This is the last letter recorded in Barringer's personal letter books.

99. J. Paul Barringer, personal communication, October 8, 1981.

100. Obituary of D. M. Barringer, *The New York Times,* December 1, 1929.

CHAPTER TWELVE. THE AFTERMATH

Epigraph: Q. A. Shaw, Report to the Directors, Meteor Crater Exploration and Mining Company, December 1930, *Barringer Papers.*

1. Magie had purchased Holsinger's seventy-five nonassessable shares in the Standard Iron Company; and, in addition, he and his daughter, Henrietta Magie, each owned ten shares of Meteor Crater Exploration and Mining Company stock. Thomson had subscribed for 100 shares of the exploration company stock, while Olivier had been content to invest in two shares. *See* "List of Stockholders," Meteor Crater Exploration and Mining Company, dated April 23, 1929, *Barringer Papers.*

2. Bridgman received his Nobel Prize in 1946.

3. D. M. Barringer, Jr., to H. N. Russell, December 6, 1929, *Barringer Papers.*

4. Q. A. Shaw to B. Barringer, December 12, 1929, *ibid.*

5. H. N. Russell to Q. A. Shaw, January 7, 1930, *ibid.*

6. H. N. Russell to E. P. Adams, November 29, 1929, *ibid.*

7. H. N. Russell to D. M. Barringer, November 30, 1929, *ibid.*

8. H. N. Russell to Q. A. Shaw, November 30, 1929, *ibid.*

9. D. L. Webster, "The Mass of the Meteorites at Meteor Crater," nine-page typescript, dated November 29, 1929, *Lowell Observatory Archives.*

10. F. R. Moulton to Q. A. Shaw, December 11, 1929, *ibid.*

11. F. R. Moulton to Q. A. Shaw, December 12, 1929, *ibid.*

12. Q. A. Shaw to Directors, Meteor Crater Exploration and Mining Company, December 16, 1929, *ibid.*

13. C. Cranz, *Lehrbuch der Ballistik,* Berlin, 1925.

14. E. P. Adams to Q. A. Shaw, December 3, 1929, *Barringer Papers.*

15. E. P. Adams, "Atmospheric Resistance to a Falling Meteor, II," eight-page typescript, dated December 18, 1929, *Lowell Observatory Archives. See also* E. P. Adams to Q. A. Shaw, December 21, 1929, *Barringer Papers.*

16. D. L. Webster to Q. A. Shaw, January 3, 1930, *Barringer Papers.*

17. E. Thomson to Q. A. Shaw, December 28, 1929, *ibid.*

18. H. N. Russell to Q. A. Shaw, January 7, 1930, *ibid.*

19. F. R. Moulton, "III—The Arizona Meteorite," twenty-one-page typescript, with three pages of corrections, dated January 20, 1930, *ibid.* Moulton submitted this to Shaw on January 8 [see Moulton to Shaw that date], but he subsequently found some minor numerical errors which delayed copying and distribution of the report to the later date.

20. Q. A. Shaw to B. Barringer, December 11, 1929, *ibid.*

21. Q. A. Shaw to F. R. Moulton, January 31, 1930, *ibid.*

22. P. W. Bridgman to Q. A. Shaw, January 31, 1930, *ibid.* This letter is nine typewritten pages.

23. F. R. Moulton, "Second Report on the Arizona Meteorite," November 21, 1929, *Lowell Observatory Archives*, p. 2 (Foreword), 124.

24. F. R. Moulton to Q. A. Shaw, February 17, 1930, *Barringer Papers*.

25. E. P. Adams to P. W. Bridgman, February 16, 1930, *ibid.*

CHAPTER THIRTEEN. METEOR CRATER—THE LATER YEARS

Epigraph: F. R. Moulton, "Second Report on the Arizona Meteorite," 127-page typescript, dated November 20, 1929, *Lowell Observatory Archives*, 125.

1. F. R. Moulton, Foreword, "Second Report on the Arizona Meteorite," November 20, 1929, *Lowell Observatory Archives*.

2. W. F. Magie, "Physical Notes on Meteor Crater, Arizona," *Proceedings of the American Philosophical Society*, 49(1910): 41–48.

3. F. R. Moulton, "The Arizona Meteorite," August 24, 1929, *Lowell Observatory Archives*.

4. F. R. Moulton, "Second Report," *op. cit.*, 88–93.

5. H. N. Russell to E. P. Adams, November 29, 1929; and P. W. Bridgman to Q. A. Shaw, January 31, 1930, *Barringer Papers*.

6. E. M. Shoemaker, "Collision of astronomically observable bodies with the Earth," *Geological Society of America Special Paper 190*, 1982, 1–13. See also D. J. Roddy, S. H. Schuster, K. N. Kreyanhagen, and D. L. Orphal, "Computer Code Simulations of the Formation of Meteor Crater, Arizona, Calculations MC-1 and MC-2," *Proceedings of the 11th Lunar and Planetary Science Conference*, 1982, 2275–2308.

7. E. P. Öpik, "Researches on the Physical Theory of Meteor Phenomena. I. Theory of the Formation of Meteor Craters," *Publications of the Astronomical Observatory of the University of Tartu*, 28(1936): 3–12. Öpik's estimate for the "mass affected" at Meteor Crater was 10^9 tons, and his total energy allowance was 8.8×10^{15} foot-pounds (11.8×10^{22} ergs, or 2.56 megatons).

8. C. C. Wylie, "Calculations on the probable mass of the object which formed Meteor Crater," *Popular Astronomy*, 51(1943): 97–99; and "Second note on the probable mass of the object which formed Meteor Crater," *ibid.*, 158–61.

9. L. LaPaz, "Remarks on four notes recently published by C. C. Wylie," *ibid.*, 339–43.

10. R. B. Baldwin, *The Face of the Moon* (Chicago: The University of Chicago Press, 1949), 141, 224; and *The Measure of the Moon* (Chicago: The University of Chicago Press, 1963), 160.

11. L. LaPaz, "The Barringer (Arizona) Meteorite Crater," in "Report of the meeting of

the Southwestern Division, American Association for the Advancement of Science," *Science*, 111(1950):679.

12. J. S. Rinehart, "Some observations on high speed impact," *Popular Astronomy*, 58(1950):458–64.

13. N. Rostoker, "The formation of craters by high-speed particles," *Meteoritics*, 1 (1953):11–27.

14. J. S. Rinehart, "Recent findings at the Arizona Meteorite Crater," *Foote Prints* (Foote Mineralogical Company), 30(1958):15–22.

15. J. J. Gilvarry and J. E. Hill, "The impact of large meteorites," *Astrophysical Journal*, 124(1956):610–22; and J. E. Hill and J. J. Gilvarry, "Application of the Baldwin Crater Relation to the scaling of explosion craters," *Journal of Geophysical Research*, 61(1956): 501–11.

16. E. J. Öpik, "Meteor impact on solid surface," *Irish Astronomical Journal*, 5(1958): 14–33.

17. E. M. Shoemaker, "Impact Mechanics at Meteor Crater, Arizona," *U.S. Geological Survey Open File Report*, 1959; and "Penetration Mechanics of High-Velocity Meteorites, Illustrated by Meteor Crater, Arizona," *Report of the 21st International Geological Congress*, Copenhagen, 1960, Part SVIII, 418–34.

18. R. B. Baldwin, *The Measure of the Moon*, op. cit., 178.

19. E. M. Shoemaker and S. Kieffer, "Guidebook to the Geology of Meteor Crater, Arizona," prepared for the 37th annual meeting of the Meteoritical Society, Flagstaff, Arizona, August 4, 1974.

20. *See* note 6.

21. E. M. Shoemaker, personal communication, September 14, 1983.

22. J. E. Dick to G. M. Colvocoresses, March 25, 1930, *Barringer Papers*.

23. W. R. Nelson to G. M. Colvocoresses, March 25, 1930, *ibid*.

24. J. J. Jakosky to G. M. Colvocoresses, June 9, 1930, *ibid*.

25. J. J. Jakosky to G. M. Colvocoresses, January 9, 1930, *ibid*.; *see also* itemized "Expense of Geophysical Survey for Meteor Crater Co. by International Geophysics, Inc., November 3, 1930," *ibid*. IGI's fee was $2,500 and twenty-five shares each of preferred and common stock, plus actual expenses.

26. "Geophysical Survey of Meteor Crater," report by International Geophysics, Inc., October 13, 1930, *Lowell Observatory Archives*. *See also* J. J. Jakosky, C. W. Wilson, and J. W. Daily, "Geophysical examination of Meteor Crater," *The Mining Journal*, 14(1931):5–6, 29; "Geophysical examination of Meteor Crater, Arizona," *Transactions of the American Institute of Mining and Metallurgical Engineers*, 97(1932):63–98; and J. J. Jakosky, "Geophysical Methods Locate Meteorite," *Engineering and Mining Journal*, 133(1931):392–93.

27. J. J. Jakosky to G. M. Colvocoresses, October 13, 1930, *Barringer Papers*.

28. J. J. Jakosky to G. M. Colvocoresses, October 18, 1930, *ibid*.

29. J. J. Jakosky to G. M. Colvocoresses, October 13, 1930, *ibid*. This is Jakosky's transmittal letter with his report which, from other documentary evidence, was not sent until several weeks after this date.

30. G. M. Colvocoresses to Q. A. Shaw, November 24, 1930, *ibid.*

31. H. Lundberg to Q. A. Shaw, January 8, 1931, *ibid.*

32. W. R. Rooney to Meteor Crater Exploration and Mining Company, May 4, 1931, *ibid.*

33. G. M. Colvocoresses, "Discussion and Cost Estimate of Proposed Treatment of Ore from Meteor Crater," fourteen-page mimeographed typescript, dated May 7, 1931, *ibid.*

34. G. M. Colvocoresses to F. R. Dravo, June 23, 1931, *ibid.*

35. J. B. Girand to G. M. Colvocoresses, October 2, 1931, *ibid.*

36. Q. A. Shaw, Report to the Stockholders, Meteor Crater Exploration and Mining Company, July 15, 1931, *ibid.*

37. G. M. Colvocoresses to C. O. Lampland, August 7, 1931; *Lowell Observatory Archives.*

38. H. L. Fairchild, "Nature and Fate of the Meteor Crater Bolide," *Science,* 72(1930): 463–67.

39. D. M. Barringer, Jr., "The Barringer Meteorite," *ibid.,* 73(1931):66–67.

40. F. S. Dellenbaugh, "Meteor Butte," *ibid.,* 38–39.

41. C. R. Longwell, "Meteor Crater Not a Limestone Sink," *ibid.,* 234–35.

42. G. M. Colvocoresses to C. O. Lampland, October 10, 1931, *Lowell Observatory Archives.*

43. G. M. Colvocoresses to C. O. Lampland, October 21, 1931, *ibid.*

44. G. M. Colvocoresses to C. O. Lampland, January 15, 1932, with enclosed copy of report on "Drill Hole #1," *ibid.*

45. G. M. Colvocoresses to Q. A. Shaw, January 11, 1932, *Barringer Papers.*

46. G. M. Colvocoresses to Q. A. Shaw, February 9, 1932, *ibid.*

47. G. M. Colvocoresses to C. O. Lampland, February 18, 1932, *Lowell Observatory Archives.*

48. G. M. Colvocoresses, "Notes re: STAR DUST," five-page mimeographed typescript, dated May 17, 1932, *Barringer Papers.*

49. L. Hecksher, Report to the Stockholders, Meteor Crater Exploration and Mining Company, October 17, 1932, *ibid.*

50. D. M. Barringer, Jr., to C. O. Lampland, March 14, 1933, *Lowell Observatory Archives.*

51. G. M. Colvocoresses to stockholders, Meteor Crater Exploration and Mining Company, November 11, 1933; *Barringer Papers.* There were eighty-one stockholders at this time.

52. H. Lundberg, "Some Geophysical Data on Meteor Crater in Arizona," abstract, *Bulletin of the Geological Society of America,* 49(1938):1953; and "Explaining the Crater by Geophysical Methods," *The Sky,* 2(1938):18–19, 29.

53. H. H. Nininger, *Arizona's Meteorite Crater* (Denver: American Meteorite Laboratory, 1956), 49, 51.

54. D. M. Barringer, Jr., to "Members of the Committee on Matters Concerning the Barringer Meteorite Crater," November 10, 1951, *Lowell Observatory Archives.*

55. N. Harding, "A Gravity Investigation of Meteor Crater," master's thesis, University of Wyoming, 1954 (copy in the *Lowell Observatory Archives*).

56. D. Hager, "Crater Mound (Meteor Crater), Arizona, a Geologic Feature," *Bulletin of the American Association of Petroleum Geologists,* 37(1953):821–57.

57. R. G. Brereton, "Aeromagnetic Survey of Meteor Crater, Arizona," *Annals of the New York Academy of Sciences,* 123(1965):1175–81.

58. R. D. Regan and W. J. Hinze, "Gravity and Magnetic Investigations of Meteor Crater, Arizona," *Journal of Geophysical Research,* 80(1975):776–88.

59. J. S. Rinehart, "Distribution of Meteoritic Debris About the Arizona Meteorite Crater," *Smithsonian Contributions to Astrophysics,* 2(1958):145–60; and "Recent findings at the Arizona Meteorite Crater," *op. cit.*

60. D. M. Barringer, Jr., "From What Direction Did the Meteorite Come?" *Foote Prints,* 30(1958):23–28.

61. E. Thomson to D. M. Barringer, March 4, 1911, and March 23, 1920, *Barringer Papers*. Thomson, however, thought that the earth's motion would carry the meteorite eastward into the southeast quarter of the crater.

62. C. H. Roach, G. R. Johnson, J. G. McGrath, and T. S. Sterrett, "Thermoluminescence investigations at Meteor Crater, Arizona," *U.S. Geological Survey Professional Paper 450-D,* 1962, 98—103.

63. H. D. Ackerman, R. H. Godson, and J. S. Watkins, "A Seismic Refraction Technique Used for Subsurface Investigations at Meteor Crater, Arizona," *Journal of Geophysical Research,* 80(1975):765–75. This study found that the velocity of shock waves in the unfractured country rock around the crater was 3.1 km/sec (1.94 miles), or essentially the value Moulton had assumed in his third report.

64. E. M. Shoemaker, personal communication, September 24, 1983.

65. D. Hager, *op. cit.*; see also J. J. Jakosky, "Geophysical Survey of Meteor Crater," *op. cit.,* 15.

66. E. Blackwelder, "The Age of Meteor Crater," *Science,* 76(1932):557–60.

67. J. J. Jakosky, "Geophysical Survey of Meteor Crater," *op. cit.,* 15ff.

68. H. H. Robinson, "The San Francisco Volcanic Field, Arizona," *U.S. Geological Survey Professional Paper 76,* 1913.

69. F. M. Brown, "The Age of Meteor Crater," *Science,* 77(1933):239–40. Some years later, an even more recent date of A.D. 1064–67 was determined from tree-ring dating; and in the past few years Shoemaker and his U.S.G.S. colleagues found that volcanic activity at Sunset Crater continued until about A.D. 1250.

70. H. H. Nininger, *op. cit.,* 65.

71. J. D. Buddhue, "The Age of the Barringer Meteorite Crater," abstract of paper presented to the 1961 Meteoritical Society meeting at Nantucket, Massachusetts (unpublished). I am indebted to Dr. Shoemaker for a copy of this abstract.

72. J. J. Jakosky, "Geophysical Survey of Meteor Crater," *op. cit.,* 9. Buddhue here added an extra foot to Jakosky's section, making it 36'8".

73. D. Hager, *op. cit.*

74. G. E. Foster, *The Meteor Crater Story* (Winslow: Meteor Crater Enterprises, Inc., 1964).

75. E. Shoemaker, "Asteroidal and Comet Bombardment of the Earth," *Annual Review of Earth and Planetary Science,* 11(1983):461–94.

76. S. Sutton, "Thermoluminescence (TL) Dating of Shock-Metamorphosed Rock From Meteor Crater, Arizona: Shock Threshold for TL Resetting and Post-Impact Temperature of the Crater Floor," paper submitted to the 1983 Nininger Competition (Meteoritical Society), August 30, 1983.

77. S. Sutton, "Thermoluminescence Measurements on Shock-Metamorphosed Sandstone and Dolomite From Meteor Crater, Arizona: Thermoluminescence Age of Meteor Crater," preprint of paper submitted to the *Journal of Geophysical Research,* April 1984.

78. N. H. Darton, "Crater Mound, Arizona," abstract, *Bulletin of the Geological Society of America,* 56(1945):1154.

79. E. Blackwelder, "Meteor Crater, Arizona," *Science,* 104(1946):38–39. *See also* F. R. Moulton, "Some Quantitative Aspects of the Fall of Meteors," abstract *Popular Astronomy,* 39(1931):17–18; and *Astronomy* (New York: The Macmillan Company, 1931).

80. E. Blackwelder to H. S. Colton, January 21, 1946, *Harold Sellers Colton Papers,* Museum of Northern Arizona, Flagstaff, Arizona; hereafter cited as the *Colton Papers.*

81. F. C. Leonard, editor's note, "Contributions to the Society for Research on Meteorites," *Popular Astronomy,* 54(1946):152–53.

82. F. C. Leonard, "The Name of the Barringer Meteorite Crater of Arizona," *Popular Astronomy,* 58(1950):469.

83. H. L. Fairchild, "The Origin of Meteor Crater (Coon Butte), Arizona," *Bulletin of the Geological Society of America,* 18(1907):493–504.

84. E. Blackwelder to H. S. Colton, March 11, 1946, *Colton Papers.*

85. E. Blackwelder to H. S. Colton, September 25, 1946, *ibid.*

86. E. Blackwelder to H. S. Colton, January 8, 1946, *ibid.*

87. H. S. Colton to N. B. Drury, January 14, 1946; and to S. P. Osborn, January 22, 1946, *ibid.*

88. E. Blackwelder to N. B. Drury, February 6, 1946, copy in *ibid.*

89. S. P. Osborn to H. S. Colton, March 11, 1946, *ibid.*

90. E. Blackwelder to H. S. Colton, March 11, 1946, *ibid.*

91. H. S. Colton to E. Blackwelder, March 16, 1946, *ibid.*

92. R. W. Barringer, "Publicity Release," five-page mimeographed typescript, undated, *Lowell Observatory Archives.* Gilbert had, in fact, suffered a stroke in 1909 which, he had noted, "permanently lowered" his general physical condition; *see* S. J. Pyne, *Grove Karl Gilbert—A Great Engine of Research* (Austin: University of Texas, 1980), 214.

93. E. Blackwelder to H. S. Colton, September 25, 1946, *Colton Papers.*

94. H. S. Colton to E. Blackwelder, September 30, 1946, *ibid.*

95. R. W. Barringer, *op. cit.*

96. J. A. Russell, "Report on the 18th Meeting of the Society," *Meteoritics,* 1(1953):353–57.

97. D. Hager, "Crater Mound ('Meteor Crater'), Arizona: Is Its Origin Geologic or Meteoric?" abstract, *Contributions to the Meteoritical Society,* 4(1949): 223–24.

98. D. Hager, "Crater Mound (Meteor Crater), Arizona, a Geologic Feature," *op. cit.*

99. E. Blackwelder and D. Hager, "Crater Mound-Meteor Crater," *Bulletin of the American Association of Petroleum Engineers,* 37(1953): 2577–80 (critique and reply).

100. *Ibid.*

101. H. H. Nininger, "Impactite Slag at Barringer Crater," *American Journal of Science,* 252(1954): 277–90.

102. D. Hager, "Notes on Crater Mound in Answer to Some Points Raised by H. H. Nininger," *ibid.,* 695–97.

103. J. D. Buddhue, "A sieve analysis of crushed sandstone from Canyon Diablo, Arizona," *Popular Astronomy,* 58(1948): 190.

104. H. H. Nininger, "Reply," *American Journal of Science,* 252(1954): 697–700.

105. D. Hager, "Additional Notes on Crater Mound," *Bulletin of the American Association of Petroleum Geologists,* 40(1956): 161–62.

106. E. M. Shoemaker and K. E. Herkenhoff, "Upheaval Dome Impact Structure, Utah," abstract, *15th Lunar and Planetary Science Conference,* March 12–16, 1984, Houston, Texas, pp. 778–79.

107. "A Society for Meteorite Research," *Journal of the British Astronomical Association,* 43(1933): 306.

108. H. H. Nininger, "Geological Significance of Meteorites," *American Journal of Science,* 246(1948): 101–08.

109. D. M. Barringer to H. H. Nininger, February 8, 1929, *Barringer Papers.*

110. The general biographical material that follows here is from H. H. Nininger, *Out of the Sky* (New York: Dover Publications, 1952); *Arizona's Meteorite Crater, op. cit.;* and *Find a Falling Star* (New York: Paul S. Eriksson, Inc., 1972).

111. H. H. Nininger, *Arizona's Meteorite Crater, op. cit.*

112. D. Heymann, "On the Origin of the Canyon Diablo No. 2 and No. 3 Meteorites," *Nature,* 204(1964): 819–20.

113. L. LaPaz, "An announcement concerning future explorations at the Canyon Diablo, Arizona, Meteorite Crater," *Popular Astronomy,* 56(1948): 559–60. See also H. H. Nininger, *Find a Falling Star, op. cit.,* 147–49, 169.

114. H. H. Nininger, *Arizona's Meteorite Crater, op. cit.,* 81–114.

115. J. S. Rinehart, "Distribution of Meteoritic Debris About the Arizona Meteorite Crater," *op. cit.,* 152–53.

116. H. H. Nininger, "Impactite Slag at Barringer Crater," *op. cit.;* and *Arizona's Meteorite Crater, op. cit.,* 117–34.

117. H. H. Nininger, *Arizona's Meteorite Crater, op. cit.,* passim, but especially 147–52, where Nininger urges government ownership of the crater.

118. *Ibid.,* 50, 152–54.

119. L. Coes, "A New Dense Crystalline Silica," *Science,* 118(1953): 131–32.

120. E. C. T. Chao, E. M. Shoemaker, and B. Madsen, "First Natural Occurrence of Coesite," *ibid.*, 132(1960):220–22.

121. S. M. Stishov and S. V. Popova, "A New Dense Modification of Silica," *Geokimiya*, 10(1961):837; and E. C. T. Chao, J. J. Fahey, J. Littler, and D. J. Milton, "Stishovite, SiO_2, a very high pressure mineral from Meteor Crater, Arizona," *Journal of Geophysical Research*, 67(1962):419–21.

122. E. M. Shoemaker, "Impact Mechanics at Meteor Crater, Arizona," *op. cit.*; and "Penetration Mechanics of High Velocity Meteorites, Illustrated by Meteor Crater, Arizona," *op. cit.*

CHAPTER FOURTEEN. IMPACT: A COSMIC PROCESS

Epigraph: E. M. Shoemaker, "An account of some of the circumstances surrounding the recognition of the impact origin of the Ries Crater," paper submitted for publication to the Verein Rieser Kulturtage, Nördlingen, Germany, spring 1984.

1. H. H. Nininger, "The Geological Significance of Meteorites," *American Journal of Science*, 246(1948):101–08.

2. E. M. Shoemaker, "Collision of astronomically observable bodies with the Earth," *Geological Society of America Special Paper 1990*, 1982, 1–13; and "Asteroid and Comet Bombardment," *Review of Earth and Planetary Science*, 11(1983):461–94.

3. L. W. Alvarez, W. Alvarez, F. Asaro, and H. V. Michel, "Extraterrestrial cause for the Cretaceous-Tertiary extinction," *Science*, 208(1980):1095–1108. See also R. Ganapathy, "A major meteorite impact on Earth 65 million years ago: evidence from the Cretaceous-Tertiary boundary," *ibid.*, 209(1980):291–93.

4. B. F. Bohor, E. E. Foord, P. J. Modreski, and D. J. Triplehorn, "Mineralogical evidence for an impact event at the Cretaceous-Tertiary boundary," *ibid.*, 224(1984):867–69. See also R. A. Kerr, "An Impact But No Volcano," *ibid.*, 858.

5. R. Ganapathy, "Evidence for a major meteorite impact on the Earth 34 million years ago: Implication for Eocene extinctions," *ibid.*, 216(1982):885–86.

6. A. R. Alderman, "The Meteorite Craters at Henbury, Central Australia," *Mineralogical Magazine*, 23(1932):19–32; and L. J. Spencer, "Meteorite Craters," *Nature*, 129(1932):781–84.

7. L. J. Spencer, "Meteorite Craters as Topographic Features on the Earth's Surface," *Smithsonian Institution Annual Report*, 1933, 307–25 (reprinted from the *Geographical Journal*, March 1933). See also R. B. Baldwin, *The Measure of the Moon* (Chicago: The University of Chicago Press, 1963), 26–29.

8. R. B. Baldwin, *op. cit.*, 30–31, 57–58.

9. The absence of a crater, or craters, at Tunguska has since been confirmed and the object, a large meteoroid or a small comet, is believed to have exploded high in the atmosphere. See E. L. Krinov, "The Tunguska and Sikhote-Alin Meteorites," in B. M. Middlehurst and G. P. Kuiper, eds., *The Moon, Meteorites and Comets* (Chicago: The University of Chicago Press, 1963), 208–34.

10. L. J. Spencer, "Meteorites and the Craters on the Moon," *Nature,* 139(1937): 655–57.

11. J. D. Boon and C. C. Albritton, Jr., "Meteorite Craters and Their Possible Relationship to Cryptovolcanic Structures," *Field and Laboratory,* 5(1936): 1–9; and "Meteorite Scars in Ancient Rocks," *ibid.,* 5(1937): 53–64.

12. W. Branca and E. Fraas, "Das kryptovulcanishes Becken von Steinheim," *Akademie der Wissenshaften, Alb. 1,* 1–64, 1905.

13. W. H. Bucher, "Cryptovolcanic structures in the United States," *Report of the International Geological Congress, Session XVI* (1933), 1055–85, 1936.

14. W. H. Bucher, "The Largest So-Called Meteorite Scars in Three Continents as Demonstrably Tied to Major Terrestrial Structures," in *Geological Problems in Lunar Research, Annals of the New York Academy of Sciences,* 123(1965): 897–903.

15. N. L. Carter, "Basal quartz deformation lamellae—a criterion for recognition of impactites," *American Journal of Science,* 263(1965): 786–806.

16. A piece of Coconino sandstone showing part of a curved slickensided surface "rudely resembling part of a shatter cone" was found at Meteor Crater in the early 1960s by E. C. T. Chao, and was characterized by Shoemaker as "perhaps the closest approach to independent evidence for the impact origin of shatter cones." See E. M. Shoemaker, D. E. Gault, and R. V. Lugn, "Shatter cones formed by high speed impact in dolomite," in *Short Papers in the Geologic and Hydrologic Sciences,* U.S. Geological Survey, 1961, D365–368.

17. R. S. Dietz, "Geological structures possibly related to Lunar Craters," *Popular Astronomy,* 54(1946): 465–67; and "Meteoritic impact suggested by orientation of shatter cones at the Kentland, Indiana, Disturbance," *Science,* 105(1947): 42–43.

18. W. H. Bucher, "Cryptoexplosion structures caused from without or within the earth (Astroblemes or Geoblemes?)," *American Journal of Science,* 261(1963): 597–649.

19. R. S. Dietz, "Shatter cones in cryptovolcanic structures (meteorite impact?)," *Journal of Geology,* 67(1959): 496–505; Meteorite impact suggested by shatter cones in rock," *Science,* 131(1960): 1781–84; and "Cryptoexplosion Structures: A Discussion," *American Journal of Science,* 261(1963): 650–64.

20. E. M. Shoemaker, D. E. Gault, and R. V. Lugn, *op. cit.*

21. R. S. Dietz, "Shatter Cones in Cryptoexplosion Structures," in *Shock Metamorphism of Natural Materials,* B. M. French and N. M. Short, eds. (Baltimore, Mono Book Corporation, 1968), 267–85.

22. R. B. Baldwin, *The Face of the Moon* (Chicago: The University of Chicago Press, 1949), 66–113.

23. R. B. Baldwin, *The Measure of the Moon, op. cit.,* 6–105.

24. E. M. Shoemaker, "Asteroid and Comet Bombardment," *op. cit.,* 484.

25. W. Goodacre, *The Moon,* London, 1931 (published by the author).

26. H. H. Nininger, "A reply to Dr. Spencer's paper on 'Meteorites and the Craters on the Moon,'" *Popular Astronomy,* 46(1938): 107–09.

27. H. L. Fairchild, "Selenology and Cosmology," *Science,* 88(1938): 555–62.

28. R. K. Marshall, "The origin of the lunar craters," *Publications of the American Astronomical Society,* 10(1946): 167–68 (abstract).

29. R. B. Baldwin, "The meteoritic origin of lunar craters," *Popular Astronomy*, 50 (1942):365–69; and "The meteoritic origin of lunar structures," *ibid.*, 51(1943):117–27.

30. R. K. Marshall, "The origin of the lunar craters (A summary)," *ibid.*, 51(1943): 415–24.

31. J. B. Hartung, "Was the formation of a 20-km-diameter impact crater on the moon observed on June 18, 1178?" *Meteoritics*, 11(1976):187–94. The date is a Julian Date.

Astronomer Kenneth Brecher more recently has suggested that the Giordano Bruno cratering event, as well as a meteoritic "storm" on the moon detected in June 1975 by the Apollo seismic network, and possibly even the Tunguska event in 1908 in Siberia, are all manifestations of a meteoritic or cometary swarm resulting from the breakup of the well-known Comet Encke. See K. Brecher, abstract, "The Canterbury Swarm: Ancient and Modern Observations of a New Feature of the Solar System," *Bulletin of the American Astronomical Society*, 16(1984):476.

32. J. E. Spurr, *Geology Applied to Selenology—I. The Imbrium Plain Region of the Moon. II. The Features of the Moon*, 1944–1945 (published by the author).

33. J. Green, "Hookes and Spurrs in Selenology," in *Geological Problems in Lunar Research, op. cit.*, 373–402.

34. R. S. Dietz, "The Meteoritic Impact Origin of the Moon's Surface Features," *Journal of Geology*, 54(1946):359–75.

35. R. B. Baldwin, *The Face of the Moon, op. cit.*

36. T. L. MacDonald, "Studies in Lunar Statistics," *Journal of the British Astronomical Association*, 41(1931):172–83, 228–39, 288–90, 367–79; and 42(1931):291–94.

37. J. J. Gilvarry and J. E. Hill, "The Impact Theory of the Origin of Lunar Craters," *Publication of the Astronomical Society of the Pacific*, 68(1956):223–29.

38. E. M. Shoemaker, "Penetration Mechanics of High-Velocity Meteorites, Illustrated by Meteor Crater, Arizona," *Report of the International Geological Congress, XXI Session*, 1960, Part XVIII, 418–34.

39. E. M. Shoemaker, "Interpretation of Lunar Craters," in Z. Kopal, ed., *Physics and Astronomy of the Moon* (New York: Academic Press, 1962), 283–359.

40. R. B. Baldwin, *The Measure of the Moon, op. cit.*, 173–74.

41. C. W. Mead, J. Littler, and E. C. T. Chao, "Metallic spheroids from Meteor Crater, Arizona," *The American Mineralogist*, 50(1965):667–81; R. Brett, "Opaque minerals in drill cuttings from Meteor Crater, Arizona," *U.S. Geological Survey Professional Paper 600-D*, 1968, D-179-180; and P. J. Blau, H. J. Axon, and J. I. Goldstein, "Investigation of the Canyon Diablo metallic spheroids and their relationship to the breakup of the Canyon Diablo meteorite," *Journal of Geophysical Research*, 78(1973):363–74.

42. E. M. Shoemaker, personal communication, September 24, 1983.

43. R. B. Baldwin, "The Origin of Lunar Features," in *Geological Problems in Lunar Research, op. cit.*, 543–46.

44. N. A. Kozyrev, "Spectroscopic proofs for the existence of volcanic processes on the moon," in Z. Kopal and Z. K. Mikailov, eds., *The Moon* (New York: Academic Press, 1962). This volume contains papers from Symposium No. 14 of the International Astronomical Union held in Leningrad, Russia, December 1960.

45. J. C. Greenacre, "The 1963 Aristarchus Events," *ibid.*, 811–17.
46. R. B. Baldwin, "The Origin of Lunar Features," *op. cit.*, 546.
47. Moon issue, *Science,* 167(1970):449–804 (reprinted as *Apollo 11 Lunar Conference* (Washington, D.C.: American Association for the Advancement of Science, 1970).

PHOTO CREDITS

1 Courtesy of NASA.

2 From *Neue Annalekten*, 1834.

3 From James Nasmyth and James Carpenter, *The Moon: Considered as a Planet, a World, and a Satellite,* London: John Murray, 1874, pp. 102, 104.

4 The Mary Lea Shane Archives Of The Lick Observatory, The University Library, University of California, Santa Cruz.

5 From *Foote Prints,* vol. 30, no. 2 (1958), p. 6. Published by Foote Mineral Co., 18 W. Chelton Ave., Philadelphia, Pennsylvania, in "The Biography of a Crater" by Norman P. Grentieu.

6 G. K. Gilbert 1785, U.S. Geological Survey.

7 G. K. Gilbert 736, U.S. Geological Survey.

8 G. K. Gilbert 784, U.S. Geological Survey.

9 Portraits 205, U.S. Geological Survey.

10 From the *Bulletin of the Philosophical Society,* Washington, D.C., 1893, p. xii.

11 From the *Bulletin of the Philosophical Society,* Washington, D.C., 1893, p. xii.

PHOTO CREDITS   425

12 From the *Bulletin of the Philosophical Society,* Washington, D.C., 1893, p. xii.
13 The Mary Lea Shane Archives Of The Lick Observatory, The University Library, University of California, Santa Cruz.
14 By permission, Princeton University Library.
15 D. M. Barringer, 1909.
16 D. M. Barringer, 1909.
17 D. M. Barringer, 1909.
18 D. M. Barringer, 1909.
19 By permission, Stanford University Archives No. 1457.
20 By permission, Smithsonian Institution, No. 84-10782.
21 D. M. Barringer, 1909.
22 By permission, AIP Niels Bohr Library, Stanford University.
23 Portraits 100, U.S. Geological Survey.
24 By permission, Princeton University Library.
25 By permission, Princeton University Library.
26 The Mary Lea Shane Archives Of The Lick Observatory, The University Library, University of California, Santa Cruz.
27 The Mary Lea Shane Archives Of The Lick Observatory, The University Library, University of California, Santa Cruz.
28 T. J. J. See, "The Capture Theory of Cosmical Evolution" in *Researches on the Evolution of the Stellar Systems,* vol. 2, Lynn, Massachusetts: Thomas P. Nichols and Sons, 1910. Plate XII, Fig. a, p. 342.
29 Carter Observatory, Wellington, New Zealand.
30 Carter Observatory, Wellington, New Zealand.
31 From *The New Zealand Journal of Science and Technology,* vol. 7, no. 3 (1924), pp. 129–42.
32 Courtesy of J. Paul Barringer.
33 Courtesy of Harvard University Archives.
34 Courtesy of J. Paul Barringer.
35 From D. M. Barringer Papers, Princeton University Archives.
36 University of Rochester Photograph No. 8413641.
37 University of Chicago Archives.

38 By permission, American Philosophical Society.

39 By permission, AIP Niels Bohr Library, Stanford University.

40 By permission, AIP Niels Bohr Library, Stanford University.

41 By permission, AIP Niels Bohr Library, Stanford University.

42 By permission, Harvard University Archives.

43 Courtesy of H. H. Nininger.

44 Portraits 1403, U.S. Geological Survey.

45 From Ralph B. Baldwin, *The Measure of the Moon,* University of Chicago Press, 1963, p. 72.

46 Courtesy of Robert S. Dietz.

47 Courtesy of Ralph B. Baldwin.

48 From Ralph B. Baldwin, *The Measure of the Moon,* University of Chicago Press, 1963, p. 141.

49 From E. M. Shoemaker, *Penetration Mechanics of High Velocity Meteorites,* International Geological Congress, XXI Session, Norden, 1960, pt.18, Structure of the Earth's Crust and Deformation of Rocks, p. 418, Copenhagen, 1960.

Index

Académie Royale des Sciences de France, 19
Academy of Natural Sciences of Philadelphia, 86, 139, 158, 178, 235–236
Accretion theory, 21–22, 24–25, 31–32, 49, 128, 156–157, 160, 187, 203, 353
Adams, Edwin Plimpton, 283–291, 293–295, 298–300, 317; first report, 283–286, 290, 294–295, 299, 300, 304; second report, 303–309, 313
Adiabatic compression, 304–305
Aerial photographs, 176, 230, 241, 275, 346
Aerolite, 25, 85, 90, 95, 112, 118, 229, 325. *See also* Meteorite
Agassiz, George Russell, 1, 223–224, 227, 231–233, 235, 241, 247–248, 257–259, 261
Ainsle, Captain Maurice, 211
Air resistance, on meteor, 108, 129, 136–137, 152, 199, 221, 267–268, 270–273, 275–279, 281–294, 297–301, 303–309, 331, 359
Albritton, Claude C., Jr., 348–350, 357
Alexander, H. H., 83–85, 110, 120, 169–170, 215, 218–219, 226, 232, 254
Alphonsus, lunar crater, 364
Alpine valley, lunar, 64
Alvarez, Luis, 345
American Association for the Advancement of Science, 31, 49, 106, 220
American Astronomical Society, 265, 353, 364
American Institute of Mining Engineers, 167, 171
American Meteorite Museum, 328, 340–341
American Museum of Natural History, 76, 106, 170, 172, 254
American Philosophical Society, 140, 259, 340–341
"Annular glaciers," 28–29

427

428 INDEX

Antoniadi, E. M., 69
Aristarchus, lunar crater, 365
Aristillus, lunar crater, 193
Arizona sandstone, 309, 311
Armijo, Mathias, 32
Ashanti crater, 347
Asteroid, 21, 32, 67, 212, 345–346, 354, 359
Astronomy (Moulton), 354
Astronomy: An Introduction (Baker), 354
Atmospheric retardation, 267, 270–271, 275–279, 281, 331. See also Air resistance
Aubrey limestone, 42–43, 46, 97
Aubrey sandstone, 43, 83, 115

Bacubirito meteorite, 125, 271
"Bakelite," 171
Baker, Marcus, 41–42, 45–48, 50
Baker, Robert H., 354
Baldwin, Ralph B., 320–321, 352–353, 354, 357–360, 362–363, 365
Ball, Sir Robert S., 69
Ballistics, 62, 69, 140–141, 168–169, 180, 182, 197, 212, 268–269, 274, 277, 280–284, 293, 303–304, 307, 319. See also Bomb analogy
Barbour, Thomas, 218, 223–224, 227, 237, 249, 254–255, 259, 290
Barford, Einar, 225–226, 231–233, 236
Barnes, Will C., 141–143
Barr, Edward, 364–365
Barringer Crater Company, 336, 341. See also Standard Iron Company
Barringer Meteorite Crater, 336
Barringer, Brandon (D. M. Barringer's son), 236, 239, 247, 277, 281–282
Barringer, Daniel Moreau: biographical, 5, 74, 238, 296–297; cluster theory of, 133, 135–140, 150–152, 171, 272. See also Cluster theory; commercial interest in Coon Mountain, 78, 84–85, 102, 110, 131, 133, 143, 146, 151, 163–176, 214–227, 230–231; conflict with editor of *National Geographic Magazine*, 250–257, 260; deterioration of support from allies, 243, 273, 286–289, 292–295; disagreement with Tilghman, 103–104, 133; early interest and claims at the crater, 73–80; efforts to attract investors, 102–103, 111, 119, 127, 131, 133, 138, 140–141, 151, 163–173, 180–181, 214–219, 220–225, 227, 237–240, 242–243, 246–248, 273; efforts to control contradictory ideas, 174–176, 221, 243–244, 283, 287–289, 294–295; financial problems, 268, 270–271; investigation of Coon Mountain, 75–89, 103, 109–110, 134–137, 148, 168, 177–180, 239, 248–249, 257–260, 273; objections to new meteoric hypothesis, 219–222; objections to the volcanic steam explosion hypothesis, 81, 84, 90–94, 115–116, 144, 150, 158–159, 172; obtaining financing, 146, 153, 230, 260–262, 289, 295; on impact origin of meteor crater, 75–76; response to his advocacy of theories, 162, 174, 221; response to mass and velocity estimates of Moulton, 262, 264, 271–281, 283–292, 294–296; restriction of access to Coon Mountain, 173–174; role in lunar controversies, 157–160, 180–184, 216, 219, 242, 259–260; works by: *The Law of Mines and Mining in the United States* (Barringer and Adams), 75; *Minerals of Commercial Value* (Barringer), 75; first paper on Coon Mountain in 1906, 87–100, 104–107, 161, 253; second paper on Coon Mountain in 1909, 133–137, 161, 191, 253; third paper on Meteor Crater in 1914, 158; fourth paper on Meteor Crater in 1924, 178–179, 183
Barringer, Daniel Moreau (Reau), Jr., 217, 224, 229, 250–251, 256–259, 298; biographical, 215; investigation of Odessa crater, 234–236; lecture and articles, 241, 243–246, 256; president of Standard Iron Company, 325; refutation of "sink-hole" theory, 229, 336; study of Coon Mountain, 327–330
Barringer, Paul B. (D. M. Barringer's cousin), 110, 119, 129

Barringer, Richard (D. M. Barringer's son), 181–182, 335–336
Beer, Wilhelm, 12–13, 15, 56, 358
Bennett, Thomas D. (D. M. Barringer's brother-in-law), 77
Bennitt, E. J. (D. M. Barringer's brother-in-law), 89, 110, 151, 216–217, 224, 238; confidant of D. M. Barringer, 85, 87, 102, 104, 129, 160, 168, 173, 180; initial involvement with Coon Mountain, 77–78, 80; negotiations for Holsinger's shares, 163; obtaining patents, 80; silica brick scheme, 164
Bergeron, Jules, 27
Bibbins, Arthur B., 233–234, 236
Bickerton, Alexander William, 185, 194–196, 204–206; partial impact theory, 205
Big splash theory, 156–162
Bilateral symmetry, 82, 94, 134, 159, 168, 329, 350, 352, 357
Bird, J. Malcolm, 181, 183
Birt, W. R., 27
Black scale, mantle of, 44
Blackwelder, Eliot, 331–337
Blake, William P., 108
"Blue Sky" commission, 230
"Blue Sky" laws, 222–227, 231–233, 236, 238
Bolide, 70, 116, 192, 212, 275–276, 349, 356
Bomb analogy, 197–198, 200, 202, 210, 219, 259, 320, 358. *See also* nuclear explosion analogy
Boon, John D., 348–350, 357
Boon and Albritton hypothesis, 348–350
Boone and Crockett Club, 74–75, 247
Boutwell, William D., 230–231, 250–251, 253–255
Branner, John Casper, 114–115, 129, 166; distrust of U.S. Geological Survey staff, 101, 105, 119, 131–132; mentor of D. M. Barringer, 75, 173; on meteoritic impact theory, 106; opinions about G. K. Gilbert, 101, 105–106; role in Coon Mountain controversies, 115, 131–133, 144, 218; study of metamorphosed sandstone, 135, 150

Brereton, Roy G., 328
Bridgman, Percy Williams, 5, 298, 313–317, 319, 321
British Astronomical Association, 69, 71, 154, 185, 190, 192, 202, 205, 208
British Museum, 107, 150, 283, 346
Brown, Ernest W., 283, 286
Brückner, Eduard, 149
"Bubble or blister" analogy, 9–10, 14, 17, 22, 27, 56, 66
Bucher, Walter H., 349–351
Buddhue, John David, 332–333, 338

"Caldera," 14, 67, 149, 155
Campbell, William Wallace, 5, 231, 270, 356; attempt to visit Coon Mountain, 173–176; on explosion after meteoritic impact, 179–180, 182, 240; on meteoritic theory, 173–176, 240; on volcanic analogy, 173, 185, 199–202, 356; restriction on publication, 174–176, 179
Campi Phlegraei, Italy, 14–15
Campo del Cielo craters, 347
Cañon Diablo Iron Mining Company, 142
Canyon Diablo irons, 34, 76, 92, 94, 116–117, 120, 122, 135, 141, 144, 150, 168–169, 179, 241–242, 340. *See also* Meteoritic iron
Canyon Diablo meteorite, 95, 325, 340
Canyon Diablo siderites, 88, 136
Cape York iron, 125
Capture theory, 189–190
Carnegie Geophysical Laboratory, 165, 242, 324
Carnegie Institution, 121, 146, 215–216, 242, 254, 283, 320
Carpenter, James, 15–18, 23, 27, 185
Celestial mechanics, 63, 266
Celestial Mechanics (Moulton), 266
Central cones, 203. *See also* Central peaks
Central peaks, 16, 28–29, 58, 63–64, 157–160, 182–183, 188–191, 197, 201–203, 331–332, 356–357
Chacornac, J., 27
Chaldni, Ernst Florens Friedrich, 20
Chamberlin, Thomas C., 105; biographi-

Chamberlin, Thomas C. (*continued*)
cal, 71; planetesimal hypothesis, 128, 156–157, 160, 194, 212, 216, 353; volcanic analogy of lunar craters, 71, 154, 260
Chambers, George F., 27
Chao, Edward C. T., 149, 343
Chinguetti meteorite, 279
Cholnoky, Eugene De, 148, 150
Circularity of craters, 4, 61–63, 134, 158, 180, 182, 185, 192, 195–196, 198, 206–208, 210, 213, 219, 254, 259, 329, 350, 353, 356–357. *See also* Lunar craters, circularity of
Clavius, lunar crater, 69
Clearwater Lakes crater, 351
Clerke, Agnes M., 13
Cluster theory, 133, 135–140, 152, 179, 263, 265–267, 276, 278, 280, 292, 299–301, 306, 308–309, 311–312, 322
Coconino sandstone, 43, 82–83, 145, 243, 249, 258, 323, 326, 339, 343
Coesite, 343, 347, 350–352
Colton, Harold S., 334–336
Colvocoresses, George M., 216–226, 228, 232–234, 239, 243, 281, 285; biographical, 216; deteriorating support of Barringer, 287; drilling at Coon Mountain, 223, 249, 257–260, 262, 273, 326; initial investigation of Coon Mountain, 217–218; Meteor Crater Exploration and Mining Company, 218–219, 223–225; on geophysical studies at Coon Mountain, 321–324; report of, 248; role in Coon Mountain controversies, 218, 220, 221–225, 287, 306, 323–324, 331; soliciting investments, 214, 219, 221, 224, 232–233, 247–248, 325, 327
Combined meteoritic and volcanic hypothesis, 52, 206, 209
Comet, 136–140, 152, 179, 267, 345
Commonwealth silver mine, 73, 75–76, 111
Compressional wave, at impact, 309, 315
Condensation of volatilized materials, 186, 220–221, 341–343, 362–363

Controversy, among scientists over crater origin, 100–101, 348–363
Coon Butte, *See* Coon Mountain
Coon Crater, 47. *See* Coon Mountain
Coon Mountain: absence of volcanic material at, 33, 40; chronology of, 79, 96, 107, 218, 265, 322, 331–333; commercial control of, 340–341, 343. *See also* Standard Iron Company; D. M. Barringer, commercial interest in Coon Mountain; description of: 1, 33, 81–84, 90, 134, 176–180, 330, 337–338; composition of, 43–44, 81–84, 90–96, 120–121, 169, 177, 179, 244, 246; dimensions of, 43, 45, 48, 50, 83, 100, 110; faulting at, 337–338; stratigraphy of, 42–43, 48, 82–83, 93–94, 268, 333; effort to acquire for public ownership, 333–336; origin of, 2, 39–40, 45, 49–53, 108–109, 141, 148–149, 229, 325–328, 331–333, 336–339, 349; *See* Meteoritic impact theory; presumed absence of iron in, 43, 106, 130, 144; priority of publication of impact origin, 87, 131–132, 174; publicity of, 140–141, 146–149, 153, 171, 175, 180–181, 227–228, 230–231, 242, 244–245, 250–251, 275, 335; recent investigations of, 331–333, 343, 351, 361–362, 364–365; restrictions of access or publication, 173–176, 179, 340–341, 343
Copernicus, 27, 71, 156, 208–213, 350
Craft, 32, 142
Craters. *See also* Lunar craters: absence of more terrestrial impact, 199, 201, 203–204, 210, 212–213, 346, 348; central peaks, 197, 201–202, 213, 364. *See also* Central peaks, circularity of. *See* Circularity of craters; comparison of terrestrial and lunar topography, 16, 57–58, 154–160, 180–184, 199–204, 206, 210, 213, 344–346, 352–354, 356, 358, 364; depth-diameter relationships, 358, 360–361; depth of, 43, 45, 83, 100, 110, 211, 267–269, 271, 281, 347, 354, 358, 361; diameter of, 43, 208–210, 213, 271, 358, 361; explosion analogy, 98, 108, 146, 197–198, 201–

203, 210, 219, 259, 280, 320, 331, 358; formation of, 2, 15–16, 18, 32, 46, 91, 130, 157–158, 160, 204, 206–209, 213, 241, 281, 292, 316, 329, 346–348, 352, 361; impact, 157–158, 160–162, 193–197, 200, 206–210, 213, 241, 344, 346–348, 352, 364; terrestrial cryptoexplosive: Clearwater Lakes, 351, Vredefort, 350–351, terrestrial impact: 4, 184, 346–348, 352, 358; distribution of, 344; Aouelloul, 279; Ashanti, 347; Campo del Cielo, 347; Coon Mountain, 51; Haviland, 340, 353; Henbury, 346–347; Odessa, 184, 233–236, 258–259, 272, 331, 346, 362; Ösel, 184, 347; Upheaval Dome, 339; Wabar, 346–347, 362; terrestrial nuclear explosion: 318, 321, 346, 361; Sedan, 321; Teapot Ess, 321, 361; terrestrial volcanic: Etna, 58, 107, 191; Kobandai, 51, 146; Krakatoa, 51; Lonar Lake, 52; Vesuvius, 51; volcanic, 156, 199, 209, 213, 364–365
Crater Mining Company, 173, 176, 178–179, 229
Crater Mound. *See* Coon Mountain
"Craters of elevation" theory, 14, 16
Crommelin, A. C. D., 212–213, 355
Cryptovolcanic structures: formation of, 349–351; Clearwater Lakes, 351; Flynn Creek, 350; Kentland Cone, 351; Sierra Madera Dome, 350; Vredefort Ring, 350. *See also* Meteoritic impact theory
"Cushion of air" concept, 272. *See also* Air resistance

Dana, James Dwight, 56, 337; volcanic analogy, 14–15, 17, 30, 58, 68; ebullition theory, 14–15, 18, 23
Darton, Nelson Horatio, 146, 161, 252, 336; role in Coon Mountain Controversies, 143–145, 152, 159–160, 333–334; volcanic steam explosion hypothesis, 143–145
Darwin, Sir George Howard, 29, 56, 128, 157
Davidson, Reverend M., 193–195
Davis, William Morris, 147–150, 231; critique of Gilbert's volcanic hypothesis, 66–67, 72; visit to Coon Mountain, 147–148
Day, Arthur L., 165, 215, 283
De Beaumont, J. B. A. L. L. Élie, 14, 21
Dellenbaugh, Frederick S., 325–326
Der Mond (Beer and Mädler), 12–13
Dialogue Concerning the Two Chief World Systems (Galileo), 8
Diamonds, at Coon Mountain, 31, 34, 76, 120, 127, 142, 147, 153–154, 163, 230
Dick, James E., 321–322
Dietz, Robert S., 351, 356–357
Dixon, S. G., 86–88
Dodd, Captain A. W., 188, 191
Douglas, James S., 131, 215–216
Drilling, at Coon Mountain, 82–89, 103, 109–110, 133–134, 152, 176–179, 214, 248–249, 257–260, 275, 326
Dravo, Frank R., 259, 324
Drury, Newton B., 334–335
"Dynamic similarity," principle of, 293

Ebert, H., 29, 56, 60
Ebullition theory, 15, 18
Echols, William E., 283–284
Ellipticity of craters, 62
Energy expenditure, of meteorite, 134, 140, 185, 266–267, 269, 271, 281–282, 285, 288, 293, 297, 301–303, 306–316, 318–321, 342–343, 361, 363
Ericsson, John, 28–29
Erosion, 43, 201, 218, 348
Etna, 58, 107, 191
"Expansion on solidification," 17, 185
Exploration of crater, 32–35, 41–46, 81–91, 109–113, 167–168, 176–179, 257–260, 268, 273–274, 321–323, 326–327, 340–341
Explosion analogy. *See* Bomb analogy; Point source explosion analogy; Nuclear explosion analogy
Explosion on impact, 4, 187, 196–198, 206–209, 212, 219–220, 240–241, 341, 361–363. *See also* New meteoric hypothesis
Explosive meteoritic impact theory, 196–198, 206–209, 212, 219, 357–363, 365

Exposition du Système du Monde (Laplace), 22

Fairchild, Herman Leroy, 2, 5, 135, 218, 332, 334; impact origin of Coon Mountain, 114–118, 231, 353, 363; "percussion cap" theory, 114–115; role in Coon Mountain controversies, 255–257, 325; visit to Coon Mountain, 114; volcanic analogy, 115–117

Farrington, Oliver C., 113–114, 161–162, 231, 283

Faye, Hervé, 29, 56

Field Museum of Natural History, 113, 161, 231, 283

Fisher, Willard F., 263, 270–272

"Flaming drop" theory, 94–96, 113–114, 119, 122

Fletcher, Lazarus A., 107–108, 283

Fletcher, S. L., 193

Flynn Creek disturbance, 350

Foote, Arthur E., 31–35, 39–40, 76, 81, 94, 113, 123, 154; investigation of Coon Mountain, 31–35, 123

Fountain, E. O., 202–203, 210

François Cementation Process, 259–260

Frankland, Edward, 29

Friction, 269

Fulgurite glass, 121, 159

Fusion of impacted materials, 92, 125, 130, 135, 137, 141, 153, 155, 188, 249, 269–271, 281, 342–343

Galileo, 8–9, 19

General Astronomy (Jones), 354

Geological Society of America, 106, 114–115, 145, 236, 333, 364

Geology (Chamberlain and Salisbury), 71, 105, 154

Geomorphology, 37

Geophysical survey, of Coon Mountain, 269, 274, 288, 291, 306, 316, 321–324, 326–330, 333

Gervase of Canterbury, 355

Geyser theory, 141, 148–149

Gifford, Algernon Charles, 184, 194, 352; biographical, 204–205; circularity, 206–208, 213; crater diameter, 208, 213; crater formation, 204, 206–208, 213; impact dynamics, explosion, 196, 204–208, 210, 213, 219, 348, 361; mass-velocity relationship, 207–208; new meteoric hypothesis, 185, 204–208, 213, 219. *See also* new meteoric hypothesis; role in the new meteoric theory controversies, 185, 210–213, 219–220; velocity of ejection, 208; volatilization, 207–208, 213

Gilbert, Grove Karl, 31–32, 35–72, 101, 104–106, 154, 356; biographical, 35–39, 114; Coon Mountain investigation, report on, 50–52; criticism of, 39, 90, 114, 159, 256–257; critique of "moonlet" hypothesis, 66–72, 156, 186, 207, 256–257, 356; critique of work by, 52, 90, 96; hollow-rim volume hypothesis, 41–42, 60; impact cratering, mechanics of, 63, 195; lunar crater origin, investigation of, 48; magnetic attraction hypothesis, 41–42; major works, 37; meteoritic impact hypothesis, 31–32, 35, 40–41, 51, 55, 59–60, 105, 120, 156, 186, 256–257; "moonlet" hypothesis, 61–67, 157, 186, 188, 356; role in Coon Mountain controversies, 3, 35, 39, 89–90, 96, 101, 104–106, 144, 251–257; scientific method of, 38–40, 50; study of moon, 54–67, 71–72, 154, 156–157; visit at Coon Mountain, 41–47; volcanic analogy, 57–58, 156; volcanic steam explosion hypothesis, 101, 107, 109, 112, 120, 124, 143–147, 162, 172, 222, 242–243, 251, 256–257, 335

Gilvarry, J. J., 320, 360

Giordana Bruno, lunar crater, possibly formed 1178 AD, 355

Glacial analogy, 28–29, 44, 52, 56, 64, 66, 210

Glacial hypothesis, 56, 60, 210, 213

Goodacre, Walter, 154–155, 190–191, 209–210, 212, 352

Grand Canyon, 36

Gravitation, 10, 65, 70, 266, 272, 285, 290, 354

INDEX   433

Gravity studies, of Coon Mountain, 168, 328–329
Green, Nathaniel E., 29
Greenacre, James, 364–365
Grimaldi, lunar crater, 208
Grosvenor, Gilbert H., 230–231, 251, 253, 255
Gruithuisen, Franz von Paula, 188, 359; habitation of moon, 11–13; impact theory, 21–24
Guggenheim mining, 166–167
Guild, F. N., 108–109

Hager, Dorsey, 229, 328, 331–333, 336–339
Hale, George Ellery, 146, 166, 173, 199
Halley, Edmund, 19–20
Halley's Comet, 138–140
Hammond, John Hays, 228, 246
Hannay, J. B., 29, 68
Harding, Norman, 328–330
Harris, John F., 224, 227
Hart, O. A., 167, 169, 171, 218
Harvard Club, 241
Heat generation, 125, 136, 140, 198, 213, 249, 269–270, 292, 311–312, 314–315, 342–343, 347, 359–360. See also Meteoritic impact theory, heat generation
Hecksher, August, 243, 246, 248
Henbury craters, 346–347
Henkel, F. W., 190
Herschel, Sir John F. W. (W. Herschel's son), 29; "moon hoax," 11–13; volcanic analogy, 14
Herschel, William, 7, 10–11, 13, 21, 319
Hill, J. E., 320, 360
Holden, Edward S., 47, 65–66
Holsinger, Samuel J., 109–111, 129, 133, 163, 332; chronology of Coon Mountain, 332; death of, 143, 150–151; discovery of largest siderite, 150–151; drilling, 78–80, 82–84, 89, 110; exploration of Coon Mountain, 76–80, 82–84; informed Barringer about Coon Mountain, 73–74, 76; meteor siting, 94–95; mining claims for Coon Mountain, 76–81; pessimism of, 79, 86–87, 109; title defense, 142–143
Hooke, Robert, 17, 27, 206; "bubble or blister" analogy, 9–10, 14, 22, 66
Hovey, Edmund O., 106–107, 172, 218
Howell, Edwin E., 52, 84, 116
Hull, Arthur H., 232–233
Huston, M. B., 276–277
Hutchinson, W. Spencer, biographical, 242, 261; reassessment of Coon Mountain enterprise, 261–263, 268–270, 272–274, 279; "stamp mill factor," 281, 302, 314
Hydrodynamic flow, 360–361
Hydrodynamic theory, 319–320, 360–361
Hyginus, lunar crater, 13, 208
Hyginus N, lunar crater, 13, 26

Impact analogy, 352
Impact origin, 3–4, 59, 155, 184, 187, 203–204, 234–236, 250, 271–272, 278–279, 284, 286, 292, 331, 339, 343–353, 357–363, 365–366. See also Coon Mountain, origin of; Moon, craters of, impact origin of
Impact process, 3–4, 362; earth, 344–345; effect on biological evolution, 3, 345–346, 359; importance in the solar system, 344; probability of terrestrial impact events, 345
Impact revolution, 346–366
Impact theory, 54, 116, 120, 155, 157–158, 160–162, 203–204, 213, 234–235, 241, 243, 252, 343, 360–363; consensus on, 318. See also Meteoritic impact theory; Moon, impact theory
Impact and steam explosion theory, 79, 125–126, 130, 135, 152, 240–241, 341, 343
"Impactite," 342–343
Ingalls, Albert G., 244–245
International Geophysics, Inc., 322–323
International Nickel Company, 102, 216, 324
Introduction to Astronomy (Moulton), 71, 259

Introduction to Celestial Mechanics (Moulton), 189
Iron oxide, at Coon Mountain, 44, 81, 107, 116, 137, 168
Iron shale, 81–82, 92–93, 113, 116, 119–122, 179, 339–340, 347
Ives, Herbert E., 185, 197–202, 210, 352, 361

Jakosky, J. J., 322–324, 326, 329–332
Jennings, Sidney Johnston, 170–172, 225–226, 232; assumes control of operations at Coon Mountain, 176–179; exploration of crater, 167–168; publication restrictions of, 175–176, 179
Johnson, Alba B., 224, 232
Johnson, Willard D., 40, 44, 252; volcanic steam explosion hypothesis of, 123–124, 147–150
Jones, H. Spencer, 354
Judd, John W., 30

Kaibab limestone, 42, 97, 330, 332, 337, 342, 351
Kentland cones, 351
Kepler, Johannes, 19
Keyes, Charles R., 145
Kilauea, volcano, 14, 366
Kinetic energy, 266–267, 277, 281, 302, 308, 310–312, 316, 361–362. *See* Meteoritic impact theory, kinetic energy
Klein, Herman J., 13, 26
Klemin, Alexander, 283, 286
Knyahinya, Hungary, crater, 124
Kobandai, volcano, 51, 146
Koenig, G. A., 34, 76, 113
Kometen ünd Meteoren (Meyer), 192
Kozyrev, N. A., 364
Krakatoa, volcano, 46, 51, 146
Kulik, L. A., 250, 347
Kunz, George F., 160, 172

L'Aigle meteoritic fall, France, 20
Laccolite, 37, 40
Laccolith, 37, 353, 355
Lake Bonneville (Gilbert), 37

Lampland, Carl Otto, 239, 283, 286, 295, 325, 326–327
"Landslip phenomena," 16
Lane, Albert C., 106
LaPaz, Lincoln, 320, 336, 341
Laplace, Pierre Simon, 21–22, 24, 69, 157, 194
Larmor, Sir Joseph, 304, 307
"Lava lakes" analogy, 23. *See also* Ebullition theory
Laws of Mines and Mining in the United States, The (Barringer and Adams), 75
Lease agreements, 171, 173, 218, 247–248, 336
Lechateliérite, 112, 249, 326, 337, 343, 346–347, 357
Leonard, Frederick C., 334
Leonid meteor swarm, 121, 267, 279
Lepper, G. H., 202–203
Limestone, arenaceous, 271
Limestone sink hypothesis, 52, 108. *See also* sink-hole hypothesis
Linné, lunar crater, 13, 26
Livermore, Robert, 261–263, 265, 268, 270, 272–273, 279, 296
Locke, Richard Adams, 11–12
Lockyer, J. Norman, 128
Lohrmann, Wilhelm Gotthelf, 12–13
Lonar Lake, India crater, 52
Longwell, Chester R., 326
Lowell Observatory, 364–365
Lowell, Percival, 223, 248
Lunar analogy, 154–160, 180–184, 199–200, 200–203, 213, 344–345, 352–354, 356, 358, 364
Lunar craters, 7–30, 57–58, 260, as named: Alphonsus, 364; Aristarchus, 365; Aristillus, 193; Clavius, 69; Copernicus, 27, 71, 156, 208–213, 350; Giordana Bruno, 355; Grimaldi, 208; Hyginus, 13, 208; Hyginus N, 13, 26; Linné, 13, 26; Moretus, 200; Theophilus, 208; Timocharis, 357; Tycho, 28–29, 156, 200; Wargentin, 355; circularity of, 4, 15, 17, 22–23, 25, 28, 55, 61–63, 158, 180–182, 185–186, 189, 191, 193, 195–196, 198–199,

206–210, 213, 219, 242, 347, 353, 356–357; depth of, 211, 354, 358, 361; diameter of, 208–210, 358, 361; height of rim of, 358; impact origin of, 3–4, 55–56, 59–63, 69, 154–158, 160–161, 180–213, 215–216, 219, 242, 246, 259–260, 344, 346–348, 352–363, 365; theories of formation, 4, 7–30, 61–68, 70–72, 185–186, 190–191, 195–199, 201–203, 206–213, 260, 344, 347–348, 352, 354–359, 361–363, 365
"Lunar planispheres," 11
Lunar tide theory, 29, 56–57, 68, 200
Lundberg, Hans, 217, 323, 326, 328
Lyons, Robert Edwards, 169

"Maars," 51, 58–59, 98, 146, 154
MacDonald, Thomas L., 358
Mädler, Johann Heinrich, 12–13, 15, 56, 358
Magie, William Francis (Billy), 78, 167–173, 287–289; deteriorating support for D. M. Barringer, 287, 293–294; energy estimate for Coon Mountain, 319–320; magnetic experiments of, 136, 140, 162; mass computations, 141; meteoritic impact theory, 181, 244; report of, 302; role in controversies, 231–232, 243–245, 254, 271–272, 287, 292–293, 299, 302, 311, 314; source of scientific data for Barringer, 221, 223, 231–232, 240, 243–245, 275, 287–288, 292–293
Magnetic attraction hypothesis, 136; of Baker, 45, 50; of Gilbert, 41–42, 45–46, 50; of Tilghman, 96–97
Magnetic iron oxide, 92
Magnetism, study of, 269, 321–324, 328–329. See also Geophysical survey
Magnetite, 83–84, 93, 97, 119
Mallet, John W., 87, 89, 104, 119, 131; analysis of meteoritic material, 83, 85, 111–112, 120, 135, 150, 169; biographical, 74–75; on volatilization, 129–130
Manning, Dr. G. F., 95–96

Manual of Geology (Dana), 30
Maps, of Coon Mountain, 48, 134, 245, 328; of the moon, 9–13, 364
Mare Crisium, lunar, 69
Mare Imbrium, lunar, 64–65, 71, 354
Mare Serenitatis, lunar, 65, 69
Mare Tranquilitatis, lunar, 65
Mars, 364–365
Marshall, Roy K., 353–356
Mass and energy, estimates of, 319–322
Mass estimation. *See* meteorite, estimates of size or mass of
Mass extinctions, caused by meteorites, 3, 345–346, 359
Mass-velocity relationship, 128–129, 140–141, 188, 195–196, 198, 207–210, 263, 265–271, 276–278, 280–282, 284–288, 291–293, 300–301, 303, 305–315, 319–321, 347, 353, 359–361
"Metallic spheroids," 92, 339, 341–342, 362–363
Material in crater, 107
Maxwell, J. E., 195
Maydenbauer, A., 27
McClenahan, Howard, 254–255
Meineke, Franz, 154–156
Merrill, George Perkins, 105, 118–135, 140, 143, 159–162, 172, 180, 216, 218, 234, 236, 250, 331, 363; impact and steam explosion theory, 125, 130, 135, 144, 152, 175; investigations of Coon Mountain, 118–126, 256; meteoric hypothesis, 100, 121; on explosion of Coon Mountain meteorite, 187, 219–220, 241, 341; on lechateliérite, 112; on metamorphosed sandstone, 120–121, 123, 135, 150, 169; on the impact origin of Coon Mountain, 101–102, 124, 155, 215, 231, 252; "plums in a pudding" hypothesis, 122; role in Coon Mountain controversies, 5, 123–126, 130–131, 161, 251–252, 254; visit to Coon Mountain, 44, 119, 131, 149; volatilization, 187, 219, 341; volcanic steam explosion theory, 123
Metamorphosed sandstone, 345–347, 350,

436   INDEX

Metamorphosed sandstone (*continued*) 352, 357. *See also* Coesite; Shock metamorphosed rock; Lechateliérite; Stishovite
Meteor, 31, 99, 117, 142, 145, 195–196, 209, 265; "flaming drop" theory, 94–96, 119–122; origin of, 2, 19–20
Meteor Crater, name changed to, 117, 123, 146
Meteor Crater Exploration and Mining Company, 214, 218, 220, 223, 227, 237, 239, 247–248, 261, 273, 288, 302, 316, 321, 324–325, 327
Meteor Silica Corporation, 335
Meteoric iron, 286, 333
"Meteoric rain," 22, 26
Meteoric impact theory. *See* Meteoritic impact theory; Impact origin
Meteorite, 2, 19–20, 43, 69–70, 103–104, 106–109, 191, 229, 236, 259, 279, 347–350, 357–362, 365. *See also* Moon, meteorites; angle of impact, 134, 180, 182, 222, 316, 329, 331, 342; assay of meteoritic materials, 33–34, 81, 83–85, 113, 118, 168–170, 177–178, 215, 231, 262–263, 268, 340; composition of, 32–35, 52, 76, 81–85, 92–93, 113–114, 116–118, 120–122, 135–136, 138, 153, 168–170, 179, 267–268, 280, 300–301, 312, 325–326, 340; dimensions and dynamics of, 263–272, 274, 276–281, 285, 288, 290–293, 298–302, 305–306, 308–316, 319–323, 342–343, 359–362; estimation of size or mass of, 45, 79, 98–99, 108, 125, 127, 134, 141, 147, 153, 159, 175, 196, 223, 238–239, 263–271, 274, 277–279, 281–282, 284–288, 291–293, 299, 301–303, 306–308, 311–316, 318–321, 323, 363; fate of, 41, 84, 92–93, 95, 97–98, 123, 125–130, 133–137, 140–141, 144, 151–152, 161–163, 168, 170–171, 175, 177, 179–180, 192, 219, 221, 223, 227, 235, 240–246, 263, 267–270, 272–274, 276, 281, 291–292, 306–307, 310–312, 316–317, 320–323, 326, 328–330, 333, 341–343, 361–363. *See also* volatilization; origin of, 19–20, 24–25; penetration of, 45, 50, 80, 98, 108, 125, 140–141, 179–180, 263, 267–269, 271, 276–278, 280–281, 285, 292, 309–310, 312, 315–316, 319–321, 330; stony type, 52; terrestrial, 40, 207, 236, 348–350, 358, 360–362: Anighito, 235, Bacubirito, 271, Chinguetti, 279, Saharan, 278; trajectory of, 134, 265–267, 308–309, 316, 329–331; velocity of, 98, 108, 121, 129–130, 138, 140–141, 152–153, 159, 200, 221, 241, 263, 265–271, 274, 276–283, 285, 287–288, 290–293, 299, 300–316, 319–321, 330; volatilization of, 125–130, 137, 141, 144–145, 152–153, 155, 159, 179, 187, 198, 215, 219–222, 227, 240, 243–244, 271, 274, 281, 314, 334, 341–343
Meteorites (Farrington), 162
Meteoritic falls, 2, 20, 24–26, 124–125, 152, 250, 275–276, 331, 335, 345–348; terrestrial, 26, 124–125, 250, 331, 345–348
Meteoritic impact, 196, 250, 297, 366
Meteoritic impact origin of craters. *See* Coon Mountain, origin of; Lunar craters, impact origin of; Meteoritic impact theory
Meteoritic impact process, 345
Meteoritic impact theory, 19–20, 24–26, 79, 97, 108–109, 116–117, 123–125, 131–133, 138, 140–141, 144–145, 151–153, 155–159, 161–162, 168–169, 171–172, 174–176, 179–184, 196, 215–216, 219, 236, 243–245, 251–254, 276–286, 310–312, 314, 318–321, 343–344, 364–366. *See also* Moon, meteoritic impact theory; Impact theory; bilateral symmetry. *See* Bilateral symmetry; central peaks. *See* Central peaks; circularity. *See* Circularity of craters; compression and rebound, 348–350, 357, 359–360; criteria for recognizing impact craters, 346–347, 350–352; "cryptovolcanic" or "crypto-

INDEX   437

explosive" structures, 346, 349–351, 356, 358; current controversy, 349, 351–353, 356–362; depth-diameter relationships of craters, 358, 360–361; dynamics of impact, 91–92, 98, 123–125, 128–130, 134–135, 138, 140–141, 152–153, 157–159, 179–180, 182, 191–192, 196, 198, 213, 241, 246, 249, 254, 263–272, 276–286, 288, 290–293, 298–312, 314–317, 319, 329–330, 342–343, 348–349, 353, 357–363. *See also* Moon, meteoritic impact theory, dynamics of impact; erosion. *See* Erosion; explosion effects, 357–363. *See also* Explosion on impact; heat generation, 60, 63–64, 70–71, 186, 188–189, 198, 207, 213, 303, 346–347, 359–362; hydrodynamic flow, 360–361; kinetic energy, 198, 209, 212, 347, 361–362; mass-velocity relationship. *See* Mass-velocity relationship; mixing of materials at impact, 91, 93–94, 316, 360; momentum of meteorite, 269, 300, 303, 308–309; penetration, 211, 284, 360–361. *See also* Meteorite, penetration of; percussion effects, 357, 360; postulation of, 21–22, 24, 59, 185, 348, 357; shatter cones, 350–351, 356; shock induced rarefaction and cratering, 360–363; shock melting, 363; shock propagation, 361–362

Meteoritic iron, 31–35, 42–43, 52, 73, 76–77, 81, 83, 92–95, 113, 119, 122, 127, 133, 135–136, 140, 142, 144–145, 152, 154, 159, 168–169, 177, 179, 242, 244–245, 274, 328–329, 333, 339–340, 342, 346–348, 362–363

Meteoritic material in crater, 83–84, 92–93, 107, 115–116, 125–126, 134, 137, 153, 161, 172, 178–179, 228, 316, 322, 326–327, 330, 333, 338, 341

Meteoritical Society, 332, 334, 336, 340

Meteoritics, 279

Meteors (Olivier), 314

Meydenbauer, A., 55–56

Meyer, M. Wilhelm, 192

Micrographia (Hooke), 9, 22, 27

Miller, Ephraim, 67–68
Miller, John A., 140, 157, 180, 283, 286, 294
Miller, Roswell, III, 328, 330
Minerals of Commercial Value (Barringer), 75
Mining, 75, 334–336; claims, 32, 76–81, 142–143; patents, 77, 79–81; prospects of, 3, 32, 84, 127, 130, 133, 138, 140–141, 151, 153, 163–164, 166–173, 177, 179, 192, 214–215, 221, 223, 227–228, 239–240, 244, 261, 263, 268, 270, 273, 282, 285, 297, 299, 312, 315–316, 322–324, 326–327
Mixing of materials at impact, 91, 93–94, 316, 360
Moenkopi sandstone, 43, 93, 330, 333, 343
Montezuma's Well, 108
Moon, alleged habitability of, 11–12, 158; atmosphere of, 12, 355–356; chronology of, 27, 65, 181–183, 354; craters of, *see* Lunar craters; evolution of, 25; hoax, 11–13; impact theory, 10, 54, 62, 64–66, 70, 156–158, 160–161, 182–193, 195–198, 201–213, 216, 344, 346–348, 352–357, 359, 360–362, 365; Mare: 204, 206, 354, 356, 358, Crisium, 69, Imbrium, 64–65, 69, 71, Serenitatis, 65, 69, Tranquilitatis, 65; meteorites, 24–25, 31–35, 65, 69–70, 193, 196, 207, 219, 259, 353, 358–359, diameters of, 25, 359–360, frequency of, 25; meteoritic impact theory, 3, 4, 15, 21–27, 54–55, 59–60, 65, 69–71, 128–129, 155–158, 161, 180–183, 185–186, 196, 198, 201, 203, 206, 215–216, 246, 260, 344, 347–348, 352–363, 365. *See also* Impact theory, dynamics of impact, 188, 191, 195–198, 201, 206–213, 219, 347–348, 354–355, 357, 359–363; new meteoric hypothesis, 185–186, 195–198, 201, 205–213, 219, 348; nomenclature of, 9, origin of, 55–56, 157, 187–188, 203, 209, 216, 353–355, 364, fission hypothesis, 354, planetesi-

438 INDEX

Moon (continued)
  mal hypothesis, 157, 160, 212, 216, 223, 353; photography of, 13, 356–357; ray systems of, 348, 353, 355, 357; valleys of, 64; vegetation on, 193; volatilization, 128–129, 185–186, 188, 198, 207–209; volcanic analogy, 3–4, 7–11, 14–16, 22–23, 25–28, 30, 55, 57–58, 66, 68, 71, 154–156, 158, 173, 184–185, 190–191, 197, 199–200, 202, 206, 209–211, 213, 242, 246, 260, 352–353, 355–357, 359, 364–365; volcanic theory, 49, 59, 66, 117, 155, 193, 199–200, 202, 206, 210, 213, 242, 246, 260, 353–354, 358, 364–366. See also Volcanic analogy; volcanoes on, 13, 16, 18, 67, 70, 319, 364; water on, 12, 16, 22, 28–29, 57, 210
Moon and the Condition and Configuration of Its Surface, The (Neison), 15
"Moon Crater," 47
Moon, The (Goodacre), 352
Moon, The (Pickering), 68
Moon: Considered as a Planet, a World, and a Satellite, The (Nasmyth and Carpenter), 15–18, 23, 27
Moon: Her Motions, Aspect, Scenery, and Physical Condition, The (Proctor), 15, 18–19, 21–23, 29
Moonlet, 55, 62–63, 67, 70, 157, 186, 188, 356
"Moonlet" hypothesis, 55–56, 61–67, 70–72, 157, 186, 188, 199, 203, 207, 356
"Moonspots," 68
Moore, C. F., 168–170, 172, 177–178, 232–233
Moretus, lunar crater, 200
Moulton, Forest Ray, 5, 190, 284–299, 310–316, 333, 341, 359; on circularity and angle of impact, 182; on cluster theory, 263; planetesimal hypothesis of, 128, 156–157, 160, 189, 194, 212, 216, 353; role in Coon Mountain controversies, 238, 302–303, 305, 315–316, 321, 325–326; first report, 263–268, 270, 273–274, 278–280, 285, 287–288, 293, 295, 319; second report, 264, 290–293, 306, 313–314, 318; third report, 264, 306–313; final report, 265; works by: *An Introduction to Astronomy* (Moulton), 71, 259, *Astronomy* (Moulton), 354, *Celestial Mechanics* (Moulton), 266, *Introduction to Celestial Mechanics* (Moulton), 189
Mount Wilson Solar Observatory, 100-inch Hooker reflector, 180–181, 199
Mulder, M. E., 191–192

Nasmyth, James, 15–18, 23, 27, 185
National Geographic Magazine, 230–231, 251, 254–255, 257
National Geographic Society, 252–254, 257, 260
Nebular hypothesis, 21, 24, 157, 194. See also Laplace
Neison, Edmund, 15, 56
New meteoric hypothesis, 185–186, 195–198, 201, 204–213, 219–222, 240, 348. See also A. C. Gifford
New York Sun, and moon hoax, 11–12
Newsom, J. F., 166–168
Nininger, Harvey Harlow, 328, 332, 337–343, 345, 353, 362
Noble, William, 69
North American Mines, Inc., 247, 261–262, 273
Nuclear explosion analogy, 318, 321, 346, 361
Nuclear explosion crater, 318, 321, 361

Odessa Crater, 184, 233–236, 258–259, 272, 331, 346, 362
Old and New Astronomy (Proctor), 23
Olivier, Charles P., 250, 295, 298–299, 314; biographical, 220; investment in Coon Mountain enterprise, 248; role in Coon Mountain controversies, 254, 257, 274–280, 282–284, 286, 289
Olympus Mons, volcano, 366
Öpik, Ernst J., 185, 196–197, 319–320
Osborn, Henry Fairfield, 170, 254
Osborn, Sidney P., 334–335
Ösel craters, 250, 347
Ösel island, Estonia, 250

About the Author
William Graves Hoyt was a full-time research associate at Lowell Observatory from 1981 until his death in 1985. With the publication of two earlier books by the University of Arizona Press, *Lowell and Mars* (1976) and *Planets X and Pluto* (1980), the author gained credence as a scholarly writer among scientists as well as historians. Formerly a journalist by profession, he first worked as a reporter for the *New York World-Telegram* and later as managing editor for the *Arizona Daily Sun,* Flagstaff. He joined the public information staff at Northern Arizona University in 1966 and was named director in 1975, a position he held until his resignation to move to the Observatory in 1981.